T0221537

Core Concepts in Supramolecular Chemistry and Nanochemistry

Core Concepts in Supramolecular Chemistry and Nanochemistry

Jonathan W. Steed,
Durham University, UK

David R. Turner,
Monash University, Australia

Karl J. Wallace,
University of Southern Mississippi, USA

BICENTENNIAL
BICENTENNIAL
1807
⊛WILEY
2007
BICENTENNIAL
BICENTENNIAL

John Wiley & Sons, Ltd

Other Wiley Editorial Offices

John Wiley & Sons, Inc., 111 River Street, Hoboken, NJ 07030, USA

Jossey-Bass, 989 Market Street, San Francisco, CA 94103-1741, USA

Wiley-VCH Verlag GmbH, Boschstr. 12, D-69469 Weinheim, Germany

John Wiley & Sons, Australia Ltd, 33 Park Road, Milton, Queensland 4064, Australia

John Wiley & Sons, (Asia) Pte Ltd, 2 Clementi Loop #02-01, Jin Xing Distripark, Singapore 129809

John Wiley & Sons, Canada Ltd, 6045 Freemont Blvd, Mississauga, Ontario, L5R 4J3

Wiley also publishes its books in a variety of electronic formats. Some content that appears in print may
not be available in electronic books.

Anniversary Logo Design: Richard J. Pacifico

Library of Congress Cataloging in Publication Data

Steed, Jonathan W., 1969–
    Core concepts in supramolecular chemistry and nanochemistry / Jonathan W. Steed,
    David R. Turner, Karl J. Wallace.
       p.   cm.
    Includes bibliographical references and index.
    ISBN 978-0-470-85866-0 (cloth : alk. paper) — ISBN 978-0-470-85867-7 (pbk. : alk. paper)
    1. Supramolecular chemistry.   2. Nanochemistry.   I. Turner, David R.
    II. Wallace, Karl J.   III. Title.
    QD878.S73 2007
    547′.7—dc22   2007001274

British Library Cataloguing in Publication Data

A catalogue record for this book is available from the British Library

ISBN-13: 978-0-470-85866-0 (HB)
ISBN-13: 978-0-470-85867-7 (PB)

The front cover depicts the Minoan Phaestos disc, ca. 1600 BC. This disc bears hieroglyphic characters,
separately impressed by means of punches and arranged in a spiral. Like many aspects of the molecular
world, the characters have yet to be deciphered. Photo courtesy of Iain Forbes, Department of Archaeology,
Cambridge University, UK.

Typeset in 10/12pt Palatino by Integra Software Services Pvt. Ltd, Pondicherry, India

To Ben and Joshua

Contents

Preface

Supramolecular Chemistry is now a mature and highly vigorous field. In 2005 alone, some 2532 scientific papers used the word 'supramolecular' in their titles, keywords or abstracts! The term 'supramolecular' has origins at least to Webster's Dictionary in 1903, but was first applied in the modern sense by Jean-Marie Lehn in 1978 as the '. . . chemistry of molecular assemblies and of the intermolecular bond'. Lehn shared the 1987 Nobel Prize in Chemistry with Charles Pedersen and Donald Cram for their pioneering work in the field in the late 1960s and subsequent decades. Since that time, chemists have attained an astonishing degree of control over the 'non-covalent bond' and have used these techniques to synthesise a plethora of beautiful and intricate functional structures with dimensions on the nanometre scale. More recently, this ability to 'synthesise-up' nanoscale architectures and components has given rise to the field of 'nanochemistry' – the preparation and manipulation of molecular structures on length-scales of ca. 1–500 nm. The boundaries of nanochemistry and supramolecular chemistry are highly subjective although they are somewhat distinct areas. The modern explosion in nanochemistry is very much based, however, upon the fundamental understanding of intermolecular interactions engendered by supramolecular chemists. It thus makes sense for this book to provide a 'one-stop' brief introduction which traces the fascinating modern practice of the chemistry of the non-covalent bond from its fundamental origins through to its expression in the emergence of nanochemistry.

Both supramolecular chemistry and nanochemistry are now featuring ever more strongly in undergraduate and postgraduate degree courses throughout the world. The amount of each discipline which is taught is highly variable but is often a relatively small component of the undergraduate curriculum. The need for a concise introductory book that could serve as a basis for supramolecular chemistry courses of varying lengths was recognised by Jerry Atwood and one of us (JWS) in 1995. Andy Slade at Wiley (UK) has been a great believer in the concept and in 2000 Steed and Atwood published the very successful *Supramolecular Chemistry*, a book that has since even made it into a Russian-language edition. To Andy's dismay, however, this 'concise introduction' weighed in at over 700 pages. It turned out that there was a lot to cover! Five years later in 2005, Geff Ozin and Andre Arsenault did the same thing for nanochemistry, producing an extremely comprehensive overview of research in the field. Andy never gave up the idea of the concise textbook, however, and the idea rumbled around a South Kensington pub one evening while the three present authors were all working together in London. Since then, we have all moved institutions and it has taken three years and a great deal of e-mails between three continents to bring the book to fruition but we hope that it will have been worth the wait. In this book, we have tried to provide a topical overview and introduction to current thinking in supramolecular chemistry and to show how supramolecular concepts evolve into nanochemical systems. By definition, this book is not comprehensive and we apologise in advance to the many fine researchers whose work we could not include. The examples we have chosen are those that best illustrate the fundamental concepts and breadth of the field. In order to highlight important (and readable!) entries into the supramolecular chemistry literature, we have chosen to adopt a system of 'key references' which are marked by a 'key symbol' at the start of most major sections. Key references are chosen predominantly from the secondary or review literature to give the interested student an up-to-date and, above-all, focused entry into the research literature for any subsection of the material which catches their interest (or is assigned as homework!). It is hoped in this way to guide the reader to the most useful or influential work as quickly as possible without the often bewildering effect that a mass of more or less obscure citations to the primary literature may have. Additional citations are given to provide useful further reading.

Finally, no book is written without the help and support of very many people. We would particularly like to thank Andy Slade at Wiley (UK) for championing the concept for this book and for many pleasant lunches! We are very grateful to Drs Stuart Batten, Mark Gray, Gregory Kirkovits, Ian van der Linde, Craig Forsyth, Anand Bhatt, Leigh Jones and Kirsty Anderson for their constructive criticism and helpful comments and suggestions. Thanks to Dr Kellar Autumn for his useful comments on Chapter 5. DRT wishes to thank his family for their unwavering support, his friends in both England and Australia and especially Jodie for always being there when needed. KJW would like to thank his partner Terri Tarbett for her endless love, support and patience throughout the last couple of years.

Jonathan W. Steed, *Durham, UK*
David R. Turner, *Melbourne, Australia*
Karl J. Wallace, *Mississippi, USA*

About the authors

Jonathan W. Steed was born in Wimbledon, UK in 1969. He obtained his B.Sc. and Ph.D. degrees at University College, London, working with Derek Tocher on coordination and organometallic chemistry directed towards inorganic drugs and new metal-mediated synthesis methodologies. He graduated in 1993, winning the Ramsay Medal for his Ph.D. work. Between 1993 and 1995, he was a NATO postdoctoral fellow at the University of Alabama and University of Missouri, working with Professor Jerry L. Atwood, where he developed a class of organometallic supramolecular hosts for anions. In 1995, he was appointed as a Lecturer at King's College, London where he built up a reputation for supramolecular chemistry, including anion binding and sensing, and crystal engineering studies using strong and weak hydrogen bonds. In 1998, he was awarded the Royal Society of Chemistry Meldola Medal and was promoted to Reader in 1999. In 2004, he was appointed as Reader in Inorganic Chemistry at the University of Durham and was elected FRSC in 2005. Dr Steed is co-author of the textbook *Supramolecular Chemistry* (2000) and more than 200 research papers. He has published a large number of reviews, book chapters and popular articles, as well as a major edited work, the *Encyclopedia of Supramolecular Chemistry* (2004). He has been an Associate Editor of the *New Journal of Chemistry* since 2001.

David R. Turner was born in London, UK in 1979. He obtained his M.Sci. in Chemistry at King's College, London where he became interested in crystal nucleation and organometallic anion sensors. He stayed on to do a Ph.D. with Jonathan Steed at King's College and at Durham University, on urea-functionalised anion receptors, including tripodal organic host species and molecular tweezers. His work also involved aspects of crystal engineering and solid state phenomena involving transition metal/ureido systems. He graduated in 2004. In January 2005, he changed countries and disciplines to begin a post-doctoral position at Monash University, Melbourne, Australia with Professor Peter Junk and Professor Glen Deacon, working on the synthesis and structural characterisation of novel lanthanoid – pyrazolate complexes. In January 2006, he was awarded an Australian Research Council post-doctoral fellowship in collaboration with Dr Stuart Batten at Monash University. His current research is focused on the synthesis and control of lanthanoid-containing coordination networks targeting systems with novel magnetic properties, in addition to pursuing his interest in hydrogen bonding networks. Dr Turner is the co-author of 20 scientific papers and is co-lecturer of the metallo-supramolecular course at his current university.

Karl J. Wallace was born in Essex (a true Essex boy!), UK in 1978. He obtained his B.Sc. at the University of the West of England, Bristol in 1999, where he developed an interest in inorganic chemistry and coordination polymers. He then completed a Ph.D. at King's College, London (2003), working with Jonathan W. Steed on the synthesis and binding studies of hosts for small molecule recognition. In 2003, he moved to the laboratories of Eric V. Anslyn at the University of Texas at Austin, USA as a post-doctorial fellow, synthesizing molecular 'scaffolds' for applications as practical sensor devices. In 2006, he was appointed as an Assistant Professor in Inorganic and Supramolecular chemistry at the University of Southern Mississippi, USA, where his research interests are in supramolecular chemistry, particularly molecular recognition and the synthesis of molecular sensors and devices.

1
Introduction

1.1 What is supramolecular chemistry?

As a distinct area, supramolecular chemistry dates back to the late 1960s, although early examples of supramolecular systems can be found at the beginning of modern-day chemistry, for example, the discovery of chlorine clathrate hydrate, the inclusion of chlorine within a solid water lattice, by Sir Humphrey Davy in 1810 (see Chapter 4, Section 4.4). So, *what is supramolecular chemistry*? It has been described as 'chemistry beyond the molecule', whereby a 'supermolecule' is a species that is held together by non-covalent interactions between two or more covalent molecules or ions. It can also be described as 'lego™ chemistry' in which each lego™ brick represents a molecular building block and these blocks are held together by intermolecular interactions (bonds), of a reversible nature, to form a supramolecular aggregate. These intermolecular bonds include electrostatic interactions, hydrogen bonding, π–π interactions, dispersion interactions and hydrophobic or solvophobic effects (Section 1.3).[†]

> **Supramolecular Chemistry:** The study of systems involving aggregates of molecules or ions held together by non-covalent interactions, such as electrostatic interactions, hydrogen bonding, dispersion interactions and solvophobic effects.

Supramolecular chemistry is a multidisciplinary field which impinges on various other disciplines, such as the traditional areas of organic and inorganic chemistry, needed to synthesise the precursors for a supermolecule, physical chemistry, to understand the properties of supramolecular systems and computational modelling to understand complex supramolecular behaviour. A great

---

[†] Note that interactions with units of energy should not be confused with forces which have units of Newtons.

Core Concepts in Supramolecular Chemistry and Nanochemistry Jonathan W. Steed, David R. Turner and Karl J. Wallace

deal of biological chemistry involves supramolecular concepts and in addition a degree of technical knowledge is required in order to apply supramolecular systems to the real world, such as the development of *nanotechnological* devices (Chapter 5).

Supramolecular chemistry can be split into two broad categories; *host–guest chemistry* (Chapter 2) and *self-assembly* (Chapter 3). The difference between these two areas is a question of size and shape. If one molecule is significantly larger than another and can wrap around it then it is termed the 'host' and the smaller molecule is its 'guest', which becomes enveloped by the host (Figure 1.1(a)). One definition of hosts and guests was given by Donald Cram, who said *The host component is defined as an organic molecule or ion whose binding sites converge in the complex...The guest component is any molecule or ion whose binding sites diverge in the complex.*[1] A *binding site* is a region of the host or guest that is of the correct size, geometry and chemical nature to interact with the other

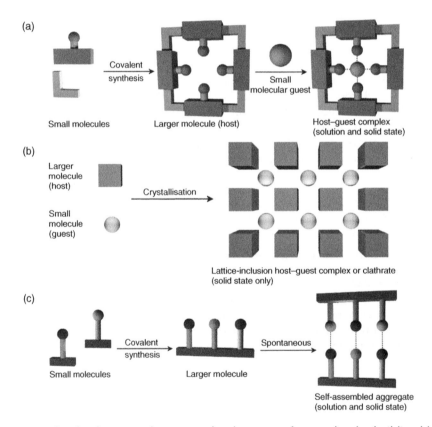

Figure 1.1 The development of a supramolecular system from molecular building blocks (binding sites represented by circles): (a) host–guest complexation; (b) lattice inclusion; (c) self-assembly between complementary molecules.

species. Thus, in Figure 1.1(a) the covalently synthesised host has four binding sites that converge on a central guest binding pocket. Host–guest complexes include biological systems, such as enzymes and their substrates, with enzymes being the host and the substrates the guest. In terms of coordination chemistry, metal–ligand complexes can be thought of as host–guest species, where large (often macrocyclic) ligands act as hosts for metal cations. If the host possesses a permanent molecular cavity containing specific guest binding sites, then it will generally act as a host both in solution and in the solid state and there is a reasonable likelihood that the solution and solid state structures will be similar to one another. On the other hand, the class of solid state *inclusion compounds* only exhibit host–guest behaviour as crystalline solids since the guest is bound within a cavity that is formed as a result of a hole in the packing of the host lattice. Such compounds are generally termed *clathrates* from the Greek *klethra*, meaning 'bars' (Figure 1.1(b)). Where there is no significant difference in size and no species is acting as a host for another, the non-covalent joining of two or more species is termed *self-assembly*. Strictly, self-assembly is an equilibrium between two or more molecular components to produce an aggregate with a structure that is dependent only on the *information* contained within the chemical building blocks (Figure 1.1(c)). This process is usually spontaneous but may be influenced by solvation or templation effects (Chapter 3) or in the case of solids by the nucleation and crystallisation processes (see Chapter 4, Section 4.5).

Nature itself is full of supramolecular systems, for example, deoxyribonucleic acid (DNA) is made up from two strands which self-assemble *via* hydrogen bonds and aromatic stacking interactions to form the famous double helical structure (see Chapter 3, Section 3.2.4). The inspiration for many supramolecular species designed and developed by chemists has come from biological systems.

Host–Guest Chemistry: The study of large 'host' molecules that are capable of enclosing smaller 'guest' molecules *via* non-covalent interactions.

Self-Assembly: The spontaneous and reversible association of two or more components to form a larger, non-covalently bound aggregate.

Binding Site: A region of a molecule that has the necessary size, geometry and functionalities to accept and bind a second molecule *via* non-covalent interactions.

> **Clathrate:** A supramolecular host–guest complex formed by the inclusion of molecules of one kind in cavities of the crystal lattice of another.

1.2   Selectivity

For a host–guest interaction to occur the host molecule must posses the appropriate binding sites for the guest molecule to bind to. For example, if the host has many hydrogen bond donor functionalities (such as primary and secondary amines) then the guest must ideally contain an equal number of hydrogen bond acceptor sites (such as carboxylates), which are positioned in such a way that it is feasible for multiple interactions between host and guest to occur (Section 1.3.2). Alternatively, if the host has Lewis acid centres then the guest must possess Lewis base functionalities. A host that displays a preference for a particular guest, or family of guests, is said to show a degree of *selectivity* towards these species. This selectivity can arise from a number of different factors, such as *complementarity* of the host and guest binding sites (Section 1.2.2), *preorganisation* of the host conformation (Section 1.2.3) or *co-operativity* of the binding groups (Section 1.2.3).

> **Selectivity:** The binding of one guest, or family of guests, significantly more strongly than others, by a host molecule. Selectivity is measured in terms of the ratio between equilibrium constants (see Section 1.2.5).

1.2.1   The Lock and key principle and induced-fit model

 Behr, J.-P. (Ed.), *The Lock-and-Key Principle: The State of the Art 100 Years On*, John Wiley & Sons, Ltd, Chichester, UK, 1995.

Emil Fisher developed the concept of the *lock and key principle* in 1894, from his work on the binding of substrates by enzymes, in which he described the enzyme as the lock and the substrate as the key; thus, the substrate (guest) has a complementary size and shape to the enzyme (host) binding site. Figure 1.2 shows a schematic diagram of the lock and key principle; the key is exactly the correct size and shape for the lock. However, the lock and key analogy is an overly simplistic representation of a biological system because enzymes are highly flexible and conformationally dynamic in solution, unlike the concept of a 'rigid lock'. This mobility gives rise to many of the properties of enzymes, particularly in substrate binding and

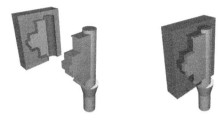

Figure 1.2 The lock and key principle, where the lock represents the receptor in which the grooves are complimentary to the key, which represents the substrate.

catalysis. To address this limitation, Daniel Koshland postulated that the mechanism for the binding of the substrate by an enzyme is more of an interactive process, whereby the active site of the enzyme changes shape and is modified during binding to accommodate the substrate (Figure 1.3). An *induced fit* has occurred and as a consequence the protein backbone or the substrate binding site itself changes shape such that the enzyme and the substrate fit more precisely, *i.e.* are more mutually complementary. Moreover, substrate binding changes the properties of the enzyme. This binding-induced modification is at the heart of many biological 'trigger' processes, such as muscle contraction or synaptic response (see Chapter 5, Section 5.3.4).

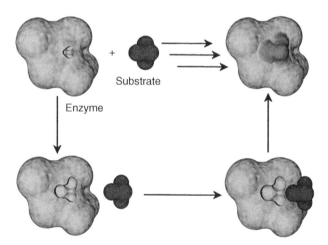

Figure 1.3 The induced-fit model of substrate binding. As the enzyme and substrate approach each other, the binding site of the enzyme changes shape, resulting in a more precise fit between host and guest.

1.2.2 Complementarity

Complementarity plays an important role in biological and supramolecular systems, for example, in the function of enzymes. An enzyme is generally a lot larger than its substrate and only a small percentage of the overall structure is involved in the binding; this region is known as the *active site* of the enzyme. The three-dimensional structure of an enzyme folds itself into a conformation whereby the active site is arranged into a pocket or cleft, which is somewhat complementary in size and shape, and is functionally compatible with the substrate. The enzyme and substrate recognise each other due to this match in size and shape and bind *via* complementary binding sites within this pocket or cleft.

In general, in order to achieve strong, selective binding, the binding site of the host must not only be complementary to the guest in terms of size and shape (*cf.* the lock and key and induced-fit models) but the binding sites on both partners must also be chemically complementary. For example, in coordination chemistry Lewis acids and bases are used to form complexes by the donation of electrons by the Lewis base to the Lewis acid. In the Lewis theory of acids and bases, the species can either be *hard* or *soft*, defined in terms of the polarisability of their electron density. Hard acids/bases are non-polarisable and soft acid/bases are polarisable. As a general rule, hard-to-hard and soft-to-soft complexes are the most stable, displaying complementarity between like species. For example, the hard alkali-metal cations are bound more strongly by the harder oxygen atoms of the crown ethers than the softer nitrogen atoms of azamacrocycles (see Chapter 2, Section 2.3.3).

> **Complementarity:** Both the host and guest must have mutual spatially and electronically complementary binding sites to form a supermolecule.

1.2.3   Co-operativity and the chelate effect

 Hancock, R. D., 'Chelate ring size and metal ion selection', *J. Chem. Edu.*, 1992, **69**, 615–621.

A frequently heard saying is that 'the whole is greater than the sum of its parts'. In other words, a team pulling together has greater effect than the sum of many individual efforts. This concept can be easily applied to supramolecular chemistry. A host species with multiple binding sites that are covalently connected (*i.e.* acting as a 'team') forms a more stable host–guest complex than a similar system with sites that are not joined (therefore acting separately from each other). This *co-operativity* between sites is a generalisation of the *chelate effect* in coordination chemistry, derived from the Greek word *chely*, meaning a lobster's claw.

> **Co-operativity:** Two or more binding sites acting in a concerted fashion to produce a combined interaction that is stronger than when the binding sites act independently of each other. The sites are *co-operating* with each other. In the case of binding two guests, co-operativity also represents the effect on the affinity of the host for one guest as a result of the binding of the other.

> **Chelate Effect:** The observation that multidentate ligands (by extension, hosts with more than one binding site) result in more stable complexes than comparable systems containing multiple unidentate ligands, a result of *co-operativity* between interacting sites.

In terms of classical coordination chemistry, Figure 1.4 shows schematically the difference between a metal ion coordinated to six unidentate ligands, such as ammonia, and one coordinated to three bidentate ligands, such as ethylene-diamine (*en*, $NH_2CH_2CH_2NH_2$). The nature of the ligand–metal dative bond is almost identical in both cases (*via* nitrogen atom lone pairs), yet the ethylene-diamine complex is 10^8 times more stable than the corresponding hexamine complex, as seen from the equilibrium constant (Figure 1.4). Indeed, in practice ethylenediamine readily displaces ammonia from a nickel ion.

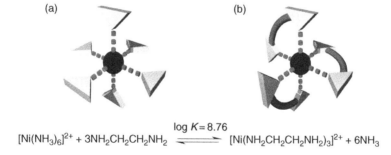

$$[Ni(NH_3)_6]^{2+} + 3NH_2CH_2CH_2NH_2 \xrightleftharpoons{\log K = 8.76} [Ni(NH_2CH_2CH_2NH_2)_3]^{2+} + 6NH_3$$

Figure 1.4 A metal ion surrounded by (a) six unidentate ammonia ligands and (b) three bidentate ethylenediamine ligands. The system with bidentate ligands is more stable, an example of the chelate effect. Triangles represent the ligand interaction sites and the sphere represents a metal ion, such as Ni^{2+}.

The enhanced stability of chelating ligands comes from a combination of entropic ($\Delta S°$) and enthalpic ($\Delta H°$) factors that lower the total complexation free energy ($\Delta G°$), as follows (where T is the temperature in Kelvin):

$$\Delta G° = \Delta H° - T\Delta S° \tag{1.1}$$

In the example shown in Figure 1.4, six unidentate ligands are replaced by three bidentate ligands. During this displacement, a greater number of molecules become free in solution (four species before and seven after). This increase in the number of free molecules gives more degrees of freedom in the system and therefore gives an increase in entropy. The $[Ni(en)_3]^{2+}$ complex is also kinetically stabilised since the bidentate ligands are harder to remove as they have two points of contact with the metal that must be simultaneously broken in order to remove the ligand. The $\Delta G°$ values for the reactions of ammonia and ethylenediamine with Ni^{2+} are -49.2 and $-104.4\,kJ\,mol^{-1}$, respectively.

One common chelating ligand is ethylenediaminetetraacetic acid (H_4EDTA) (**1**). This ligand is able to coordinate to a vast range of metals in a hexadentate manner utilising the four deprotonated acid groups and two nitrogen lone pairs. The six interaction sites of $EDTA^{4-}$ arrange themselves in such a way as to form an octahedral array around the central metal atom. As just one $EDTA^{4-}$ fully saturates the metal coordination sites, the resulting complex is extremely stable (*e.g.* the Al^{3+} complex has a log K value of 16.3). Figure 1.5 shows an X-ray crystal structure of the complex of $EDTA^{4-}$ ligating an aluminium cation. The hexadentate nature of the ligand can clearly be seen as it wraps around the central guest atom. The EDTA ligand is used extensively in metal analysis applications, such as measuring the Ca^{2+} and Mg^{2+} content of urine.

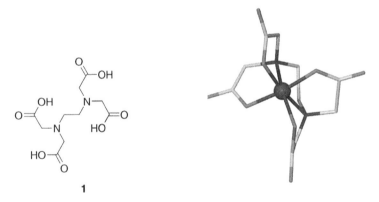

1

Figure 1.5 A host–guest complex of $EDTA^{4-}$ binding an aluminium cation, where the ligand forms an octahedral geometry around the metal ion.

The stability of metal chelate complexes is also significantly affected by the size of the *chelate ring*. A chelate ring is a ring consisting of the guest metal, two donor atoms and the covalent backbone connecting these donors. Figure 1.6 shows a chelating *podand* (a term applied to any flexible acyclic host capable of wrapping around a guest) with a six-membered chelate ring highlighted. The two nitrogen donor atoms and the metal centre account for three of the ring members; the remaining three are from the C_3 chain bridging the nitrogen atoms.

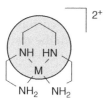

Figure 1.6 A chelating podand, with a six-membered chelate ring highlighted.

The number of members within a chelate ring has an effect on the binding of the guest. If the ring is too small, then the ring will be strained, thus making binding unlikely on enthalpic grounds. The optimum ring geometry for large metal cations is a five-membered chelate ring (Figure 1.7(a)) such as those formed in ethylenediamine complexes. Five-membered rings are particularly stable with large metal cations, such as K^+, as the donor atoms present a larger space for binding. Six-membered rings, on the other hand, are more stable with smaller guests such as Li^+, as the donor atoms result in more limited space to bind the metal (Figure 1.7(b)). As the chelate ring size becomes increasingly large, the chelate effect diminishes, as there is increasing loss of entropy associated with the greater conformational flexibility of the ring. A larger ring requires a larger backbone separating the donor atoms, which becomes less rigid with increasing length. A precise match between optimum chelate ring sizes and metal ionic radii also depends on the orbital hybridisation of the donor atoms.

(a)   (b)

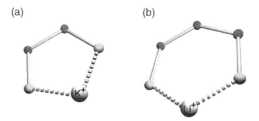

Figure 1.7 Schematic representations of (a) five-membered and (b) six-membered chelate rings (metal–ligand interactions are shown as dashed bonds).

In energy terms, the co-operativity arising from the chelate effect (or more generally from the interaction of a guest with two binding sites, A–B) with a bidentate host can be expressed in terms of the overall binding free energy, ΔG_{AB}° which is equal to the sum of the intrinsic binding free energies of each component A and B (ΔG_A^i and ΔG_B^i), plus a factor arising from the summation or connection of A and B (ΔG^s), as follows:[2,3]

$$\Delta G_{AB}^{\circ} = \Delta G_A^{i} + \Delta G_B^{i} + \Delta G^{S} \tag{1.2}$$

The intrinsic binding energy represents the energies that these groups impart to the rest of the molecule assuming that there are no unfavourable strain or entropy components introduced into the binding by the linking of the group with the rest of the molecule *i.e.* Eq. (1.3) (and similarly for component B):

$$\Delta G_A{}^i = \Delta G_{AB}{}^\circ - \Delta G_B{}^\circ \qquad (1.3)$$

We can thus write Eq. (1.4) which shows that the connection energy is equal to the sum of the separate affinities of the isolated ligands A or B minus the binding free energy of the connected molecule:

$$\Delta G^S = \Delta G_A{}^\circ + \Delta G_B{}^\circ - \Delta G_{AB}{}^\circ \qquad (1.4)$$

The above equation can be used to give an empirical measure of the co-operativity, since the equilibrium constants for the binding of A, B and A–B by a host can be measured and related to the Gibbs free energy *via* Eq. (1.1). If ΔG^S is negative, then the binding sites A and B exhibit unfavourable negative co-operativity. A positive value for ΔG^S implies a favourable positive co-operativity.

The chelate effect represents co-operativity between individual binding sites or ligating groups. Co-operativity is also possible when a host binds two guest species. Again, there are two types of co-operativity, either positive or negative. *Positive co-operativity* is when the presence of the first species *increases* the receptor's affinity for the second species. Often this process involves a structural change, *i.e.* an *induced fit* (Section 1.2.1), and occurs in many biological systems and is part of the *allosteric effect* observed in enzymes. An allosteric effect occurs when the binding of a guest at one site is influenced by the binding of another guest at a different site on the same molecule. When the two guests are the same, this is termed a *homotropic effect* and when they are different it is called a *heterotropic effect*. For example, the binding of one molecule of O_2 to one of the four myoglobin units in haemoglobin increases the O_2 affinity of the remaining three myoglobin sub-units, aiding both O_2 absorption in the lungs and O_2 decomplexation in tissues such as muscle. *Negative co-operativity* is the reverse of positive co-operativity and it is believed that there are very few examples of negative co-operativity occurring in nature. The presence of binding co-operativity (either positive or negative) in any system is indicated by a sigmoidal shape to the binding curve and may be subjected to strict, well-defined tests.[4] (The binding curve is a plot of the variation in some observable property such as spectroscopic absorbance as a function of added guest concentration.) Formally, a multiequilibrium system exhibits positive co-operativity if the ratio of the equilibrium constants, $K_{m+1} : K_m$, is higher than the value calculated from Eq. (1.5). A non-co-operative (statistical) system has a value equal to that calculated by this equation, while a lower value means negative co-operativity:

$$\frac{K_{m+1}}{K_m} = \frac{m(t-m)}{(m+1)(t-m+1)} \qquad (1.5)$$

where m is the number of occupied binding sites in species G_mH_t and t is the total number of sites (G, guest; H, host). The K-values are the equilibrium constants for the formation of the relevant species.

1.2.4 Preorganisation

 Cram, D. J., 'Preorganization – from solvents to spherands', *Angew. Chem., Int. Ed. Engl.*, 1986, **25**, 1039–1134.

We have already seen that complexes containing a chelating ligand, with multiple interaction sites that are covalently connected, have increased stability compared to similar non-chelating systems due to co-operativity between the sites. Introducing an element of *preorganisation* to a host can further enhance this stability. A preorganised host is one that has a series of binding sites in a well-defined and complementary geometry within its structure and does not require a significant conformational change in order to bind to a guest in the most stable way possible. This can be achieved by making a host that is rigid, with a preformed cavity that is already of the correct size to accept the potential guest species and with the appropriate interaction sites already in place. This arrangement is most frequently accomplished by using a host that contains one or more large rings, *macrocycles*, within its structure. Such rings are either rigid or have relatively restricted conformational freedom. The increased stability of ring-based host complexes compared to acyclic analogues has been traditionally referred to as the *macrocyclic effect* and is really just an example of the preorganisation principle.

> **Preorganisation:** A host is said to be preorganised when it requires no significant conformational change to bind a guest species.

> **Macrocyclic Effect:** Host systems that are preorganised into a large cyclic shape form more stable complexes as there is no energetically unfavourable change in conformation in order to bind a guest.

Figure 1.8(a) shows a podand binding to a metal cation. For binding to occur, the host must undergo a conformational change to adapt its shape and binding site disposition to that of the potential guest. Figure 1.8(b) shows the binding of the same guest by a macrocyclic host. This ring is already of the correct geometry to bind the guest and therefore does not have to change shape in order for the binding to take place.

(a)

(b)

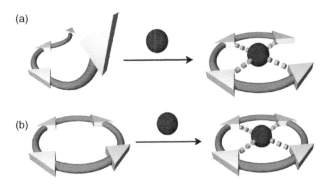

Figure 1.8 (a) A podand is not preorganised and must undergo a change in conformation in order to bind a guest destabilising the complex. (b) A macrocycle that is preorganised for a specific guest does not need to change conformation significantly for binding to occur.

Macrocyclic hosts show enhanced guest binding because of both entropic and enthalpic factors (Eq. (1.1)). Entropically, the binding of a podand results in the loss of many degrees of freedom from the system as the 'floppy' molecule must rigidify as it wraps around the host. This decreases the entropy of the system, meaning that the ΔS of binding is negative and the ΔG of the binding process becomes more positive and unfavourable. A free macrocyclic host does not have such conformational freedom and so the change in entropy between the free and binding host is much less and hence more favourable than that of an analogous podand host. Unfavourable enthalpic contributions from the binding of a podand come from bringing mutually repulsive donor groups into close proximity as the conformation changes. The free podand in solution will minimise its energy by tending to adopt the conformation with the maximum possible distance between repulsive groups, but when binding a guest such groups are brought closer together and the repulsions are overcome by the favourable interaction enthalpy between the binding sites. The macrocyclic host has the donor groups placed into the correct conformation during the synthesis, meaning that energy does not need to be expended during binding, therefore lowering the ΔG of the binding process. Figure 1.9 shows a polyamine podand and a related macrocycle, both of which are capable of binding metal cations such as Zn^{2+} and Cu^{2+}. The macrocyclic host is capable of binding guests 10 000 times more strongly than the podand as a

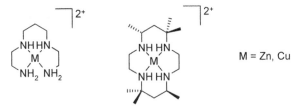

Figure 1.9 Polyamide acyclic and macrocyclic host complexes. The macrocycle displays enhanced binding compared to the podand due to the macrocyclic effect.

consequence of the macrocyclic effect. A further enthalpic effect comes from the negation of repulsions within the macrocycle when a guest binds. The binding sites within a macrocycle, usually electron lone pairs for metal guests, are all pointing towards each other, producing an unfavourable interaction. When a guest is bound to these sites, the unfavourable interactions are reduced in favour of the favourable binding interactions.

Additional enthalpic consequences of binding by macrocyclic ligands concern the desolvation of the host prior to guest binding. The donor sites of a macrocycle are less accessible to solvent molecules than those of a podand as they are generally orientated towards the interior of a cavity. This conformation prevents some solvent molecules from reaching them (Figure 1.10(b)). Podands can be fully solvated as they are flexible, with the donor sites well-separated (Figure 1.10(a)). When a podand binds to a guest, more host–solvent interactions must be broken before the guest is able to bind and therefore a greater amount of energy is required for the binding to occur.

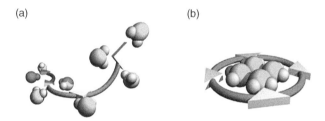

(a)   (b)

Figure 1.10 A podand (a) is fully solvated in solution as it is flexible and the donor sites are easily accessible and (b) macrocycles are often not fully solvated as the solvent molecules would have to be packed in close proximity in the centre of the host.

The macrocyclic effect can be taken one step further by synthesising *macrobicycles* (Figure 1.11). Such species can provide a three-dimensional array of interactions so that a guest is 'more surrounded' by the host. A simple macrocycle leaves the top and bottom of the guest accessible to the bulk environment, whereas a bicyclic host isolates the guest.

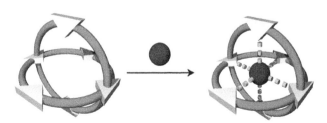

Figure 1.11 A macrobicycle is more rigid and preorganised than a macrocycle (Figure 1.8), hence resulting in stronger guest binding.

1.2.5 Binding constants

Connors, K. A., *Binding Constants: The Measurement of Molecular Complex Stability*, John Wiley & Sons, Ltd, Chichester, UK, 1987.

The binding of a guest by a host species, or the interaction of two or more species by non-covalent bonds, is an equilibrium process. The equilibrium constant for a binding process is called the *binding constant* or *association constant*. The equilibrium that exists for a simple 1:1 host–guest system is shown in Scheme 1.1. The binding constant is calculated by Eq. (1.6), using the concentrations of the species present at equilibrium: host (H), guest (G) and the resulting complex (H·G). The final value, K, has units of mol dm^{-3} or M^{-1}.[‡] These values can range from near zero to very large and so for convenience a log scale is utilised and values are commonly seen quoted as log K. Binding constants are calculated from experimental data (from titrations monitored by NMR, UV–Vis or fluorescence spectroscopy, for example), which supply information about the position of the equilibrium.

$$\text{Host} + \text{Guest} \;\rightleftharpoons\; \text{Host} \cdot \text{Guest}$$

Scheme 1.1   The equilibrium between a host–guest complex and the free species.

$$K = \frac{[\text{H}\cdot\text{G}]}{[\text{H}][\text{G}]} \tag{1.6}$$

> **Binding Constant, K:** The equilibrium constant for the interaction of a host with one or more guests. The *binding constant* provides a quantitative representation of the degree of association and is also called the association constant.

Frequently, host–guest complexes do not form exclusively in a straightforward 1:1 ratio. In such cases, there is more than one binding constant as subsequent guests bind to the host. Multiple equilibria of this type are described by stepwise binding constants for each guest as it binds, and an overall binding constant for the final complex which is termed beta (β). The definition of the overall binding constant is shown in Scheme 1.2.

[‡] Formally binding constants are defined as ratios of activities, which are dimensionless. After all, it is not possible to take a logarithm of a unit! Chemists thus make the approximation that concentrations are very similar to the activities.

$$H + G \rightleftharpoons H \cdot G \qquad K_1 = \frac{[H \cdot G]}{[H][G]}$$

$$HG + G \rightleftharpoons H \cdot G_2 \qquad K_2 = \frac{[H \cdot G_2]}{[H \cdot G][G]}$$

$$HG_2 + G \rightleftharpoons H \cdot G_3 \qquad K_3 = \frac{[H \cdot G_3]}{[H \cdot G_2][G]}$$

$$H + 3G \rightleftharpoons H \cdot G_3 \qquad \beta_3 = \frac{[H \cdot G_3]}{[H][G]^3}$$

$$\beta_3 = K_1 \times K_2 \times K_3$$

Stepwise binding constants (K_1 for first event, *etc.*)   Overall binding constant (β) for a 1:3 host–guest complex

Scheme 1.2 Derivation of stepwise and overall binding constants for a 1:3 host–guest complex.

1.2.6   Kinetic and thermodynamic selectivity

One of the most important factors in the design of host–guest systems is to ensure that a host has a preference for the target guest species above all other possible guests. The host must be able to discriminate between species and hence show a good degree of *selectivity* for the desired guest. There are two kinds of selectivity that may come about; thermodynamic and kinetic.

Thermodynamic selectivity is the ratio of the binding constants for a host binding two different guests (Eq. (1.7)). The relationship between the binding constant of any given supramolecular complex is directly related to the change in free energy during the association process by Eq. (1.8), where R is the gas constant (8.314 J mol^{-1} K^{-1}), T is the temperature (K) and ln K is the natural logarithm of the binding constant. The energy of association can be controlled to a certain extent when the host system is designed, by applying design principles such as the chelate and macrocyclic effects (Sections 1.2.2 and 1.2.3). The correct selection of supramolecular interactions between the two species is also of great importance (Section 1.3). This means that thermodynamic selectivity can be enhanced through rational changes to the design of the host.

$$\text{Selectivity} = \frac{K_{\text{GUEST 1}}}{K_{\text{GUEST 2}}} \tag{1.7}$$

$$\Delta G = -RT \ln K \tag{1.8}$$

Kinetic selectivity is based on a very different principle to thermodynamic selectivity. The word 'kinetic' implies that there is a time-element involved. Kinetic selectivity is usually found in the context of catalytic or enzyme-based processes, whereby a guest (substrate) is transformed upon binding. The rate at which competing substrates are transformed is the determining factor for kinetic selectivity, with the enzyme or catalyst being selective for the fastest-reacting substrate. To cater for a reacting guest, enzyme binding sites are not rigidly preorganised as they have to change to be complementary to the substrate at any given time along the reaction profile. Strong binding would slow down the exchange rate at the enzyme active site and therefore reduce the activity of the enzyme. Enzymes are usually selective for the transition state of a given substrate transformation, adopting a strained geometry, referred to as the *entatic state*. It is this strained geometry that lowers the activation energy for the substrate reaction and gives the enzyme its catalytic properties.

1.2.7 Solvent effects

Smithrud, D. B., Sanford, E. M., Chao, I., Ferguson, S. B., Carcanague, D. R., Evanseck, J. D., Houk, K. N. and Diederich, F., 'Solvent effects in molecular recognition', *Pure Appl. Chem.*, 1990, **62**, 2227–2236.

So far, we have looked at the interactions between a host and its guest(s) as if they were isolated from any other influences. This is not the case in real systems as there are competing interactions from other potential guests and surrounding solvent molecules. Solvent molecules greatly outnumber the amounts of the host and guest present and therefore can have a very pronounced effect upon the dynamics and energetics of association.

When in solution, host and guest species are surrounded by solvent molecules which interact with them. In order for binding to occur, many of these interactions must be broken, which has both enthalpic and entropic consequences. This desolvation process is shown in a simplified way in Figure 1.12. Enthalpically, energy must be expended to break the solvent–host and solvent–guest bonds. The removal of solvent molecules from the host and the guest leads to the solvent molecules having more freedom in the solution, which increases the entropy and also leads to the formation of solvent–solvent bonds. The choice of solvent can have significant consequences on the binding of a guest.

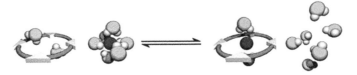

Figure 1.12 Host–guest binding equilibrium showing the desolvation of both species required prior to the binding occurring. The final complex is still solvated but overall there are more free solvent molecules present, hence increasing the entropy of the system.

Solvent effects can be understood by the way in which the individual molecules can interact with the host and the guest. Polar solvents are able to interact with host molecules *via* electrostatic interactions (Section 1.3.1). Such solvents are particularly able to inhibit binding of charged species, as the solvent dipole can interact strongly with a charged centre, thus making the solvent–host or solvent–guest interactions harder to break. Other solvents are able to disrupt the binding by means of electron-pair or hydrogen bond donation and acceptance. Many solvents display both of these properties, for example, dimethyl sulfoxide (DMSO, $OSMe_2$) acts as both an electron-pair donor and hydrogen bond acceptor by virtue of oxygen and sulfur lone pairs. The vast majority of supramolecular interactions are electrostatic in nature (Section 1.3), meaning that polar solvents often act to reduce the observed binding. For this reason it is usual for any studies to be carried out in the least polar solvent possible to reduce the competition for the host. The conditions used can help to moderate the binding process, for example, if the binding is too strong to be conveniently measured, more polar solvents can be employed to reduce the binding constant.

1.3 Supramolecular interactions

Non-covalent interactions represent the energies that hold supramolecular species together. Non-covalent interactions are considerably weaker than covalent interactions, which can range between ca. 150 kJ mol^{-1} to 450 kJ mol^{-1} for single bonds. Non-covalent bonds range from 2 kJ mol^{-1} for dispersion interactions to 300 kJ mol^{-1} for 'ion-ion' interactions. However, when these interactions are used in a co-operative manner a stable supramolecular complex can exist. The term 'non-covalent' includes a wide range of attractions and repulsions which are summarised in Table 1.1 and will be described in more detail in the following sub-sections.

Table 1.1 Summary of supramolecular interactions

Interaction	Strength (kJ mol^{-1})	Example
Ion–ion	200–300	Tetrabutylammonium chloride
Ion–dipole	50–200	Sodium [15]crown-5
Dipole–dipole	5–50	Acetone
Hydrogen bonding	4–120	(See Table 1.2)
Cation–π	5–80	K^+ in benzene
π–π	0–50	Benzene and graphite
van der Waals	< 5 kJ mol^{-1} but variable depending on surface area	Argon; packing in molecular crystals
Hydrophobic	Related to solvent–solvent interaction energy	Cyclodextrin inclusion compounds

1.3.1   Ionic and dipolar interactions

Anslyn, E. V. and Dougherty, D. A., *Modern Physical Organic Chemistry*,
University Science Books, Sausalito, CA, USA, 2006, pp. 162–168.

Ionic and dipolar interactions can be split into three categories: (i) *ion–ion interactions*, (ii) *ion–dipole interactions*, and (iii) *dipole–dipole interactions*, which are based on the Coulombic attraction between opposite charges. The strongest of these interactions is the ion–ion (Figure 1.13(a)), which is comparable with covalent interactions. Ion–ion interactions are non-directional in nature, meaning that the interaction can occur in any orientation. Ion–dipole (Figure 1.13(b)) and dipole–dipole interactions (Figure 1.13(c)), however, have orientation-dependant aspects requiring two entities to be aligned such that the interactions are in the optimal direction. Due to the relative rigidity of directional interactions, only mutually complementary species are able to form aggregates, whereas non-directional interactions can stabilise a wide range of molecular pairings. The strength of these directional interactions depends upon the species involved. Ion–dipole interactions are stronger than dipole–dipole interactions (50–200 and 5–50 kJ mol^{-1}, respectively) as ions have a higher charge density than dipoles. Despite being the weakest directional interaction, dipole–dipole interactions are useful for bringing species into alignment, as the interaction requires a specific orientation of both entities.

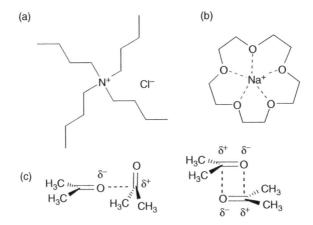

Figure 1.13   Examples of electrostatic interactions: (a) ion–ion interaction in tetrabutylammonium chloride; (b) ion–dipole interaction in the sodium complex of [15]crown-5; (c) dipole–dipole interactions in acetone.

Electrostatic interactions play an important role in understanding the factors that influence high binding affinities, particularly in biological systems in which there is a large number of recognition processes that involve charge–charge interactions; indeed these are often the first interactions between a substrate and an enzyme.

1.3.2   Hydrogen bonding

Jeffery, G. A., *An Introduction to Hydrogen Bonding*, Oxford University Press, Oxford, UK, 1997.

The *hydrogen bond* is arguably the most important non-covalent interaction in the design of supramolecular architectures, because of its strength and high degree of directionality. It represents a special kind of dipole–dipole interaction between a proton donor (D) and a proton acceptor (A). There are a number of naturally occurring 'building blocks' that are a rich source of hydrogen bond donors and acceptors (*e.g.* amino acids, carbohydrates and nucleobases). Hydrogen bond donors are groups with a hydrogen atom attached to an electronegative atom (such as nitrogen or oxygen), therefore forming a dipole with the hydrogen atom carrying a small positive charge. Hydrogen bond acceptors are dipoles with electron-withdrawing atoms by which the positively charge hydrogen atom can interact, for example, carbonyl moieties (Figure 1.14).

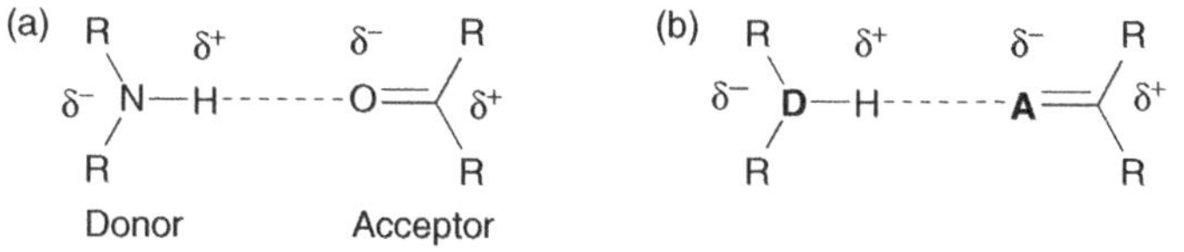

Figure 1.14   A carbonyl accepting a hydrogen bond from a secondary amine donor (a) and (b) the standard way of expressing donor and acceptor atoms (D, donor atom; A, acceptor atom).

The strength of hydrogen bonds can be very different between various systems and is not necessarily correlated with the Brønstead acidity of the proton donor. It depends on the type of electronegative atom to which the hydrogen atom is attached and the geometry that the hydrogen bond adopts in the structure. Typically, the strengths range from 4 to 120 kJ mol^{-1}, with the vast majority being under 60 kJ mol^{-1} and scales of hydrogen bond acidity and basicity have been developed.[5] The types of geometries that can be adopted in a hydrogen bonding complex are summarised in Figure 1.15.

The geometries displayed in Figure 1.15 are termed *primary hydrogen bond interactions* – this means that there is a direct interaction between the donor group and the acceptor group. There are also *secondary interactions* between neighbouring groups that must be considered. The partial charges on adjacent atoms can either increase the binding strength by virtue of attraction between

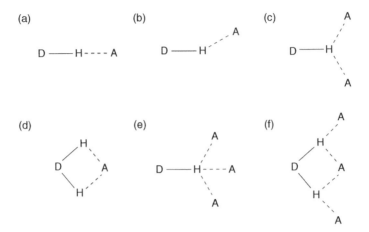

Figure 1.15 Various types of hydrogen bonding geometries: (a) linear; (b) bent; (c) donating bifurcated; (d) accepting bifurcated; (e) trifurcated; (f) three-centre bifurcated.

opposite charges or decrease the affinity due to repulsion between like charges. Figure 1.16 shows two situations in which arrays of hydrogen bond donors and acceptors are in close proximity. An array of three donors (DDD) facing three acceptors (AAA) (Figure 1.16(a)) has only attractive interactions between adjacent groups and therefore the binding is enhanced in such a situation. Mixed donor/acceptor arrays (ADA, DAD) suffer from repulsions by partial charges of the same sign being brought into close proximity by the primary interactions (Figure 1.16(b)).

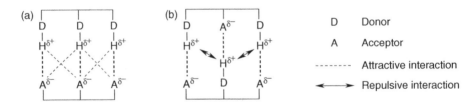

Figure 1.16 (a) Secondary interactions providing attractions between neighbouring groups in DDD and AAA arrays and (b) repulsions from mixed donor/acceptor arrays (ADA and DAD), with primary interactions shown in 'bold'.

A real-life example of hydrogen bonding is the double helix of DNA. There are many hydrogen bond donors and acceptors holding base pairs together, as illustrated between the nucleobases cytosine (C) and guanine (G) in Figure 1.17. The CG base pair has three primary interactions (*i.e.* traditional hydrogen bonds) and also has both attractive and repulsive secondary interactions.

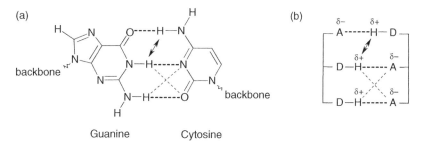

Figure 1.17 (a) Primary and secondary hydrogen bond interactions between guanine and cytosine base-pairs in DNA and (b) a schematic representation.

The geometry of a hydrogen bond and the type of donor and acceptor groups determine the strength, length and nature of the interaction. Hydrogen bond interactions can be divided into three broad categories, the properties of which are listed in Table 1.2. A strong interaction is somewhat similar in character to a covalent bond, whereby the hydrogen atom is close to the centre-point of the donor and acceptor atoms. Strong hydrogen bonds are formed between two strong bases, for example in the HF_2^- ion, which is practically linear with the hydrogen atom between the two fluorine atoms $[F \cdots H \cdots F]^-$. Moderate-strength hydrogen bonds are formed between neutral donor and neutral acceptor groups *via* electron lone pairs, for example, the self-association of carboxylic acids. Moderate hydrogen bond interactions do not have a linear geometry but are slightly bent. Hydrogen bonds commonly deviate from linearity and their angular distribution is influenced by statistical factors. A 'conical correction' for statistical effects often appears in the analysis of hydrogen bond-angle distributions, particularly from searches of the Cambridge Structural Database (see Chapter 4, Section 4.5.2). A linear hydrogen bond requires a fixed position of the hydrogen atom in relation to the acceptor, whereas non-linear hydrogen bonds have many possible positions that form a conical shape around the linear position. Larger bond angles

Table 1.2 Hydrogen bond interactions and their properties (A, acceptor; D, donor)

Interaction/property	Strong	Moderate	Weak
$D–H \cdots A$	Mainly covalent	Mainly electrostatic	Electrostatic
Bond energy (kJ mol^{-1})	60–120	16–60	< 12
Bond length (Å)			
$\quad H \cdots A$	1.2–1.5	1.5–2.2	2.2–3.2
$\quad D \cdots A$	2.2–2.5	2.5–3.2	3.2–4.0
Bond angle (degrees)	175–180	130–180	90–150
Example	HF complexes	Acids	$C–H \cdots A$
	$H_5O_2^+$	Alcohols	$D–H \cdots \pi$
	—	DNA/RNA	—

result in a larger cone, and therefore there are more possible positions for the bond to occur in. Weak hydrogen bonds are even less linear and in some cases can form perpendicular interactions, for example the C−H···π interaction between benzene rings when the C−H bonds point directly towards the conjugated system (Section 1.3.3).

The highly directional nature of hydrogen bonding interactions, together with the specific alignment of hydrogen bond donors and acceptors, has proved to be a fruitful asset for the design of supramolecular systems.

1.3.3   π-Interactions

Ma, J. C. and Dougherty, D. A., 'The cation-π interaction', *Chem. Rev.*, 1997, **97**, 1303–1324.
Hunter, C. A., Lawson, K. R., Perkins, J. and Urch, C. J., 'Aromatic interactions', *J. Chem. Soc., Perkin Trans. 2*, 2001, 651–669.

There are two main π-interactions that can be found in supramolecular systems, namely (i) cation–π interactions and (ii) π–π interactions. Cation–π interactions are well known in the field of organometallic chemistry, whereby olefinic groups are bound to transition metal centres, for example, ferrocene and Zeise's salt ($[PtCl_3(\eta^2\text{-}C_2H_4)]^-$), but these are not regarded as non-covalent interactions.[6] However, alkaline- and alkaline-earth metals also form interactions with double-bond systems, typically between 5 and 80 kJ mol^{-1}. For example, the interaction of potassium ions with benzene has a similar energy to the K^+–OH_2 interaction. The potassium cation is more soluble in water than in benzene, however, as it is not sterically possible to fit as many benzene molecules around the metal ion as water molecules (Figure 1.18).

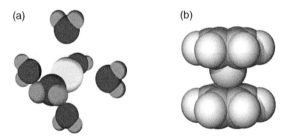

Figure 1.18   (a) Six or more water molecules can fit around K^+ whereas (b) there is space for only two benzene molecules.

The two types of π–π interactions are *face-to-face*, whereby parallel ring-systems, separated by ca. 3.5 Å, are offset and the interaction is between the centre of one ring and the corner of another (Figure 1.19(a)), and *edge-to-face*, whereby a hydrogen atom from one ring interacts in a perpendicular orientation with respect to the centre of another ring (Figure 1.19(b)). These π–π interactions arise from the attraction between the negatively charge π-electron cloud of one conjugated system and the positively charged σ-framework of a neighbouring molecule.[7]

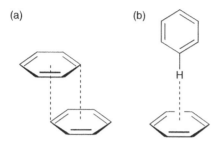

(a)   (b)

Figure 1.19 The two types of π–π interactions: (a) face-to-face; (b) edge-to-face.

The layered structure of graphite is held together by weak, face-to-face π-interactions and therefore feels 'slippery' (Figure 1.20). It is because of the slippage between layers that graphite can be used as a lubricant (albeit in the presence of oxygen). Interactions involving π-systems can be found in nature, for example, the weak face-to-face interactions between base-pairs along the length of the double helix are responsible for the shape of DNA (see Chapter 3, Section 3.2.4).

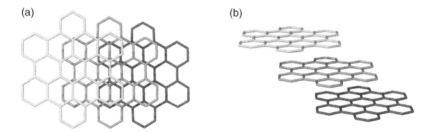

(a)   (b)

Figure 1.20 (a) Top and (b) side views of the layered structure of graphite, held together by face-to-face π-interactions.

1.3.4 van der Waals interactions

Schneider, H.-J., 'Van der Waals forces', in *Encyclopedia of Supramolecular Chemistry*, Vol. 2, Steed, J. W. and Atwood, J. L. (Eds), Marcel Dekker, New York, NY, USA, 2004, pp. 1550–1556.

Van der Waals interactions are dispersion effects that comprise two components, namely the *London interaction* and the *exchange and repulsion* interaction.

Van der Waals interactions arise from fluctuations of the electron distribution between species that are in close proximity to one another. As the electron cloud moves about a molecule's momentary location, an instantaneous dipole is formed within the molecule. This 'flickering' of electron distribution (or dipole) between two adjacent species will align the molecules such that a partial positive charge from one species will be attracted to a partial negative charge from another molecule (Figure 1.21); therefore, the two instantaneous dipoles attract one another and produce a London interaction. The strength of these interactions is dependant on the polarisability of the molecule; the more polarisable the species, then the greater the strength of the interaction. The potential energy of the London interaction decreases rapidly as the distance between the molecules increases (this depends on the reciprocal of the sixth power of the distance r – an r^{-6} dependence). These interactions are non-directional and do not feature highly in supramolecular design. However, van der Waals interactions are important in the formation of inclusion compounds (see Chapter 4, Section 4.3), in which small organic molecules are incorporated into a crystalline lattice, or where small organic molecules have been encapsulated into permanent molecular cavities.

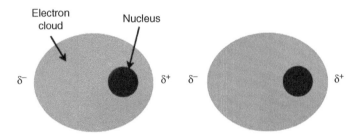

Figure 1.21 A London interaction between two argon atoms. The shift of the electron cloud around the nucleus produces instantaneous dipoles that attract each other.

In the solid state, species tend to align so there is a maximum number of interactions between each molecule, which minimises the lattice energy of the solid state structure. This close packing arrangement has been rationalised by Kitaigorodskii in a classic treatise.[8] As molecules 'grow' into a crystal, they arrange themselves so that all of the void space is occupied, to achieve the maximum interaction with their neighbours and hence the most stable lattice energy for the crystal. This close-packed arrangement is achieved by most solid state structures but there are a few examples were there is a void space, *i.e.* zeolites and channel coordination polymers (see Chapter 4, Sections 4.2 and 4.6, respectively) where the rigid framework is strong enough to withstand external forces.

1.3.5   Hydrophobic effects

 Southall, N. T., Dill, K. A. and Haymet, A. D. J., 'A view of the hydrophobic effect', *J. Phys. Chem.*, 2002, **106**, 521–533.

Hydrophobic effects arise from the exclusion of non-polar groups or molecules from aqueous solution. This situation is more energetically favourable because water molecules interact with themselves or with other polar groups or molecules preferentially. This phenomenon can be observed between dichloromethane and water which are immiscible. The organic solvent is forced away as the inter-solvent interactions between the water molecules themselves are more favourable than the 'hole' created by the dichloromethane. Hydrophobic interactions play an important role in some supramolecular chemistry, for example, the binding of organic molecules by cyclophanes and cyclodextrins in water (see Chapter 2, Sections 2.7.1 and 2.7.5, respectively). Hydrophobic effects can be split into two energetic components, namely an enthalpic hydrophobic effect and an entropic hydrophobic effect.

Enthalpic hydrophobic interactions occur when a guest replaces the water within a cavity. This occurs quite readily as water in such systems does not interact strongly with the hydrophobic cavity of the host molecule and the energy in the system is high. Once the water has been replaced by a guest, the energy is lowered by the interaction of the former water guest with the bulk solvent outside the cavity (Figure 1.22). There is also an entropic factor to this process, in that the water that was previously ordered within the cavity becomes disordered when it leaves. An increase in entropy increases the favourability of the process.

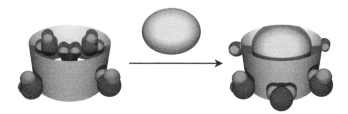

Figure 1.22   The displacement of water molecules from a hydrophobic cavity is responsible for the enthalpic hydrophobic effect.

Entropic hydrophobic interactions come about when there are two or more organic molecules in aqueous solution, the combination of which creates a hole in the water to form a supramolecular complex (Figure 1.23). There is less disruption (one hole in the aqueous phase instead of multiple holes) and hence an entropic gain, as the overall free energy of the system is lowered.

The hydrophobic effect is also very important in biological systems in the creation and maintenance of the macromolecular structure and supramolecular

Figure 1.23 Two organic molecules creating a hole within an aqueous phase, giving rise to the entropic hydrophobic effect – one hole is more stable than two.

assemblies of the living cell or the formation of amphiphilic structures such as micelles, where hydrophilic 'heads' assemble in a roughly spherical geometry and lipid bilayers where the heads meet end-to-end (see Chapter 5, Section 5.4.1).

1.4   Supramolecular design

The principles and phenomena outlined within this introductory chapter are the basic concepts upon which supramolecular chemistry is based. A union of these phenomena can lead to intricate and complex designs that form the heart of the many facets of supramolecular chemistry.

In terms of 'designer' host–guest chemistry, it is necessary to understand the nature of the target guest molecule. The host must be designed to be complementary to the guest in terms of size, shape and chemical properties (charge, hardness, acidity, *etc.*). Other factors must also be considered in the design process, such as the medium in which the binding must occur and any competing molecules which must be excluded from binding, therefore requiring a more selective host. Once all of the guest properties have been taken into consideration, the host may be designed in a specific manner, incorporating the basic phenomena outlined in this chapter, followed by a process of 'trial and improvement' based on laboratory results. Moving away from the host–guest aspect of supramolecular chemistry, the underlying principles remain the same although the systems formed are often much more complex. For example, protein tertiary structure (folding of a protein to give a three-dimensional entity by non-covalent interactions) results in a very complicated system when viewed as a whole, but the individual interactions are quite easily understood. Biological systems have provided an inspiration to chemists who design and synthesise complex supramolecular architectures capable of practical applications.

Supramolecular systems have a wide variety of uses, such as trapping molecules within solid state lattices (Chapter 4), sensing and remediation of species from solution (Chapter 2), understanding biological self-assembly (Chapter 3) and nanotechnological devices (Chapter 5). Together, these topics form the core concepts upon which supramolecular chemistry is based.

References

1. Cram, D. J., 'Preorganization – from solvents to spherands', *Angew. Chem., Int. Ed. Engl.*, 1986, **25**, 1039–1134.
2. Jencks, W. P., 'On the attribution and additivity of binding-energies', *Proc. Nat. Acad. Sci. USA*, 1981, **78**, 4046–4050.
3. Williams, D. H. and Westwell, M. S., 'Aspects of weak interactions', *Chem. Soc. Rev.*, 1998, **27**, 57–63.
4. Perlmutter-Hayman, B., 'Co-operative binding to macromolecules. A formal approach', *Acc. Chem. Res.*, 1986, **19**, 90–96.
5. Laurence, C. and Berthelot, M., 'Observations on the strength of hydrogen bonding', *Persp. Drug Disc. Des.*, 2000, **18**, 39–60.
6. Elschenbroich, C., *Organometallics: A Concise Introduction*, 3rd Edn, VCH, Weinheim, Germany, 2006.
7. Hunter, C. A. and Sanders, J. K. M., 'The nature of π–π interactions', *J. Am. Chem. Soc.*, 1990, **112**, 5525–5534.
8. Kitaigorodskii, A. I., *Organic Chemical Crystallography*, Consultants Bureau, New York, NY, USA, 1961 (originally published in Russian by Press of the Academy of Sciences of the USSR, Moscow, USSR, 1955).

Suggested further reading

Buckingham, A. D., Legon, A. C. and Roberts, S. M. (Eds), *Principles of Molecular Recognition*, Kluwer Academic Publishers, Dordrecht, The Netherlands, 1993.

Cragg, P. J., *Practical Supramolecular Chemistry*, John Wiley & Sons, Ltd, Chichester, UK, 2006.

Desiraju, G. R. (Ed.), *The Crystal as a Supramolecular Entity*, John Wiley & Sons, Ltd, Chichester, UK, 1996.

Lehn, J.-M., *Supramolecular Chemistry*, VCH, Weinheim, Germany, 1995.

Lehn, J.-M., Atwood, J. L., Davies, J. E. D., MacNicol, D. D. and Vögtle, F. (Eds), *Comprehensive Supramolecular Chemistry*, Pergamon, Oxford, UK, 1996.

Ozin, G. A. and Arsenault, A. C., *Nanochemistry*, The Royal Society of Chemistry, Cambridge, UK, 2005.

Schneider, H.-J. and Yatsimirski, A. K., *Principles and Methods in Supramolecular Chemistry*, Wiley-VCH, Weinheim, Germany, 2000.

Steed, J. W. and Atwood, J. L., *Supramolecular Chemistry*, John Wiley & Sons, Ltd, Chichester, UK, 2000.

Steed, J. W. and Atwood, J. L. (Eds), *Encyclopedia of Supramolecular Chemistry*, Marcel Dekker, New York, NY, USA, 2004.

2
Solution host–guest chemistry

2.1 Introduction: guests in solution

In Chapter 1, we discussed the type of interactions that are found in supramolecular systems and demonstrated that the stability of a host–guest complex in solution is quantified by the equilibrium constant or binding constant for the system (see Chapter 1, Section 1.2.5). In this chapter, we will examine the host–guest chemistry of anions, cations and neutral-guest species in solution. While it would be convenient to deal with these types of guest as separate topics, in reality there is a great deal of overlap, particularly with regard to solvation effects and the general design principles for creating selective hosts. Indeed, in the case of anion and cation binding, the two topics go hand-in-hand since the electrostatic charge on any ion must be balanced by a corresponding counter-ion. Thus a 'cation' or 'anion' host is always, in effect, a host for an ion pair (either contact or solvent-separated) unless the host itself bears a formal net charge (Section 2.6). In this latter case, the binding equilibrium becomes a competition reaction between the existing counter-ion and the host. The measured binding constant thus represents the competitiveness of the host over and above a specified counter-ion. In practice, the effect of the counter-ion is sometimes ignored or assumed to be negligible, particularly if weakly interacting counter-ions are used, such as NBu_4^+, PF_6^- or the very weakly interacting $B(C_6H_3(CF_3)_2)_4^-$.[1] However, there are a number of successful ion-pair binding hosts. Moreover, ion-pairing can be used to great effect in phase-transfer catalysis. In general, the selective inclusion of guests by a host molecule in solution is subject to an unprecedented level of design and control based on supramolecular principles. This allows the preparation of tailored systems for a wide range of applications, including sensing, food additives, drug delivery, imaging, biological modelling and in cosmetic therapy. We refer to discoveries of particular historical and industrial importance in each section.

Core Concepts in Supramolecular Chemistry and Nanochemistry   Jonathan W. Steed, David R. Turner and Karl J. Wallace
© 2007 John Wiley & Sons, Ltd   ISBN: 978-0-470-85866-0 (Hardback); 978-0-470-85867-7 (Paperback)

2.2   Macrocyclic versus acyclic hosts

There are two major classes of host: acyclic (*podands*) and cyclic (*macrocycles*, *macrobicycles* or *macrotricycles*). Podands are linear or branching chain species with two or more sets of guest-binding functional groups positioned on the spacer unit in such a way as to chelate a target guest species to maximise guest affinity (*cf. co-operativity*, see Chapter 1, Section 1.2.3). Podands containing several rotatable bonds generally have less intrinsic affinity for their guests than hosts that are more rigidly *preorganised* because of the unfavourable enthalpic and entropic effects associated with the change in host conformation upon binding. Podands generally have a high degree of flexibility and on binding to a guest the conformational change that occurs to produce a stable host–guest complex, may result in *allosteric effects* (see Chapter 1, Section 1.2.3) if the podand is capable of binding more than one guest.[2] In general, host flexibility is of key importance, especially in biological systems, in which recognition of a substrate results in a conformational change that may be of major significance, for example, in a protein's biochemical role (*cf. induced-fit model*, Chapter 1, Section 1.2.1). Moreover, in sensing applications, flexible hosts are associated with rapid binding and de-complexation kinetics and hence rapid sensor response. The corollary is that flexible, podand-type hosts generally exhibit lower binding constants than cyclic analogues.

Cyclic receptors have binding sites positioned in a closed-ring arrangement. As a result, cyclic systems are more preorganised and hence form more thermodynamically stable complexes because less conformational change is required upon binding. In effect, the energetic cost of adopting the cyclic arrangement needed for binding is 'paid in advance' during the irreversible, kinetically controlled synthesis of the macrocycle.

> **Podand:** An acyclic chain-like or branching host with a number of binding sites that are situated at intervals along the length of the molecule, or about a common spacer.

> **Macrocycle:** A cyclic molecule usually with nine or more atoms in the ring and generally in supramolecular chemistry containing a number of binding sites that are arranged around the closed system.

Allosteric Effect: The binding of a guest at one binding site that is influenced by the binding of a second guest, either the same species (homotropic effect) or a different kind of guest (heterotropic effect), bound elsewhere within the host.

The synthesis of podand hosts is conceptually relatively straightforward and is generally achieved using conventional synthetic methods such as ether, sulfide or amide bond formation; however, the synthesis of macrocycles is a little more challenging and is accomplished by one or both of two general procedures, *i.e.* high-dilution synthesis and template synthesis.

2.2.1   High-dilution synthesis

 Knops, P., Sendhoff, N., Mekelburger, H. -B. and Vogtle, F., 'High-dilution reactions – new synthetic applications', *Top. Curr. Chem.*, 1991, **161**, 1–36.

In the high-dilution method, small quantities of reactants are mixed together at a controlled rate in a large volume of solvent. The starting materials are separately, but simultaneously, added slowly to a large reservoir of solvent with continuous stirring over a period of time. This ensures that a low reactant concentration is sustained and that macrocycle formation is favoured over polymerisation. The addition rate of the reactants is critical and can be controlled by using either precision addition funnels or syringe pump apparatus. This method of synthesis has been extensively utilised and is employed in the commercial production of some macrocyclic polyamines, for example. A typical apparatus used for high-dilution synthesis is shown in Figure 2.1.

Why is macrocycle formation favoured when using this technique? Simply, a low concentration of the 'open-strand' reactant has a greater chance of reacting with itself to form a closed-ring system than of reacting with another molecule to form a polymer. In terms of reaction rate, intramolecular ring closure is a unimolecular process – it has a rate that is proportional to the concentration of the reactant. Intermolecular reaction leading ultimately to undesired polymer formation is a bimolecular process and its rate is proportional to the square of the concentration of a single reactant. Therefore, high dilution favours an intramolecular reaction (Figure 2.2). Richman and Atkins, who prepared a variety of amine macrocycles, for example, **2.1** (Scheme 2.1), reported a classic example of this type of reaction.[3]

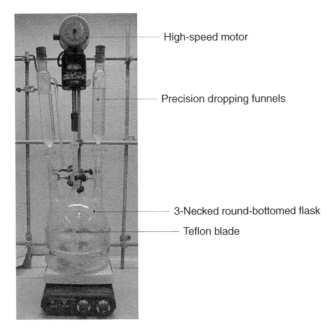

High-speed motor

Precision dropping funnels

3-Necked round-bottomed flask

Teflon blade

Figure 2.1 Typical apparatus used for high-dilution synthesis.

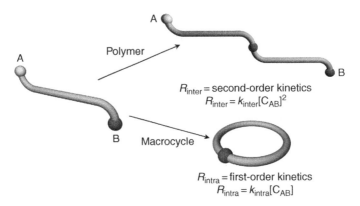

A

Polymer

A

R_{inter} = second-order kinetics
$R_{inter} = k_{inter}[C_{AB}]^2$

B

B

Macrocycle

R_{intra} = first-order kinetics
$R_{intra} = k_{intra}[C_{AB}]$

Figure 2.2 Schematic representation of the products formed in the first step of a polymerisation reaction (rate, R_{inter} (top)) or in an intramolecular macrocyclisation reaction (rate, R_{intra} (bottom)) of a linear chain molecule C_{AB}; $k_{intra} \approx k_{inter}$, where the k-values represent the rate constants.

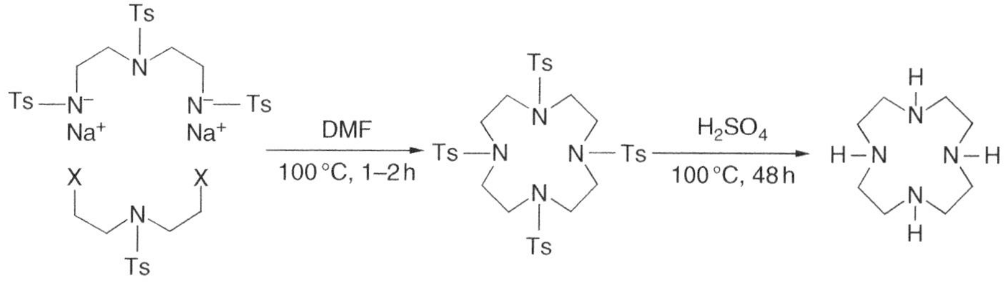

Scheme 2.1 The synthesis of 1,4,7,10-tetraaza-cyclododecane (**2.1**) *via* the Richman–Atkins cyclisation method, followed by deprotection of the secondary amine groups $(OTs=OSO_2C_6H_4CH_3)$.

2.2.2 Template synthesis

Gerbeleu, N. V., Arion, V. B. and Burgess, J., *Template Synthesis in Macrocyclic Chemistry*, John Wiley & Sons, Ltd, Chichester, UK, 1999.

The other common technique used for macrocycle formation is the use of templating species, such as metal ions, to induce cyclisation. The template brings about the organisation of the reacting components, which directs the geometry towards a particular product. The formation of macrocycles in preference to linear oligomers through the use of a template is known as the *template effect*. This reaction is demonstrated by the original synthesis of dibenzo[18]crown-6 by Pedersen in 1967 (Section 2.3.2).[4] The synthesis of cyclic molecules can occur by the use of external templates, such as metal ions, known as *exo-templates*, whereby the metal acts as a temporary centre and is removed in the final step, for example, the synthesis of cyclam (**2.2**) (Scheme 2.2(a)). Alternatively, ring expansion of a smaller ring system, known as an *endo-template*, can be used, as in the synthesis of 1,5,9-triazacyclododecane (**2.3**) (Scheme 2.2(b)). Out of the two synthetic methods the exo-template is by far the most common.

The template effect is demonstrated by the synthesis of [18]crown-6 (**2.4**), from triethylene glycol and the dichloroethyl-derivative of ethylene glycol when two different bases (either potassium carbonate or triethylamine) are employed (Scheme 2.3). A polymer (**2.5**) is formed when triethylamine is used as the base. When potassium carbonate is used in the reaction, the oxygen atoms coordinate to the K^+ ion in solution, thus acting as an exo-template. This coordination brings the reactive ends of the two glycol chains into close proximity, and thus they are able to ring-close and the cyclic product is formed. Strictly, this reaction is an example of the *kinetic template effect* in which product formation is irreversible and the reaction is kinetically controlled. The role of the template is to enhance the rate of formation of the cyclic product by stabilising the intermediate to its formation.

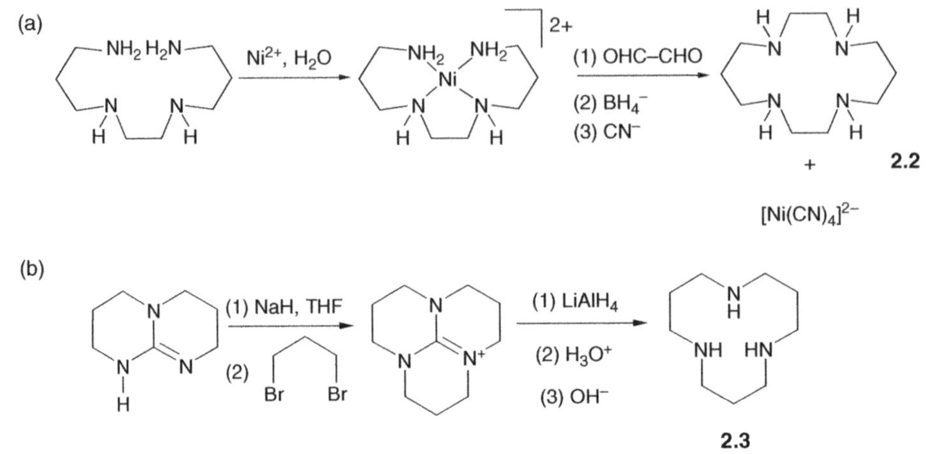
(a)
Ni2+, H2O
2+
(1) OHC–CHO
(2) BH4−
(3) CN−
+
2.2
[Ni(CN)4]2−
(b)
(1) NaH, THF
(2)
Br
Br
(1) LiAlH4
(2) H3O+
(3) OH−
2.3

Scheme 2.2 The two templated synthetic approaches to azo-macrocycles: (a) exo-template effect; (b) endo-template effect.

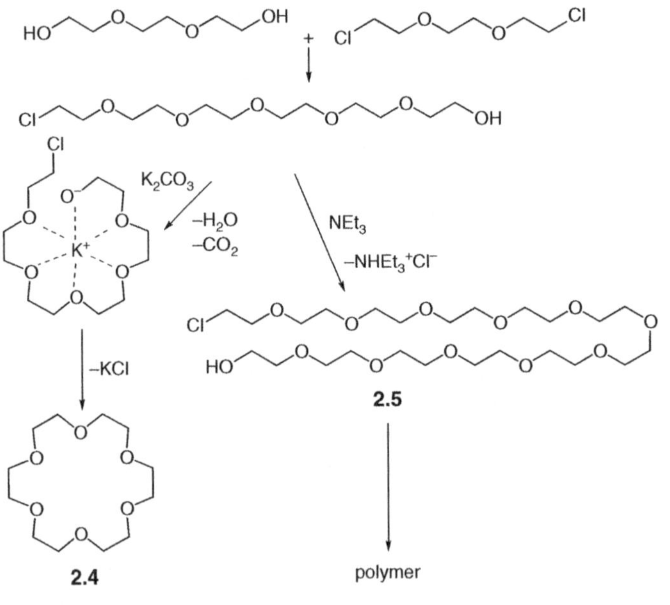
K2CO3
−H2O
−CO2
NEt3
−NHEt3+Cl−
−KCl
2.5
2.4
polymer

Scheme 2.3 The kinetic template effect in macrocycle synthesis. A polymer is formed with Et_3N, whereas the desired macrocycle is formed when K_2CO_3 is used as the template.

In contrast, the *thermodynamic template effect* involves a particular template species (usually a transition-metal ion) binding to a ligand that is complementary to itself, within an equilibrating mixture of products that are *formed without the involvement of the template*. The binding of the template thermodynamically stabilises the most complementary product (usually a macrocyclic compound). An excellent example is the preparation of phthalocyanine (**2.8**). Treatment of 1,2-dicyanobenzene with either boron trichloride or uranyl chloride results in two different-sized macrocycles (**2.6** and **2.7**, respectively) (Scheme 2.4). Macrocycles **2.6** and **2.7** are themselves only stable when the template is still present. On the removal of the template, the normal phthalocyanine (**2.8**) is formed, which is highly stable and forms many coordination complexes with a range of transition-metal ions.[5] This is also a very attractive synthetic procedure for the preparation of unsymmetrical phthalocyanines.

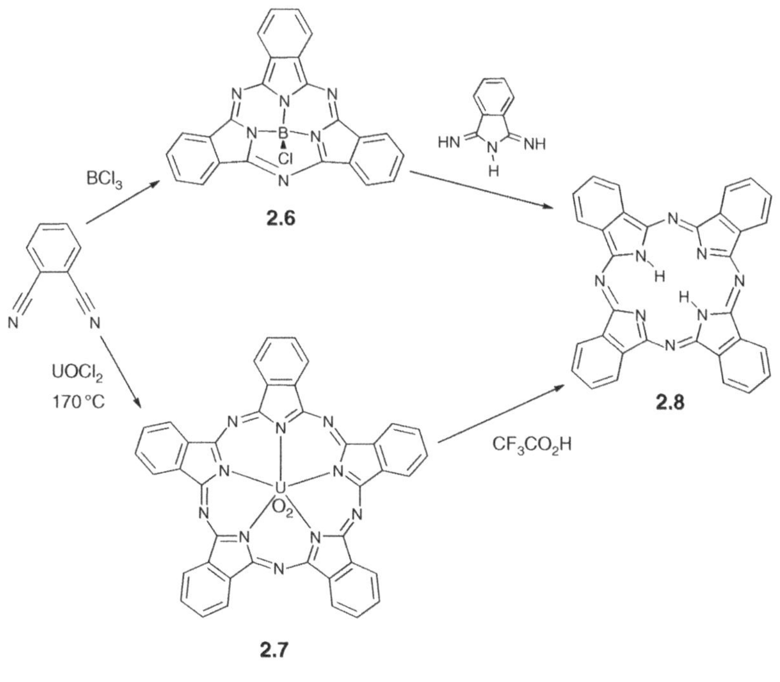

Scheme 2.4 The thermodynamic template effect for the preparation of phthalocyanine (**2.8**).

A major disadvantage of employing the template effect is the need to remove the template once the macrocycle has been formed, without destroying the product. This is not necessarily an easy task and is, in some cases, impossible. There are a number of effective demetallation strategies, depending on the stability of the metal–macrocycle complex (or more generally of the template–host complex).

For example, if the metal is labile or weakly bound to the macrocycle (as in K^+ complexes of the crown ethers), then the complex may be dissolved in an organic solvent and washed with water as the water will solvate and extract the metal salt. However, if the metal ion is bound strongly within the cavity it can be removed by using strongly coordinating ligands, such as the cyanide ion, which has been used to demetallate Cu(I)-containing catenates (see Chapter 3, Section 3.4.3) to give the very stable $[Cu(CN)_4]^{3-}$. An alternative method involves redox reactions; for example, Schiff base metal macrocycle complexes can be reduced to the corresponding metal-free amine (Section 2.3.5). Finally, it is possible to change the oxidation state of the metal template from a chemically inert configuration such as low-spin Co(III) to a more labile configuration, such as Co(II), which is easily washed out.

2.3   Cation binding

Hartley, J. H., James, T. D. and Ward, C. J., 'Synthetic receptors', *J. Chem. Soc., Perkin Trans.*, 1, 2000, 3155–3184.

2.3.1   Introduction

Cation complexes play an essential role in many biological systems; large quantities of sodium, potassium, magnesium and calcium ions, in particular, are all critical to life. For example, a concentration gradient of K^+ and Na^+ across biological cell membranes is vital to nerve signal transduction. Elucidation of the structure of a K^+ ion transport protein, along with work on Cl^- transport, resulted in the award of a share of the 2003 Nobel Prize in Chemistry to Roderick MacKinnon for his work in this area. Similarly, magnesium is a vital component in nucleotide hydrolysis and as a pigment component in photosynthesis, while calcium is intimately linked with cellular signalling and control. Cations also play important roles in the environment where build-up of toxic heavy metals, such as lead, mercury and cadmium, is a growing concern. The presence of these metals in soil results in their uptake by plants. Even in small amounts, they can be potentially concentrated in the food chain, resulting in significant doses to humans. It is of critical importance to be able to remove these ions from the environment or prevent their release in the first place. Work on ligands for heavy-metal remediation and for chelation therapy is therefore an active area of research. Metal chelate ligands are also highly important in biomedical imaging and radiotherapy applications, *e.g.* using ^{99}Tc and ^{60}Co, while, from an industrial perspective, extraction of more valuable metal ions from a mixture of species is a cost-effective way to improve many industrial processes, particularly those involving valuable platinum group metal catalysts.

2.3.2 The early years

The discovery of dibenzo[18]crown-6 (**2.9**), by Charles J. Pedersen[6] was a key step in the development of the discipline of supramolecular chemistry. Dibenzo[18]crown-6 (which comprises an 18-membered ring with six oxygen donor atoms) was discovered serendipitously as a side-product during the attempted preparation of a bis(phenol) derivative (**2.10**) (Scheme 2.5). The formation of the crown ether was a fortuitous consequence of the presence of an alkali metal-containing base and hence the operation of the template effect (Section 2.2.2). Pedersen immediately began to investigate how various metal cations bound to the new macrocycle and a range of derivatives. Subsequent work has resulted in the preparation of a wide variety of analogues, including amine derivatives (*azacrowns*) and macrocycles with thioether binding sites (*thiacrowns*). Collectively, the family of crown ethers and related heteroatom donor macrocycles have been termed the *corands*.

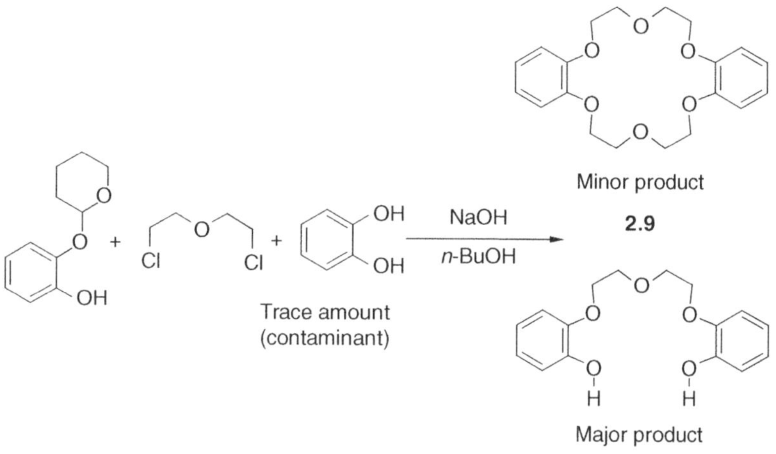

Scheme 2.5   Pedersen's first synthesis of dibenzo[18]crown-6 (**2.9**).

Shortly after the original crown ether work, Jean-Marie Lehn realised that donor atoms can be situated within a three-dimensional array to completely encapsulate the ion from the outside medium.[7] Lehn and co-workers prepared a range of bicyclic systems named 'cryptands', one of the first of which was [2.2.2]cryptand (**2.13**). Each cryptand is denoted by the number of heteroatoms that are incorporated in the bridges linking between the bridgehead atoms (normally nitrogen atoms). Therefore, [2.1.1]cryptand, [2.2.1]cryptand and [2.2.2]cryptand contain four, five and six oxygen atoms, respectively (**2.11–2.13**).

The cryptand cation-binding hosts are closely analogous to the first hosts for simple anions, the *katapinands* (**2.14**), first synthesised by Simmons and Park in 1968.[8] Katapinands are bicyclic amines that are able to bind halides. The name, 'katapinand', is derived from the Greek word meaning 'swallow up' or 'engulf'. The katapinands contain a molecular cavity determined by the chain length joining the bridgehead nitrogen atoms and, when the nitrogen atoms are protonated, become strong, positively charged hydrogen bond donors. Halide anions are bound by the *in, in*-geometry in which both NH^+ groups are orientated toward the interior of the cavity. One of the earliest examples of such katapinands was structurally authenticated by X-ray crystallography, and was found to have a chloride anion encapsulated within the cavity.

[2.1.1] Cryptand   [2.2.1] Cryptand   [2.2.2] Cryptand

2.11   **2.12**   **2.13**

Katapinands

2.14   **2.15**

In 1987, Pedersen and Lehn shared the Nobel Prize for the development of Supramolecular Chemistry with a third recipient, Donald Cram,[9] also very well-known for his work in carbanion chemistry. Cram realised that the chemical reactivity of a carbanion could be increased by the separation of a counter-cation from its anion. This separation was achieved by fully encapsulating the cation with cyclic polyethers and represents an example of the *naked anion effect*. This is where the anion has been 'stripped' away from its counter-ion, such that there is little or no association between the anion and cation. Cram and his team went about synthesising cyclic polyethers in which the oxygen atoms comprise a small cavity that is rigidly preorganised in an octahedral array. These macrocycles are termed *spherands*, of which the Li^+-selective spherand-6 (**2.15**) is one of the best-known examples.

Since the work of Pedersen, Lehn and Cram, an enormous amount of research has been conducted using macrocycles for binding cation, anion and neutral species, sensing and catalysis, as well as the synthesis of self-assembled rotaxanes, catenanes and knots (see Chapter 3). Subsequent sections give a brief overview of some of the most important classes of compound.

2.3.3   Crown ethers, lariat ethers and cryptands

It was initially proposed that there is an optimal spatial fit between crown ethers and particular cations. It is true that [18]crown-6 is selective for K^+, whereas the larger [21]crown-7 has a higher affinity for Rb^+ and Cs^+ than K^+. However, modern understanding of these systems has somewhat modified this simple 'size-fit' idea, particularly because of the flexibility of the crown ethers – their 'size' is not a constant. Table 2.1 shows the binding constants obtained for a selection of cations with various crown ethers.

Table 2.1   Binding constants obtained for various cations and a selection of crown ethers (log K, methanol, 20 °C)

Crown ether	Na^+	K^+	Rb^+	Cs^+	Ca^{2+}	NH_4^+
[12]crown-4	1.70	1.30	—	—	—	—
[15]crown-5	3.24	3.43	—	2.18	2.36	3.03
[18]crown-6	4.35	6.08	5.32	4.70	3.90	4.14
[21]crown-7	2.52	2.35	—	5.02	2.80	3.27
Benzo[18]crown-6	4.30	5.30	4.62	3.66	3.50	—

Complementarity between the crown ether and guest has been demonstrated by X-ray crystallography. Figures 2.3(a–c) show the X-ray structures of [18]crown-6 with Na^+, K^+ and Cs^+, respectively. The Na^+ complex shows that the crown ether has wrapped itself around the metal ion to maximise the electrostatic interactions to the ligand; however, the small size of the Na^+ ion means that the Na–O distances are much longer than optimum and in fact the structure of $[Na([18]crown-6)]^+$ is highly dependent on the nature of the counter-anion and degree of hydration in the solid state. The larger K^+ ion is generally found to sit within the cavity of [18]crown-6, forming a one-to-one complex with the $K^+ \cdots$ oxygen distances being approximately equal and optimum. Some 'doming' of the K^+ ion out of the macrocycle ring is noted, depending on the identity of the counter-ion. In contrast, the Cs^+ ion does not fit within the cavity but instead 'perches' above the ring. Many X-ray crystal structures show that the large Cs^+ ion is often found to bind between two crown ethers, forming a 'sandwich' complex. Solid state structures have also shown that two small metal ions, such as lithium, are able to bind within the cavity of [18]crown-6 (Figure 2.3(d)), while larger crown ethers, such as dibenzo[30]crown-10, can either wrap about K^+ in a three-dimensional fashion or encompass two Na^+ ions.[10]

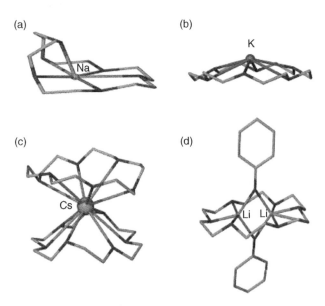

Figure 2.3 X-ray crystal structures of [18]crown-6 containing (a) Na^+, (b) K^+, (c) Cs^+ and (d) two Li^+ ions (phenolate salt). Na–O bond lengths are significantly longer than optimal.

In reality, the 'size-match' criterion does not bear up to close scrutiny. The crown ethers are highly flexible molecules and while there is some relationship between size-match and binding constants, the magnitude of log K also depends very much on the charge on the metal cation, the chelate ring size (five-membered rings favour K^+; six-membered rings favour Li^+) (see Chapter 1, Section 1.2.3) as well as the nature of the solvent, degree of electronic complementarity of the metal with the donor atoms (hard–soft acid and base theory[11]) and the nature of the counter-anion.

Lariat ethers are a class of compounds that contain a single side arm attached to a macrocycle, often by the nitrogen atom of an azacrown.[12] This side arm is flexible (as in a podand) but contributes to the envelopment of the guest cation in a three-dimensional array, as in the cryptands, while the macrocycle exerts stronger complexation than in podand systems. The *pseudo*-podand arm comes over the face of the macrocycle and binds to the cation that has been encapsulated within the host, as demonstrated in Figure 2.4.

The lariat ethers retain much of the flexibility and hence fast binding and decomplexation kinetics of podands. The three-dimensional binding serves to isolate the guest more effectively from the surrounding medium and enhance the binding constants (Scheme 2.6). While the preorganised cryptands and particularly spherands have even higher binding constants, as

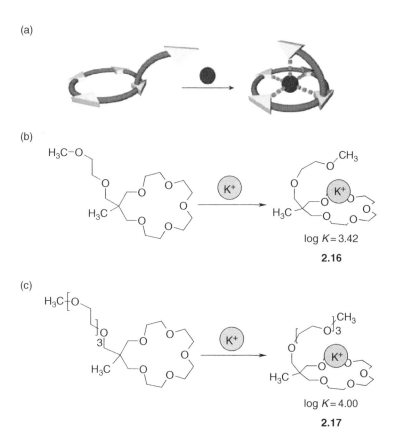

Figure 2.4 (a) Schematic representation of a macrocycle containing one side arm, (b) the side arm, stabilises the host–guest complex by binding to the cation within the cavity and (c) a longer lariat side arm attached to a carbon (binding constants were obtained in MeOH at 25 °C).

the cation is totally enveloped in a three-dimensional cage, their binding kinetics are much slower.

The lariat arm is usually introduced through substitution of a tertiary amine nitrogen atom within the macrocycle (*viz.* **2.21**), although C-backbone derivatives such as **2.17** have also been prepared. The stability of lariat ether complexes is varied; the binding constants for Group 1 and 2 elements in the Periodic Table are generally lower than those exhibited by the crown ethers, which have all oxygen donor atoms and are hence more complementary to 'hard' cations. However, they are higher than podands and analogous azacrowns. Thus, potassium is bound effectively by [18]crown-6 in methanol (log $K = 6.08$), whereas the same guest binds to the podand, pentaethylenegycol, with much lower affinity (log $K = 2.3$). The lariat ether falls in between these extremes, log $K = 4.8$, a value noticeably larger than that measured for the comparable diaza[18]crown-6 (log $K = 2.04$).

If the cation is exchanged for an ammonium ion instead of an alkali-metal, the complex stability drastically increases in favour of the lariat ether over the

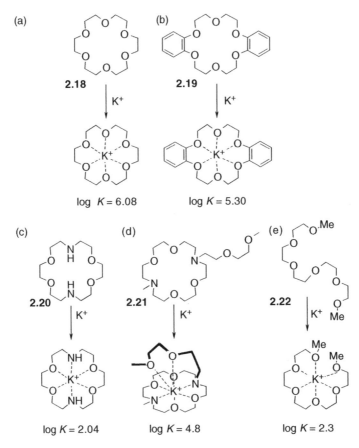

Scheme 2.6 The binding constants (MeOH, 25 °C) of a potassium ion bound to: (a) [18]crown-6; (b) dibenzo[18]crown-6; (c) diaza[18]crown-6; (d) a lariat ether; (e) the podand pentaethyleneglycol.

crown. This shift is due to the directionality of the NH bonds in NH_4^+. The interaction is now of a charge-assisted hydrogen bonding type (hydrogen bonding involving formally charged species) as opposed to the much less directional ion–dipole interaction. In solution, the binding constants for NH_4^+ complexation by [18]crown-6 (**2.18**) and the lariat ether **2.23** in methanol, are $\log K = 4.35$ and 4.75, respectively. This contrast may be explained by recognising that the ammonium ion engages in three hydrogen bonding interactions with the oxygen atoms in [18]crown-6 (or bifurcated interactions to pairs of oxygen atoms). One of the NH functionalities is not involved in binding. In the case of the lariat ether, the pendant side arm is able to come over the face of the macrocycle and hydrogen bond to the remaining NH functionality, forming a more stable complex with an additional hydrogen bonding interaction (Figure 2.5).

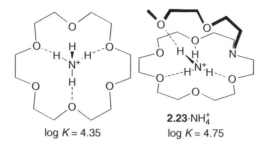

2.23·NH₄⁺
log K = 4.35   log K = 4.75

Figure 2.5 Binding constants and possible geometries for (a) [18]crown-6 (**2.18**) and (b) the lariat ether **2.23** binding NH_4^+ in methanol.

Diaza[18]crown-6 (**2.20**) can also be readily functionalised with two side arms, instead of just one as in **2.21**, to form double lariat ethers termed *bibracchial lariat ethers* (BiBLEs), such as **2.24**. This modification results in the possibility of the side arms covering both faces of the macrocycle (Figure 2.6(a)). A particularly striking example is the complexation of sodium in a BiBLE that contains two thymine groups as the side arms (Figure 2.6(b)).

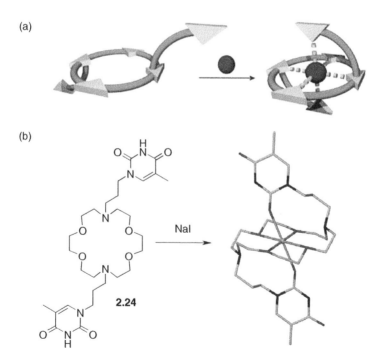

Figure 2.6 (a) Schematic showing a guest species being bound by two pendant side arms in a bibracchial lariat ether (BiBLE) and (b) complexation of sodium in the BiBLE **2.24** which contains two thymine groups as the side arms.

The lariat concept is extremely powerful. To take just one example, side arms have been appended to more complex systems, such as the members of the cyclodextrin family (Section 2.7.5), as in compound **2.25**. The fluorescent indole group of the L-tryptophane-derived lariat arm in this compound is encapsulated within the cavity of the cyclodextrin in aqueous solution by hydrophobic effects. On the addition of a guest, the lariat arm is displaced and the guest takes its place within the cavity, as illustrated in Figure 2.7.

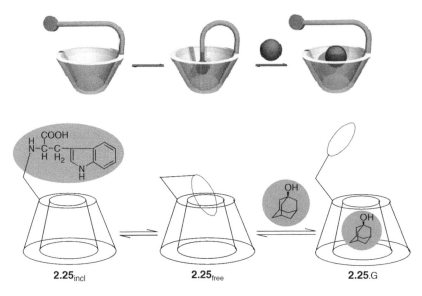

2.25$_{incl}$ **2.25**$_{free}$ **2.25**.G

Figure 2.7 A lariat arm attached to a beta-cyclodextrin scaffold. The arm occupies the central cavity until the guest displaces it.

This lariat arm displacement in **2.25** has been used as a 'fluorescent handle' to monitor host–guest complexation. Fluorescent titration shows that **2.25** is able to distinguish between chemically distinct alcohol guests and also between enantiomers and geometrical isomers, with 2-adamantanol giving the most stable complex because of a good size match with the cyclodextrin cavity and strong hydrophobic interactions.

Since the initial preparation of the cryptands (Section 2.3.2), many analogues have been synthesised using high-dilution techniques, but the most important of the cryptands is [2.2.2]cryptand, commercially available as an ion-sequestration agent under the name of Kyptofix® (**2.13**). The cavity size of the [2.2.2]cryptand is similar to the cavity size of [18]crown-6 and is also a good host for K$^+$. Due to the three-dimensional nature of the host, the cryptand encapsulates the metal ion and shields it from the outside environment. The bicyclic structure renders it more preorganised and hence the binding constant for K$^+$ is significantly greater than that of [18]crown-6. The binding enhancement arising from the preorganisation of bicyclic species was originally termed the *macrobicyclic effect* (see Chapter 1, Section 1.2.4) but is simply a manifestation of the preorganisation principle. For example, the binding constant in MeOH of the [2.2.2]cryptand (**2.13**) for K$^+$ is 10^4 times larger than [18]crown-6. The smaller cryptand, [2.2.1]cryptand (**2.12**), is selective for Na$^+$. Due to their lower flexibility and greater degree of preorganisation, cryptands display *peak selectivity*, in which binding constants are at a maximum for a particular metal ion. This is in contrast to the crown ethers, which deeply *plateau selectivity* since they often bind a number of metal ions equally well.

2.3.4   Spherands, hemispherands, cryptaspherands, heterocrowns and heterocryptands

Spherands are the most preorganised of the macrocyclic ligands and are very rigid. Their extreme preorganisation is demonstrated by the binding constant of spherand-6 (**2.15**) for Li$^+$ and Na$^+$, which are in excess of 16 and 14 log units, respectively, in water-saturated chloroform. This affinity represents a remarkably strong interaction with these highly labile ions and arises, in part, from the fact that inter-oxygen repulsions present in the free host are ameliorated when the metal ion is inserted into the spherand cavity. Compare these values to the binding affinities of alkali-metal ions for crown ethers (Table 2.1). Unlike the crown ethers and even the cryptands, the *binding activation energy* (the energy required for a metal ion to force its way into the cavity) is extremely high as a consequence of the rigid nature of the spherand. Hence binding kinetics, for *spheraplexes* (complexes of spherands) are 10^5 slower than for the crown ethers. Decomplexation of metal ions from spheraplexes is also extremely slow at 25 °C, of the order of days. Spheraplexes demonstrate how the extent of preorganisation influences the binding power of host–guest systems. To date, there have been many derivatives of spherands prepared and these include hybrids of spherands with crown ethers, podands and cryptands, to form *hemispherands*, *e.g.* **2.26** and *cryptaspherands*, *e.g.* **2.27**.

Thus far, the types of host molecules that have been discussed contain either an oxygen or nitrogen atom within the scaffold. Another common donor atom that is incorporated into the scaffold is sulfur. Oxygen, nitrogen and sulfur are the three most common heteroatoms used to prepare *heterocrowns* or *heterocryptands* (*e.g.* **2.28–2.33**). By selectively exchanging the type of atom within a crown ether scaffold for softer donor atoms there is a dramatic change in the selectivity of the host for softer metal cations. For example, Ag$^+$ is a large, polarisable, heavy-metal ion that forms more stable complexes with soft donor atoms, such as sulfur.[13]

X = CO$_2^-$
Y = OMe

2.26   **2.27**

2.28   **2.29**   **2.30**

2.31   **2.32**   **2.33**

This effect is demonstrated by comparing the structures of [Ag([18]crown-6)]$^+$ and the analogous [Ag([18]aneS$_6$)]$^+$ complexes (Figure 2.8).

The geometry around the Ag$^+$ ion in the oxygen crown ether is clearly distorted in comparison to the analogous sulfur system. The sulfur crown is more flexible and the bonding more directional and hence the geometry around the metal centre is approximately octahedral. Lariat ether and BiBLE ligands, such as **2.34**, have even been prepared with alkene functionalities that bind both soft metals, such as Ag$^+$, and also, more surprisingly, alkali-metals, such as Na$^+$ and K$^+$, using cation–π interactions in the solid state (Figure 2.9).[14]

As with the heterocrowns, heterocryptands can also be prepared. For example, K$^+$ forms a stable complex with the [2.2.2]cryptand (**2.13**). If one or more of the oxygen atoms are substituted with NMe functionalities (**2.35–2.37**), the binding affinity for the K$^+$ ion is drastically reduced. However, the corresponding affinity for softer metal ions, such as the Ag$^+$ ion, increases.

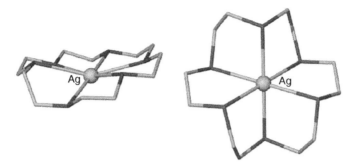

Figure 2.8 Silver(I) complexes of [18]crown-6 and [18]aneS$_6$. The hard–soft interactions between the silver and the oxygen atoms result in poor directionality. The analogous sulfur crown forms a recognisable octahedral coordination geometry around the silver ion (counter-anion not shown).

(a)   (b)

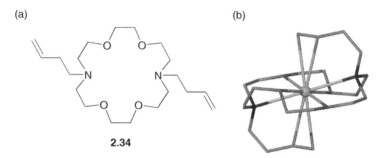

2.34

Figure 2.9 (a) The alkene BiBLE ligand **2.34** and (b) its Na$^+$ complex. The sodium ion is bound *via* cation–π interactions supported by the crown ether donor set.

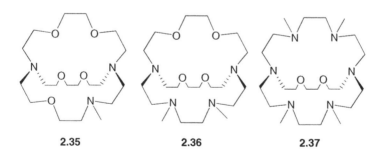

2.35   **2.36**   **2.37**

2.3.5 Schiff bases

A *Schiff base* is the name given to the imine formed when an aldehyde condenses with an amine, resulting in a loss of a water molecule. Many aliphatic imines are unstable with respect to hydrolysis (*i.e.* the formation of the Schiff base is reversible). As a result, Schiff base macrocycles are often reduced to the corresponding amine with a reducing agent, such as sodium borohydride. In contrast, many aryl imines are relatively stable due to the delocalisation of the electrons across the π-system. The imine nitrogen atom has a lone pair that can donate to a metal centre and thus this class of host molecules have been extensively used in cation binding, with early work dating well before the discovery of the crown ethers. Because imine formation is reversible, Schiff base macrocycles are generally formed by a thermodynamic-templated condensation in the presence of a metal salt (Section 2.3.5), as seen for the macrocycle and the lariat macrocycle Schiff bases **2.38** and **2.39**, respectively. Different-sized macrocyclic imines can be obtained depending on the metal used. One of the easiest ways to synthesise Schiff base complexes is to use a template Schiff-base condensation reaction between dicarbonyl compounds and diamines (Scheme 2.7, *cf.* Scheme 2.4). Schiff bases have found extensive application as models for biological systems.[15]

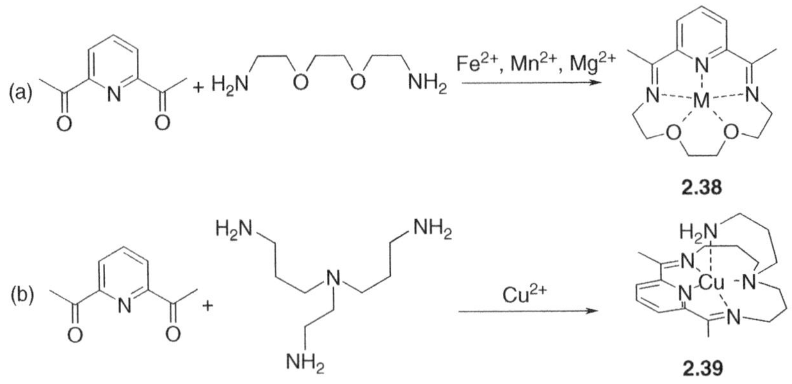

Scheme 2.7 Schiff-base condensation reactions between dicarbonyl compounds and diamines: (a) the formation of a macrocyclic Schiff base; (b) the formation of a lariat ether Schiff base.

2.3.6 Calixarenes

Gutsche, C. D., *Calixarenes*, The Royal Society of Chemistry, Cambridge, UK, 1989; *Calixarenes Revisited*, The Royal Society of Chemistry, Cambridge, UK, 1997.

The calixarenes are cyclic hosts synthesised by the condensation reaction between a *p*-substituted phenol and formaldehyde. The name comes from a Greek vase, called a *calix crater*, similar in shape to the bowl-like calix[4]arene (**2.40**). Using a number of different bases, varying thermal conditions and altering the ratio of reactants, it is possible to produce *p*-*t*-butylcalix[*n*]arenes (where *n* = the number of phenol-derived repeat units) for *n* = 4–16, with the most common macrocycles being *n* = 4 (**2.40**), 6 and 8. The tertiary butyl substituent blocks the *para* position of the phenol, hence preventing extensive cross-linking which would otherwise result in the formation of Bakelite-type phenol–formaldehyde polymers.

n = 4, 5, 6, 7, 8, ...16
R = H, *t*-butyl, etc.
Y = wide range of functional groups

upper rim

lower rim

2.40   (*n* = 4)

The calixarene framework is very versatile and many derivatives have been prepared by functionalising the groups on the 'upper' and 'lower' rims (the upper, or wide rim is where the *t*-butyl substituents are located, while the hydroxyl groups are on the lower or narrow rim). By selectively changing the framework, chemists have been able to design hosts capable of binding cations, anions, neutral species or simultaneously combining different guests, such as toluene and Na^+ (Section 2.6.4).

The *p*-*t*-butylcalix[4]arene (**2.40**) has been drawn in a 'cone' conformation in which the hydroxy groups are pointing in the same direction. In solution, there is a high degree of rotation around the CH_2 groups connecting the phenol rings, giving four possible isomers of the calix[4]arenes. At room temperature, all four are in equilibrium (Figure 2.10). Variable-temperature (VT) NMR spectroscopic experiments have been carried out on the unsubstituted tetraphenol **2.41** in a mixture of $CDCl_3$ and CD_3CN, so making it possible to 'freeze-out' the partial-cone and cone isomers on the ^1H NMR spectroscopic time-scale by lowering the temperature.

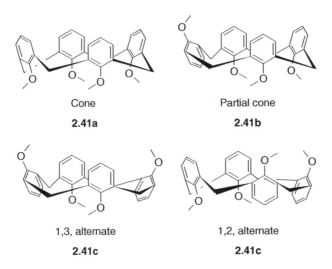

Cone

2.41a

Partial cone

2.41b

1,3, alternate

2.41c

1,2, alternate

2.41c

Figure 2.10   The four different conformers of calix[4]arenes.

2.3.7   Biological ligands: ion channels and siderphores

Interest in modelling cation transport across biological cell membranes by ion channels is significant, particularly with the 2003 Nobel Prize in Chemistry being awarded for the X-ray crystal structure determination of biological K^+ and Cl^- channels.[16] A model calixarene system for a K^+-gated ion channel has been developed by coupling two calixarenes together with ethyleneoxy chains (**2.42**) to give a bis(calixarene) 'tube'. This 'calixtube' selectively binds K^+ over any other alkali- or alkaline-earth metals; $K = 4 \times 10^4 \, M^{-1}$ in chloroform/water. This selectivity is two orders of magnitude over other metal cations. Molecular mechanics

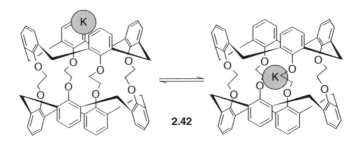

2.42

Scheme 2.8   Potassium-ion complexation by the compound **2.42**.

calculations suggest that the K^+ transport to the centre of the cavity is *via* a two-step process (Scheme 2.8). This work has led to an array of calixtubes that are able to shuttle ions through the tube.[17]

Another important natural cation-binding process is the detection and bioaccumulation of the trace element Fe(III).[18] For example, iron is essential for all microorganisms that live in aerobic conditions, which need the element for biological functions, including reduction of oxygen for synthesis of ATP, reduction of ribotide precursors of DNA and for the formation of heme-units for oxygen transport. The majority of iron found in nature is 'tied up' within a number of enzymes and proteins that are involved either in iron storage or transport. Fe(III) is virtually insoluble under physiological conditions; the solubility of $[Fe(H_2O)_6]^{3+}$ is 10^{-18} M^{-1}. For optimum growth to occur, microbes require iron concentrations in the micromolar range. When a cell becomes iron-deficient, the organism produces low-molecular-weight organic compounds, called *siderophores* (from the Greek, meaning 'iron carriers'). The latter are a class of naturally occurring, low-molecular-weight compounds that are ferric-ion-specific chelating agents, utilised by many microbes. Siderophores are unique ligands that selectively bind Fe(III) ($K_a > 10^{30}$ M^{-1}) and are able to transport the iron across the cell membrane.

Many siderophores are three-armed podands that contain hydroxamates or catechol moieties which bind to the metal ion. Siderophore iron(III) complexes are high-spin and are highly thermodynamically stable. The highest stability constant for a natural siderophore is for enterobactin (**2.43**), whose affinity for iron(III) is 10^{52} M^{-1}. The iron(III) ion is totally enveloped by the catechol arms in a six-coordinate geometry (**2.43**). Artificial, macrobicyclic siderophores have achieved binding constants of up to 10^{59}.[19]

2.43   **2.43a**

$= Fe^{3+}$

2.4   Anion binding

2.4.1   Introduction

Anion binding is now a major area of endeavour. In contrast to cation binding, however, anion binding was relatively slow to develop, even though one of the earliest anion receptors can be traced back to the work of Simmons and Park in 1968 on the *katapinands* (*cf.* **2.14**). The design of cation and anion hosts uses the same criteria, *i.e.* specificity arising from the preorganised placement of complementary binding sites. However, there are some properties of anions that make the task a little more challenging, as follows:

1. Anions come in many shapes and sizes. Anions are generally larger than cations and therefore larger hosts are required to bind them. The smallest anion, F^-, has a radius approximately comparable to the radius of K^+. Generally, cations are spherical (except for organic cations, such as ammonium ions), but anions are found in various shapes, for example, spherical (halides), linear (SCN^-), planar (NO_3^-), tetrahedral (HPO_4^{2-}) and octahedral (PF_6^-). Biologically important anions, such as nucleotides and many proteins, have much more complex shapes.

2. Anions have high free energies of solvation compared to cations of similar size and hence hosts for anions experience more competition from the surrounding medium. For example, the standard free energies of hydration, $\Delta G_{hydr}°$, for F^- and K^+ are -465 and $-295\,kJ\,mol^{-1}$, respectively.

3. Generally, most anions exist in a narrow pH window. This can be problematic for hosts that contain, for example, polyammonium functionalities within the receptor, as the host may not be fully protonated in the pH range where the anion is deprotonated.

4. Many anions are generally coordinatively saturated and they only bind *via* weak interactions, such as hydrogen bonding and van der Waals interactions.

An obvious anion binding strategy is to use a positively charged host. However, electrostatic interactions are non-directional and all anions will bind to cations and form either a solvent-separated or contact-ion pair (Section 2.6). Anions will also bind to neutral receptors since there is a difference in electrostatic charge, although the forces between anions and cations are larger. It has been argued that neutral receptors, which do not come already burdened with a competing counter-anion, offer the possibility of enhanced selectivity. To overcome the non-directional nature of electrostatic interactions, an alternative approach would be the introduction of a Lewis-acidic site (*e.g.* organoboron, mercury or tin compounds), as most anions are Lewis-basic (Section 2.4.4). The Lewis acid–base interaction is both strong and directional. The Lewis-basic nature of anions also means that they are good hydrogen bond acceptors and both hydrogen bonding and Lewis-acid receptors have been extensively studied. Anions are also polarisable and therefore van der Waals interactions play a significant role in binding, particularly when the anion has been encapsulated within the cavity of a host system and exhibits a high degree of surface-area contact. Finally, the key role of solvent cannot be overstated. As with cations, the solvent medium drastically influences the host–guest affinity (as measured by the binding constant). Markedly higher binding affinity in charged systems is observed in non-polar solvents, *e.g.* chloroform, than in highly polar solvents, such as dimethyl sulfoxide or water, in which a lot of competition between the anion and the solvent is observed. However, in the case of hydrogen bonding receptors in particular, strong solvation of the binding sites can significantly compete with anion binding in polar media.

2.4.2   Charged receptors

 Bowman-James, K., 'Alfred Werner revisited: The coordination chemistry of anions', *Acc. Chem. Res.*, 2005, **38**, 671–678.

Electrostatic interactions

The most obvious way to bind anions is to design a host that contains a positive electrostatic charge. In most modern compounds, the non-directional and non-selective nature of the electrostatic interactions are generally modulated by additional directional interactions from functionalities such as hydrogen bond donors or acceptors (Section 2.4.3). Electrostatic interactions play an important role in understanding the factors that influence high binding affinities, particularly in biological systems in which there are a large number of recognition processes that involve charge–charge interactions. Indeed, electrostatic interactions are often the first interactions between the substrate and enzyme.

Early work showed that electrostatic interactions to a tetrahedral tertiary ammonium cryptand **2.44** result in a very effective host for iodide. This work was later extended to a series of zwitterionic, overall neutral cryptands (Section 2.7), containing either BH_3 (as in **2.45**) or a pendant carboxylate functionality.

A common building motif that is used in host systems for anions is 1,4-diazabicyclo[2.2.2]octane (DABCO). The quaternised DABCO motif represents a simple method of introducing a positive charge into the framework. For example, compound **2.46** contains three DABCO substituents around a planar framework. The result is a relatively unselective cationic host based solely on electrostatic interactions and which binds large anions such as $[Fe(CN)_6]^{3-}$ (Figure 2.11).

The simple DABCO-based podand **2.46** is not preorganised and a degree of flexibility is likely in solution due to free rotation around the methylene groups.

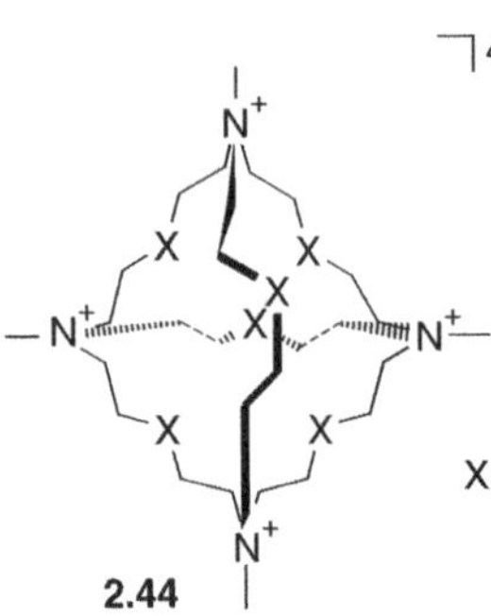

$X = (CH_2)_2$

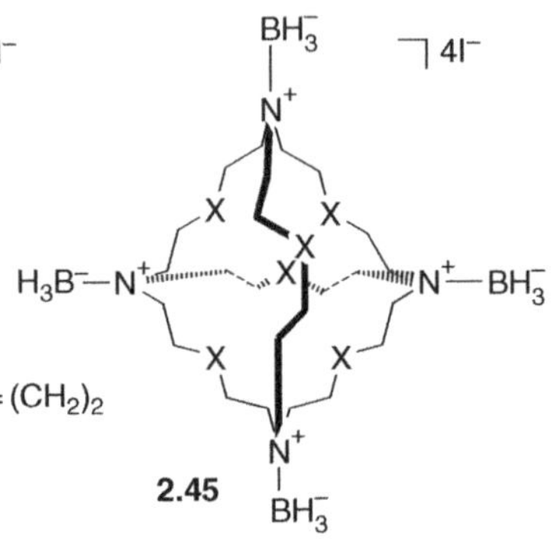

2.44   **2.45**

DABCO moiety

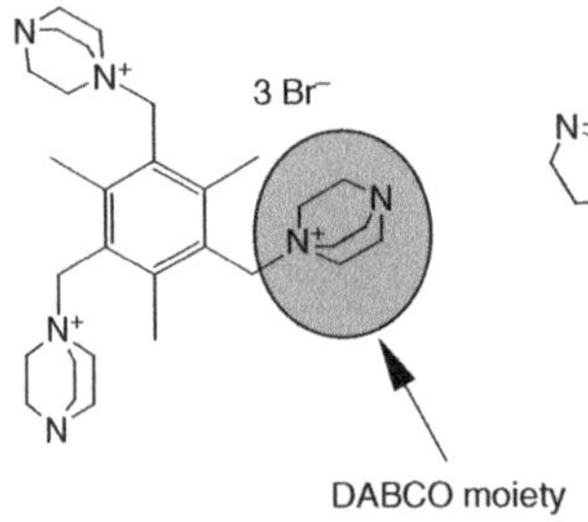

2.46 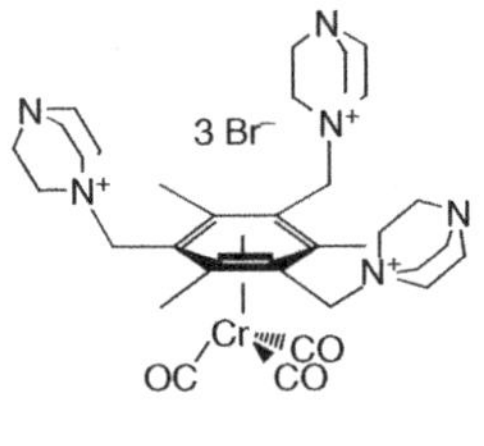   **2.47**

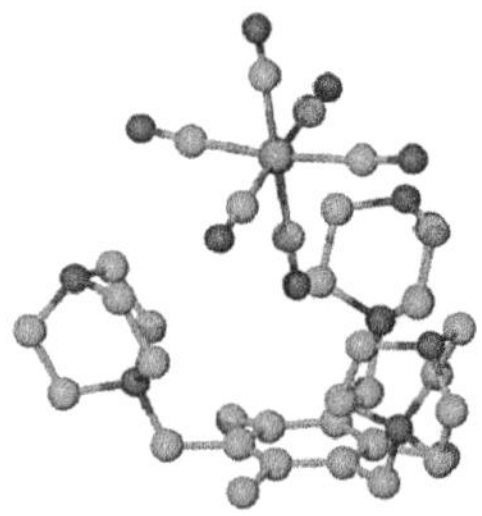

Figure 2.11 X-ray crystal structure of the 1:1 ferricyanide complex of the receptor **2.46**.

The free rotation can be controlled by the incorporation of a metal-containing moiety, such as chromium tricarbonyl, onto the aryl ring of the organic spacer group, as in **2.47**.

Electrostatic and hydrogen bonding interactions

Perhaps the most common class of anion binding host utilises more than one non-covalent interaction, in particular, a combination of electrostatic interactions and hydrogen bonding. This is readily achieved through the use of protonated nitrogen-containing macrocycles and macrobicycles (azacorands and cryptands), such as those shown in Figure 2.12. These kinds of polyammonium species have been used in the supramolecular catalysis of phosphoryl transfer chemistry (Section 2.6). The $R_2NH_2^+$ groups are complementary to the anionic phosphatyl residues. The azacrown **2.48**, for example, has binding constants (log K) of 3.4, 6.5 and 8.9 for AMP^{2-}, ADP^{3-} and ATP^{4-}, respectively. These polyamine macrocycles also bind to polycarboxylates, with similar binding constants. Monocyclic azacorands such, as **2.49**–**2.51**, are not very selective towards simple oxoanions in aqueous solution because water is highly solvating and has a high dielectric constant, and therefore competes strongly with the host. However, as the charge of the anion increases, the affinity also increases. This explains why important biological

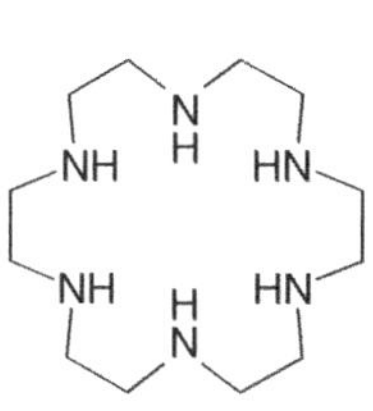

2.48

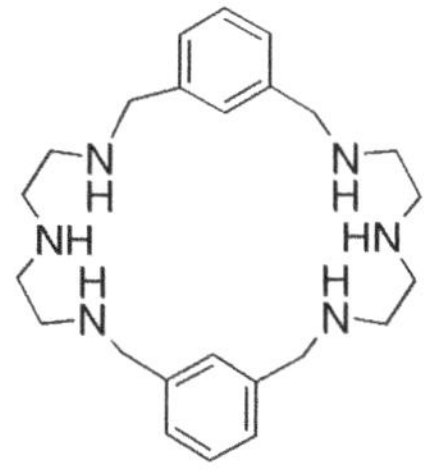

2.49

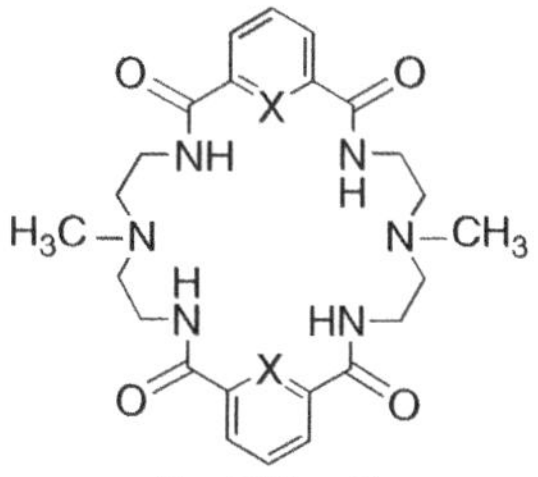

X = CH or N

2.50

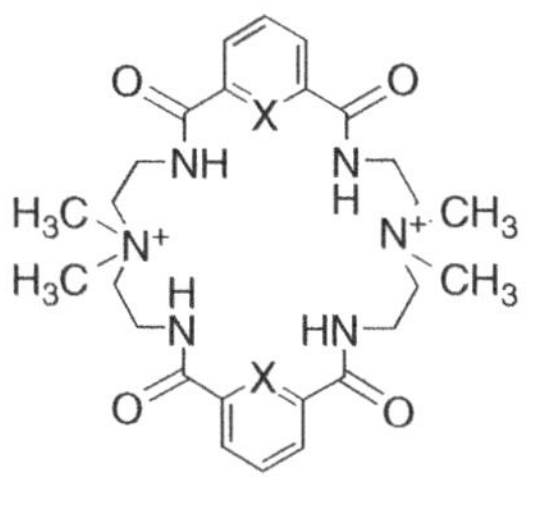

X = CH or N

2.51

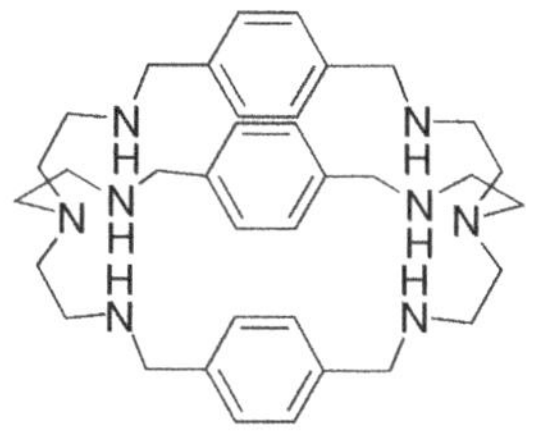

2.52

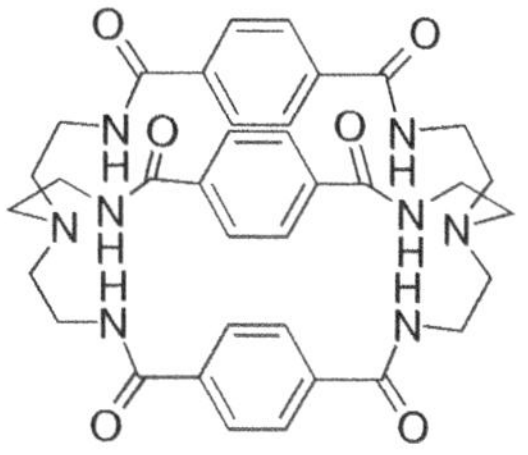

2.53

Figure 2.12 Some common aza and azooxa-polyammonium macrocycles.

anions have very large binding affinities. Bicyclic hosts are more preorganised and less strongly solvated, for example, receptors **2.52** and **2.53**, but are too small to include large biological anions. The binding affinities for the protonated receptor [**2.49**–H_4]$^{4+}$ for nitrate and sulfate are log $K = 2$ and over 4, respectively. The analogous protonated [**2.52**–H_6]$^{6+}$ has a higher binding affinity for the same two anions, log $K = 3$ and 4, respectively. The importance of hydrogen bonding together with the positive charge from the quaternised amine means that receptor **2.51** shows a higher affinity for most anions in comparison to **2.50**. For example, the affinities of **2.50** and **2.51** for dihydrogen phosphate in DMSO-d_6 are log $K = 2.92$ and 4.06, respectively (X = CH). This result demonstrates the importance of complementary non-covalent interactions (hydrogen bonding and electrostatic) and also the importance of preorganisation.

Surprisingly, introducing the stronger hydrogen bond-donor amide functionalities within the framework of the receptor, as in macrocycle **2.51** and the macrobicycle **2.53**, does not enhance the binding affinity. The binding constant of sulfate by receptor **2.51** is log $K = 3$, while the binding constant for protonated **2.53** is an order of magnitude less. This difference might be attributed to the extensive hydrogen bonding occurring between the amide oxygen atoms and adjacent amide NH groups, preventing the sulfate anion from entering the cavity, and competition from the water solvent.

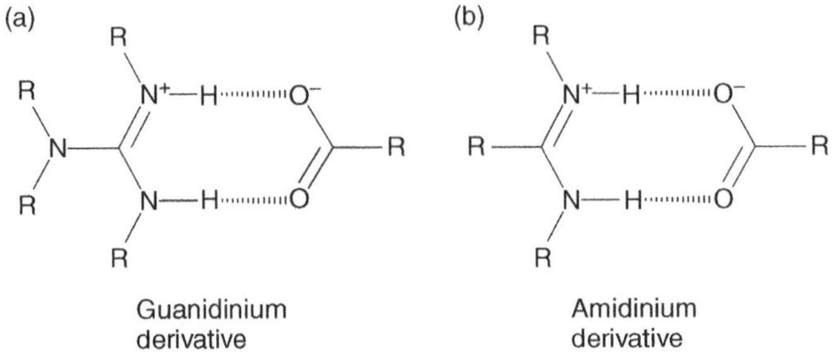

Figure 2.13 Carboxylate anion binding by (a) guanidinium and (b) amidinium derivatives.

Some of the simplest but most effective anion hosts contain guanidinium or amidinium motifs (Figure 2.13). These functionalities are highly polar and hence effective hydrogen bond donors, and consequently the guanidinium moiety, in particular, has become a very popular building block in the design of anion complexation hosts. An important property of guanidinium moieties is that they are generally water-soluble; an attractive attribute if the receptor is to be used under physiological conditions. Guanidine is readily protonated (pK_a 13.5) and is thus positively charged over a large pH range. Indeed, the guanidinium moiety is ubiquitous in natural anion binding proteins, such as carboxypeptidase A and *Yersinia* protein tyrosine phosphatase in the form of the amino acid arginine. Guanidinium-based anion receptors often exhibit strong anion complexation, even in highly competitive solvents, such as water and dimethyl sulfoxide. Indeed, even simple guanidinium-based podands, such as *m*-xylylene diguanidinium (**2.54**), have been shown to act as effective anion receptors for anions such as sulfate. Receptors **2.55a–2.55c** are simple hosts that have been found to be effective hosts for carboxylates or the phosphate group, due to the presence of double hydrogen bonding interactions.

A great deal of work has been carried out on tripodal species such as **2.56** which contains an array of guanidinium functionalities situated around an aryl core. The receptor is laden with hydrogen bond-donor moieties that are effective at

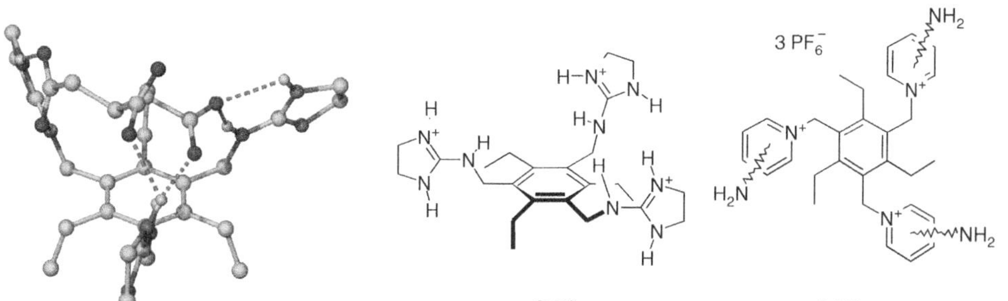

2.56

$3 PF_6^-$

2.57a = *meta*
2.57b = *para*

Figure 2.14   X-ray crystal structure of the citrate complex of **2.56**.

binding polycarboxylates, such as citrate (Figure 2.14). The host–guest interaction is then enhanced by the positive charge on the guanidinium side arms. The role of the ethyl groups is to preorganise the podand arms into a '3-up and 3-down' arrangement, producing an anion-chelating conformation in which all of the hydrogen bond donors can converge on the guest. The energetic preference for alternation about a hexasubstituted aryl ring imparts ca. 15 kJ mol^{-1} additional binding energy to the complex.

The binding of citrate by **2.56** has been investigated by a technique known as *Indicator Displacement Assay* (IDA). Binding of an indicator competes with the substrate for the same binding site *via* rapid, reversible complexation. The indicator is bound by the receptor, and exhibits a characteristic electronic absorption maximum (λ_{max}), indicative of the micro-environment of the receptor. Upon addition of a guest (citrate anion in this example), the indicator is displaced into free solution, changing the wavelength of the absorption maximum due to the enhanced solvation of the indicator (Scheme 2.9). The process is particularly effective when the indicator is chemically related to the target guest species and the sensitivity of the system can be tuned by close attention to the relative affinity of the host for indicator and analyte. In Chapter 5 (Section 5.3.6), we will explore how such assays can be used to produce electronic microarray sensors.

At neutral pH, citrate has a charge of three negative and the three carboxylate groups act as two-point hydrogen bonding-acceptor moieties. The substituted triguanidinium host is designed such that the three hydrogen bonding-donating guanidinium groups can align in a complementary way to the acid acceptor groups in citrate to give a binding constant for **2.56** with citrate, in a methanol/water mixture in an approximately neutral buffer solution, of 6.2 $\times 10^4$ M^{-1}. Host **2.56** has been shown to bind citrate in the presence of different indicators. An example is xylenol orange which has structural features similar to citrate. The free host **2.56** has an electronic absorption band (λ_{max}) at 445 nm. On the addition of citrate to **2.56** in the presence of xylenol orange, the band at 577 nm, assigned to

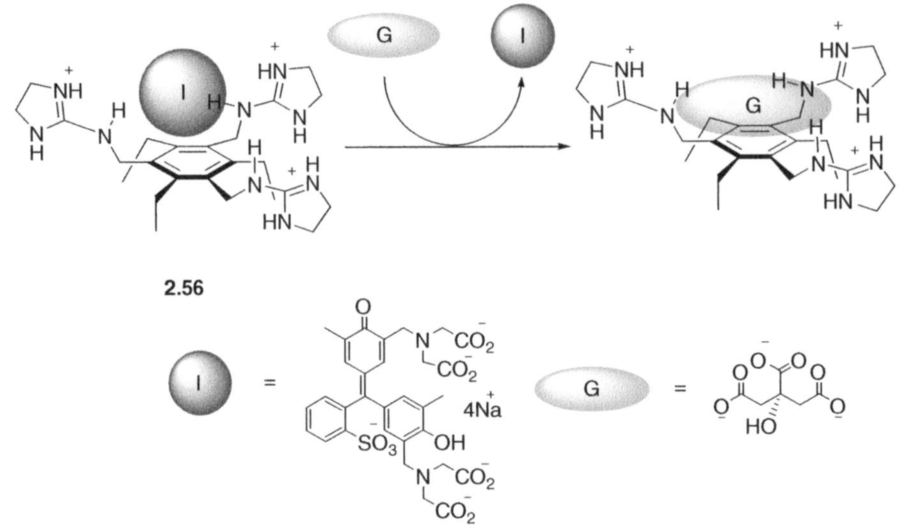

2.56

Scheme 2.9 Measurement of complexation equilibria by the indicator (I) displacement assay (IDA) technique in the trisguanidinium host **2.56** designed for citrate guest (G) binding.

the xylenol orange bound to the receptor, decreases and a new absorbance band at 445 nm appears, assigned to free xylenol orange, thus, the citrate is displacing the indicator. The inclusion of citrate by **2.56** is demonstrated by the X-ray crystal structure of the complex (Figure 2.14).[20]

A related system using a cationic receptor is based on aminopyridinium moieties distributed around a hexasubstituted triethylbenzene core (**2.57a**) (see Figure 2.14). The hydrogen bonding sites comprise amine and aryl CH groups, together with the positive charge on the pyridinium moiety. The host forms a stable chloride complex in which the anion is surrounded in a trigonal prismatic array of NH and CH donors, despite the fact that the amine is an intrinsically weaker hydrogen bond donor than guanidinium. The binding constant for chloride is ca.$10^4 \, M^{-1}$ in acetonitrile solution and there are large chemical shift changes observed for the NH proton and the adjacent pyridyl CH resonance ($\Delta\delta$ up to 2.5 ppm). Unusually for a podand, exchange between bound and free anion is slow on the ^{1}H NMR time-scale just below room temperature. The analogous *para* derivative **2.57b** does not show a convergent conformation and exhibits low anion affinity, highlighting its non-complementarity and lack of preorganisation.[21]

A very closely related compound (**2.58**) may be used as an effective host for the ammonium cation, NH_4^+, while the similar compound **2.59** binds the anion $[ZnCl_4]^{2-}$ when triprotonated, but when neutral, forms a complex with Ru^{2+} in which the metal is bound simultaneously to all three imine nitrogen atoms and in an η^6-fashion to the aryl ring.[22] These results highlight the fact that the same organisational principles apply across the spectrum of guest types with the particular affinity being determined by choice of binding sites.

2.58   **2.59**

2.4.3   Neutral receptors

Antonisse, M. M. G. and Reinhoudt, D. N., 'Neutral anion receptors: design and application', *J. Chem. Soc., Chem. Commun.*, 1998, 443–448.

While there is less intrinsic electrostatic affinity between an anion and a neutral host compared to that between an anion and a cationic receptor, neutral receptors do not automatically come accompanied by a counter-anion and so it has been suggested that they could potentially exhibit greater selectivity as a class of receptors. Of course, we have to remember that if the receptor is neutral then it will not only bind an anion but will also bind its counter-cation. If this aspect is neglected, then the identity and solvation characteristics of the counter-cation can play an important and unpredictable role in the overall host selectivity (Section 2.4.2)

Hydrogen bonding interactions

Typically, neutral anion receptors incorporate strong, multiple hydrogen bond-donor groups, such as ureas. For example, the urea group in **2.60** acts as both a binding site and also as the backbone of the dipodal system. The receptor also utilises two amide functionalities that are also excellent hydrogen bond donors. The four hydrogen bond-donor NH groups can interact with oxo-anions, such as benzoate, as shown in Figure 2.15. This receptor was designed such that all of the hydrogen bonding sites are pointing into the molecular cleft, which is complementary to benzoate. The binding constant for benzoate is high, at $1.5 \times 10^4\,\mathrm{M}^{-1}$ in acetonitrile, hence suggesting a good match between the host and the guest.

Compound **2.61** is also an example of a rigid urea-based 'molecular-cleft' host. This species was designed to bind *meta*-disubstituted aromatic guests and is also able to chelate carboxylates, phosphates and sulfonates.

The binding affinity of amide-based receptors can be increased by appending $Cr(CO)_3$ groups to the phenyl rings of the original cleft, as in **2.62**, and their organometallic derivatives bind Cl^- with $K_{assoc} > 10^4\,\mathrm{M}^{-1}$ in acetonitrile solution.

The importance of a hydrogen bonding array is clearly demonstrated by the 2, 2′, 2″-tris(aminoethyl)amine moiety (**2.63**). This particular functional group has

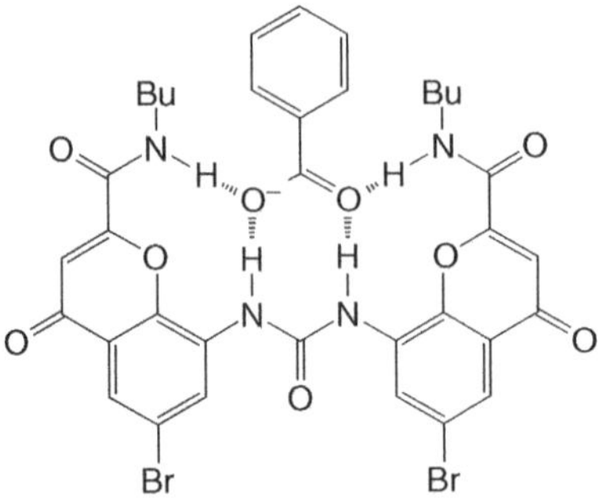

2.60

Figure 2.15   Binding of benzoate by the neutral anion receptor **2.60**.

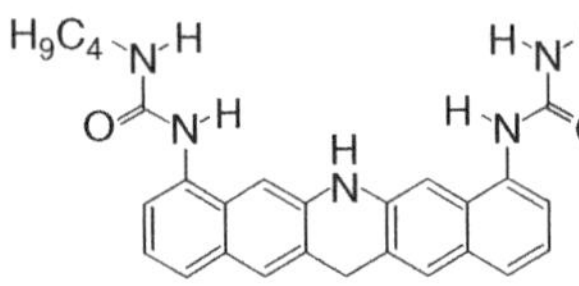

proved to be an extremely effective anion-binding motif as a result of the convergence of the NH binding sites. The naphthyl rings enhance binding in **2.63**, by preorganising the host by π-stacking interactions. The association constant for $H_2PO_4^-$ is $1.4 \times 10^4 \, M^{-1}$, which represents a 400-fold selectivity over HSO_4^-.

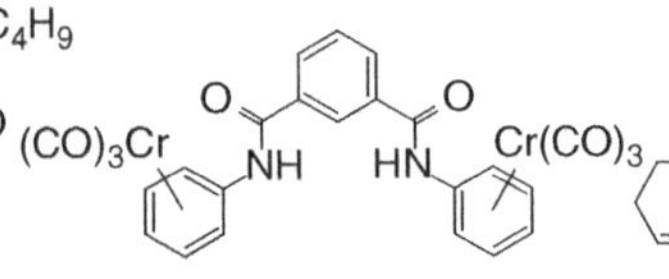

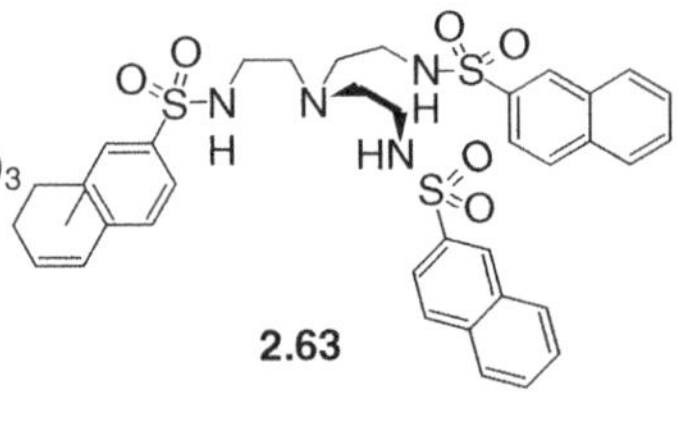

Receptors **2.64** and **2.65** (derived from *cis*-1,3,5-tris(aminomethyl)cyclohexane) have both been incorporated into membrane ion-selective electrodes. The two hosts differ dramatically in their degree of rigidity and preorganisation, and the size of the cavity that they create and consequently display very different response characteristics. The more rigid **2.65** shows Hofmeister-like selectivity while the more flexible tris(2-aminoethylamine) derivative (**2.64**) proved to be distinctly 'anti-Hofmeister'. The Hofmeister series was developed in 1888, based on the ability of ions to 'salt-out' proteins and correlates the hydrophobic tendencies of various ions.

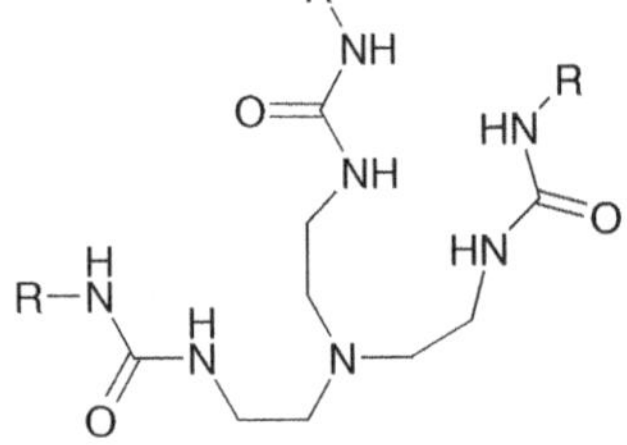

2.64 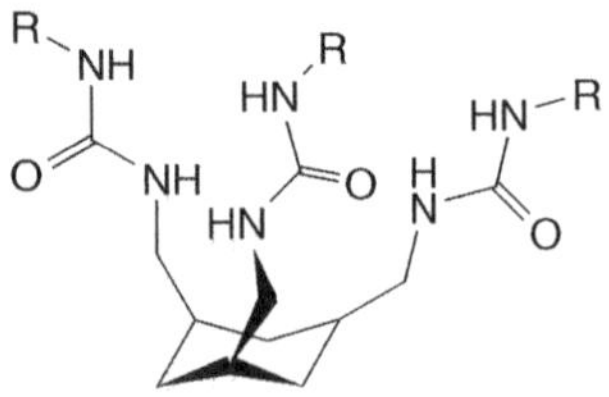   **2.65**

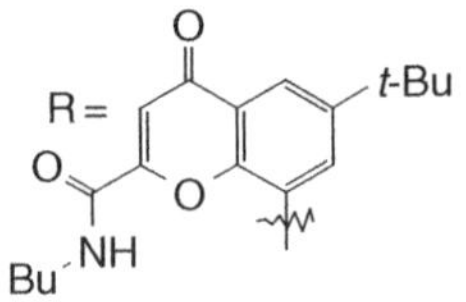

Hofmeister Series:

Strongly hydrated   Weakly hydrated
Anions: citrate^{3-} > SO$_4$$^{2-}$ > HPO$_4$$^{2-}$ > F$^-$ > Cl$^-$ > Br$^-$ > I$^-$ > NO$_3$$^-$ ≈ ClO$_4$$^-$
Cations: Al^{3+} > Mg^{2+} > Ca^{2+} > H$^+$ > Na$^+$ > K$^+$ > Rb$^+$ > Cs$^+$ > NH$_4$$^+$ > N(CH$_3$)$_4$$^+$

The effectiveness of the preorganisation imparted by the cyclohexyl core in compound **2.65** is conclusively demonstrated by the *cholapods*, such as **2.66**, based on a cholesterol framework.[23] These 'second-generation' hosts feature a number of refinements that make them extremely powerful anion-chelating agents. The array of five acidic NH protons, which somewhat resemble the tris(2-aminoethylamine) moiety (a component of many successful anion hosts), is held in a preorganised anion-binding conformation by the rigid cholesterol framework. The addition of electron-withdrawing −CF$_3$ and −NO$_2$ substituents increases the hydrogen bond acidity of the urea and sulfonamide groups. This results in extremely large affinities for halides, with **2.66b** displaying a binding constant for Cl$^-$ of 6.60×10^{10} M^{-1}. Affinities as high as 1.03×10^{11} M^{-1} have been observed for the thiourea analogue! Such large binding constants are out of the reach of determination when using NMR spectroscopy. Instead, the affinities were obtained from an extraction method in which an organic phase containing a lipophilic receptor is mixed with an aqueous phase containing substrate. The extraction constant (K_e) is then determined based on the amount of guest extracted into the organic phase at equilibrium. Once K_e is determined, then K_d can be calculated. This is a very useful strategy for the determination of very high binding constants. By varying the substrate concentration in the aqueous phase, the degree of complexation in the organic phase can be controlled and therefore saturation can be avoided (Cram's original paper is suggested reading for a more comprehensive explanation[24]).

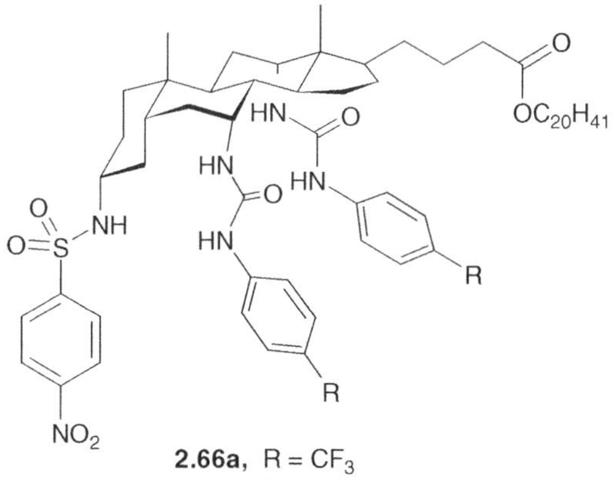

2.66a, R = CF$_3$
2.66b, R = NO$_2$

The calixpyrroles represent another well-developed class of neutral-anion receptor. Unlike other functional groups, such as amides and ureas, that often self-associate, pyrroles remain discrete and this makes them attractive functionalities for anion binding, as there is no competition from host aggregation. These compounds were first synthesised by Baeyer in the 19th Century by condensation of pyrrole and acetone in the presence of an acid but their ability to bind anions or neutral species was not recognised until the mid-1990s. The free macrocycles, such as calix[4]pyrrole (**2.67**), generally adopt a 1,3-alternate conformation in contrast to the closely related calixarenes (Section 2.3.6) but undergo a conformational change on the addition of fluoride.[25]

One of the simplest classes of pyrrole-based molecular receptors (**2.68**) binds oxo-anions in polar organic solvents. Compounds **2.68** show a clear preference for oxo-anions over halides, for example, host **2.68a** binds benzoate 18 times more strongly than chloride in acetonitrile, whereas host **2.68b** binds benzoate 51 times more strongly than chloride in a 0.5 % DMSO/H_2O solvent mixture. Dihydrogen phosphate also has some affinity for **2.68**, with $K_{assoc} = 357 \, M^{-1}$ and $1450 \, M^{-1}$ for **2.68a** and **2.68b**, respectively.

Pyrrole-containing receptors include porphyrins, extended porphyrins and calix[n]pyrroles. A variety of calixpyrroles have been used as sensors and in anion-separation technologies. An interesting example is the cryptand-like calixpyrrole (**2.69**) that binds fluoride *via* six of the nine hydrogen atoms, forming a one-to-one host–guest complex in dichloromethane and THF mixed solvent. The actual stoichometry is rather complicated as each of the pairs of clefts can bind fluoride ions, forming one-to-one, one-to-two or one-to-three stoichiometries.

| 2.67 | x = n-Bu **2.68a**
x = Ph **2.68b** | 2.69 |

2.4.4 Lewis-acid receptors and anticrowns

Wedge, T. J. and Hawthorne, M. F., 'Multidentate carborane-containing Lewis acids and their chemistry: mercuracarborands', *Coord. Chem. Rev.*, 2003, **240**, 111–128.

Lewis-acidic podands, such as the rigidly preorganised 'hydride sponge' (**2.70**) were among the earliest types of anion-binding Lewis-acidic receptors. Compound **2.70** is analogous to the H^+-binding 'proton sponge' (**2.71**) (Scheme 2.10). The compound can chelate hydride between the two Lewis-acid centres and extracts it from almost any hydride source. The host–guest interaction energy has been estimated by NMR spectroscopy to be $71\,kJ\,mol^{-1}$. The electron-withdrawing naphthalene spacer makes the boron centres more electron deficient than aliphatic trialkylboranes, which also enhances the stability of the hydride complex. Compound **2.70** also binds other anionic Lewis bases, such as fluoride and hydroxide. The key to its effectiveness as a Lewis-acid receptor is the very rigid preorganised framework. Since the preparation of this simple receptor, there has been a plethora of Lewis-acid receptors for anion recognition in supramolecular chemistry (**2.72–2.76**).

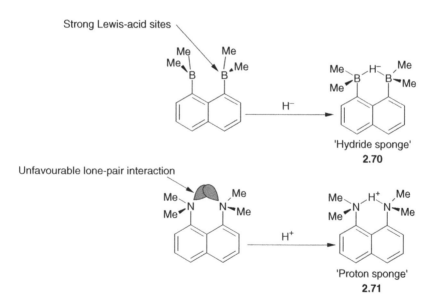

Scheme 2.10 2.10 Hydride sponge (**2.70**) and proton sponge (**2.71**).

An easily visualised example is the multidentate Lewis-acid host **2.77**, a colourimetric fluoride sensor. The receptor contains a non-innocent ferrocenyl linker between the boronate groups. Reaction of **2.77** with n-$Bu_4N^+F^-$ in CHCl$_3$

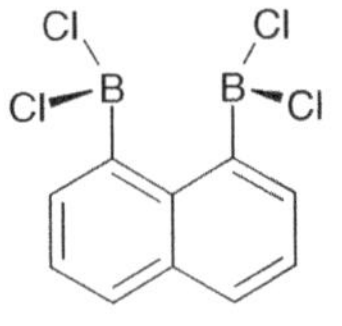

2.72

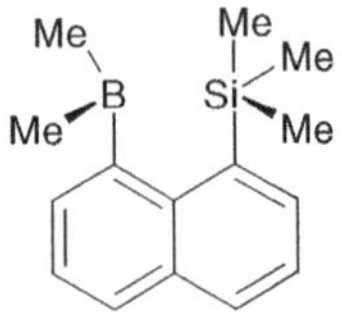

2.73

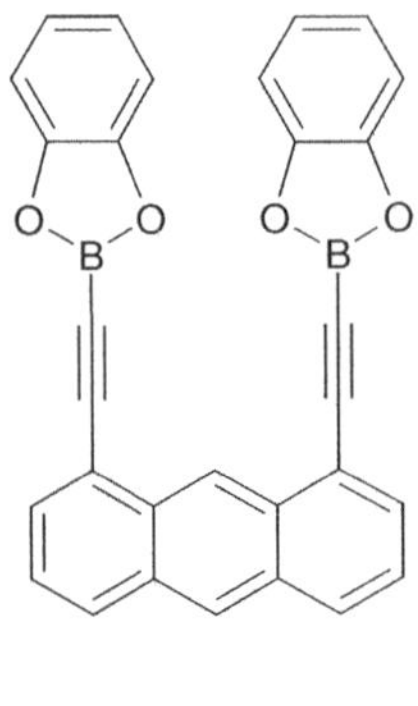

2.74

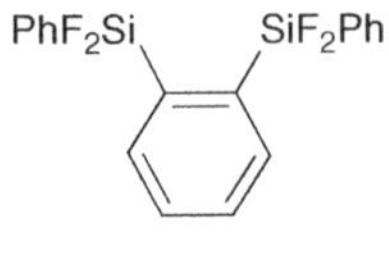

2.75

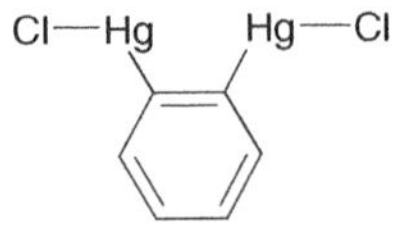

2.76

produces a marked colour change (orange to pale green) under aerobic conditions. There is no colour change under anaerobic conditions or with the addition of Cl^-, Br^-, I^-, BF_4^-, PF_6^-, $H_2PO_4^{2-}$, HSO_4^{2-} and NO_3^-, thus suggesting a selectivity for the fluoride ion alone. The mode of binding of the anion is not a true chelation as the anion is not trapped in the cleft of the receptor. Instead, complexation of the fluoride occurs in a two-to-one stoichiometry with one anion bound by each of the Lewis-acid centres (Scheme 2.11). Cyclic voltammetry of the complex shows an anodic shift of 146 mV of the oxidation potential of the Fe(II)/Fe(III) couple in the presence of fluoride and hence F^- binding allows aerobic oxidation of the ferrocene-derived centre, giving rise to the observed colour change in the presence of oxygen.

orange   green

Scheme 2.11 Proposed mode of binding between the receptor **2.77** and fluoride.

The two-to-one stoichiometry contrasts to the related cobaltocenium receptor **2.78**. Reduction of the Co(III) centre to Co(II) and re-oxidation with $Cu(OH)_2$ in non-polar solvents gives a 1:1 complex with OH^- that adopts a *syn*-conformation, with an OH^- anion chelated between the Lewis-acid groups (Figure 2.16). This behaviour, in contrast to other anions which form *anti* 2:1 complexes, may be explained by the strong Lewis basicity of hydroxide and its small size, hence allowing it to sit within the cavity formed by the two B^iPr_2 groups.

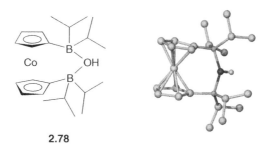

2.78

Figure 2.16 X-ray crystal structure of **2.78**·OH, showing the chelation of the hydroxide anion.

An interesting and unusual class of Lewis-acid receptors are the *anticrowns*. These receptors contain Lewis-acid centres (Si, Sn and Hg being the most common) incorporated into a macrocyclic ring system. These systems are the conceptual opposite of the Lewis-basic crown ethers. The simple tin complexes $Cl_2Sn[(CH_2)_n]_3SnCl_2$, where $n = 6$, 8, 10 and 12 (**2.79**) (Figure 2.17) have been shown by ^{119}Sn NMR spectroscopy to form complexes with halide anions in acetonitrile solution. Fluoride coordinates to both the Lewis-acidic centres in the small $n = 6$ host, whereas chloride only binds to a single Sn atom in the larger $n = 8$ homologue – the anion is too small to stretch across the larger cavity.

To increase the binding affinity of anticrowns towards anionic species, electron-withdrawing groups are attached to the scaffold, for example, **2.80** and **2.81** (Figure 2.17). These perfluorinated polymercuramacrocycles are among the most common anticrowns. The fluorine atoms drastically increase the Lewis acidity of mercury atoms. The anticrown **2.81** shows high affinity towards Cl^-, Br^-, I^- SCN^- and BH_4^-. In the solid state, the Br^- complex of **2.81** is a polymeric, polydecker bent sandwich complex, in which each Br^- anion is coordinated to six Hg atoms, three from each of two adjacent macrocycles. These complexes are very stable, for example, compound $\{[(2.81)]_2(BF_4)\}^-$ in THF at 20 °C has a binding constant of $10^7 M^{-1}$.

Probably the best known of the anticrowns are the mercuracarborands,[26] *e.g.* **2.82** and **2.83** (Figure 2.17). As with the fluorinated complexes, the carborane substituents act as elaborate electron-withdrawing groups and increase the Lewis acidity of the mercury(II) ions. Most crown ethers exhibit one-to-one host:guest stoichiometry, but most anticrowns form 2:1 complexes on size-fit grounds. One exception to this rule is the chloride complex of **2.83** in which the chloride ion binds within the centre of the host cavity with a long Hg–Cl distance of 2.94 Å.

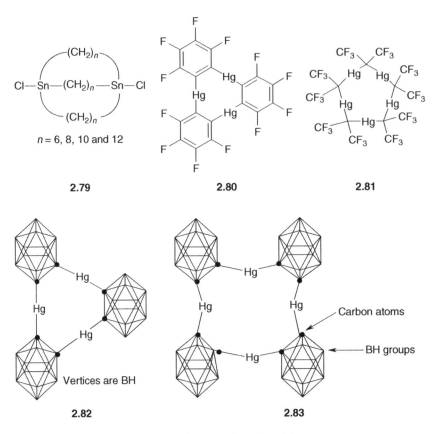

Figure 2.17 A selection of Lewis-acid receptors.

2.5 Metal-containing receptors

 Beer, P. D. and Hayes, E. J., 'Transition metal and organometallic anion complexation agents', *Coord. Chem. Rev.*, 2003, **240**, 167–189.

Anion receptors based on metal centres can be classified into three broad categories: those in which the metal plays a structural role, those in which it is a key component of the anion-binding site and those in which it acts as part of a redox, fluorescent or colourimetric reporter group. The latter types of compound will be covered in the next section, although, of course, there are examples of compounds that fall into more than one group, such as **2.77** in which the ferrocene-derived unit acts as both a colorimetric reporter and a structural element.

2.5.1 Metals as structural elements

Hosts with metals as structural elements may be further divided into inert and labile coordination complexes. In the former case, the metal–ligand bonds are long-lived on the human time scale and the host compound binds anions as a unit in exactly the same way as non-metal-containing analogues. For example, in compound **2.85**, two thiourea-derived terpyridyl ligands are held together in a well-defined way by a relatively inert (low-spin d^6) ruthenium(II) centre (Scheme 2.12). The host binds long dicarboxylates, particularly pimelate.[27]

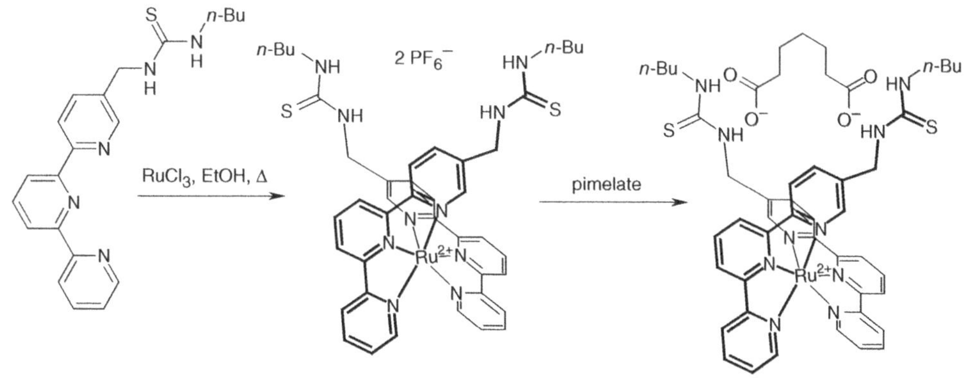

2.84 **2.85** **2.85a**

Scheme 2.12   Use of an inert Ru(II) centre to organise an anion chelate ligand.

Labile coordination compounds are not true anion hosts in the conventional sense. Instead, they fall into the category of self-assembly (Chapter 3) and are frequently templated by anions, cations or both. The ensemble may be thought of as a multi-component self-assembly of ligands that, like the urea-containing terpyridines in **2.84**, contain both anion- and cation-binding functionality. A good example is the ligand **2.86** which contains a single pyridyl functionality and a urea anion-binding group. In the presence of the labile Ag(I) cation and nitrate anion, the complex [Ag(**2.86**)$_2$(S)]NO$_3$ (S = a solvent molecule) self-assembles. The labile nature of the complex may be demonstrated by addition of excess nitrate which results initially in conversion to a dinitrate complex, [Ag(**2.86**)$_2$(S)](NO$_3$)$_2$, in which each urea group binds a single anion and, ultimately, binding of nitrate at the Ag(I) centre itself (Scheme 2.13).[28]

An illustration of the boundary between labile and inert-complexes comes from the arene ruthenium(II) derivatives **2.87** and **2.88**. In the presence of a bidentate bispyrrole ligand, complex **2.87** is inert and binds a number of anions *via* hydrogen bonding interactions to the bispyrrole NH functionalities. In contrast,

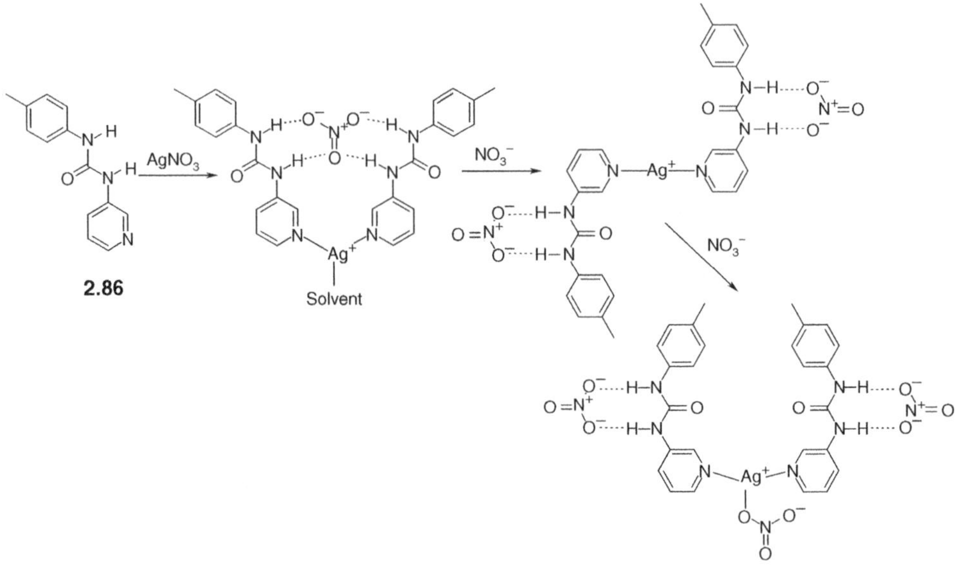

Scheme 2.13   Self-assembly of a labile nitrate-binding Ag(I) complex and its evolution in the presence of excess nitrate.

while **2.88** binds effectively to chloride, nitrate and hydrogen sulfate, the monodenate pyridyl ligands are displaced over a period of hours by Cl⁻ which moves from binding to the amine NH groups to direct coordination to the semi-labile metal centre. Compound **2.88** represents an interesting example in which one metal centre, ruthenium(II), is used as a structural element, while another, iron(II), acts as a redox reporter group.[29]

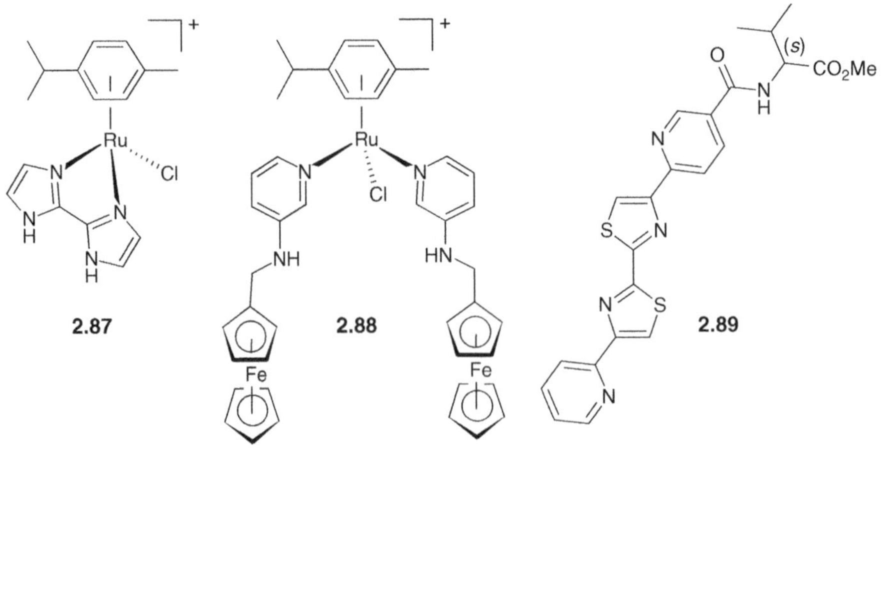

The self-assembly principle (Chapter 3) may be used to very good effect to give selective anion hosts with labile metal ions. Ligand **2.89** contains both cation and amide anion-binding functionalities. Reaction with the labile Co(II) results in the formation of a dinuclear complex, $[Co_2(\textbf{2.89})_3]^{4+}$, that can exist as a mixture of isomers, either with all of the three amino acid groups at the head of the molecule (HHH) or a 'head-to-head-to-tail' arrangement (HHT), with one pointing the other way. With $Co(ClO_4)_2$ the HHH:HHT ratio is 1:3; however, if the complex is refluxed with nitrate the system converts almost exclusively to the HHH isomer. This is because the nitrate forms a strong, three-fold hydrogen bonded interaction with the three amide NH groups of the amino acid functionalities in this isomer – in effect, the nitrate anion organises its own binding pocket in a variation of the thermodynamic template effect (see Section 2.2.2).[30]

2.5.2 Metals as electrochemical sensing elements

Anion receptors containing redox-active groups, such as ferrocene or cobaltocenium, that allow the detection of anions and cations *via* electrochemical means, are a popular area of research.[31] Ferrocene and cationic cobaltocenium moieties have been incorporated into many acyclic, macrocyclic and calix[4]arene frameworks. These receptors contain either amide or amine functionalities that are able to form hydrogen bonds to the anions, or typical cation-binding groups, such as bipyridine. Three interesting examples are the cyclophane organometallic receptor (**2.90**), tripodal receptor (**2.91**) and the molecular cleft (**2.92**).[32] Receptor **2.90** binds bromide in acetonitrile *via* electrostatic interactions only, whereas the receptors **2.91** and **2.92** utilise amide functionalities attached to the cyclopentadienyl rings on the cobaltocenium. Compare the preorganisation of the tripodal hydrogen bonding array in **2.91** and **2.93** with the *cholapods* (**2.66a** and **2.66b**).

Both receptors **2.91** and **2.92** have an affinity for dihydrogen phosphate over chloride in acetonitrile solution. For **2.91**, $K(H_2PO_4^-) = 1200$ and $K(Cl^-) = 100\,M^{-1}$, while for **2.92** $K(H_2PO_4^-) = 320$ and $K(Cl^-) = 35\,M^{-1}$. The reason for the selectivity for dihydrogen phosphate over other anions is the basicity of $H_2PO_4^-$. The electrochemical response when an anion is bound to the receptors complements the findings obtained in the NMR spectroscopy experiments. When the receptor binds to the anion, the cobalt(III) centre becomes harder to reduce. Dihydrogen phosphate produces a larger cathodic shift than other anions in cyclic voltammetry (200 and 240 mV for receptors **2.91** and **2.92**, respectively). The difference between the ferrocene-based receptors and the cobaltocenium derivatives is that the overall receptor complex is neutral for the ferrocene analogues (**2.93** and **2.94**), whereas the cobalt compounds are cations, thus the intrinsic affinity between ferrocene receptors and anions is lower. The host–guest complex

2.90　　　　　　　　　　　　　　　　**2.91**

2.92　　　　　　　　**2.93**　　　　　　　　**2.94**

only forms hydrogen bonding interactions and lacks the electrostatic interactions to aid in binding, as in receptors **2.91** and **2.92** where the weakly interacting PF_6^- anions do not compete strongly for the binding sites. Electrostatic interactions are switched on once the ferrocene has been oxidised to ferrocenium, however.

Two interesting ferrocene-derived receptors, **2.95** and **2.96**, have been synthesised as redox-switchable cation-binding cryptands. Upon oxidation of the ferrocene moiety, guest cations are expelled from the cryptand cavity due to increased electrostatic repulsion resulting in large anodic shifts in the Fe(II)/Fe(III) couple upon binding various metal ions in a 1:2 host:guest stoichiometry. The largest anodic shifts are seen for Y^{3+} and Eu^{3+} (326 and 302 mV, respectively), whereas Na^+ binding results in only a 70 mV anodic shift. The change in redox potential in systems such as this is related to the ratio of guest binding constants in the reduced and oxidised species. Large shifts result from very different values of K_{ox} and K_{red}, according to the following equation (2.1):

$$E_{1/2}(\text{host}) = E_{1/2}(\text{complex}) + \frac{RT}{nF} \ln\left(\frac{K_{ox}}{K_{red}}\right) \qquad (2.1)$$

where $E_{1/2}$ are the redox potentials for host and complex, R is the gas constant, T is the temperature (in K), n is the number of electrons transferred, F is the

Faraday constant and K_{ox} and K_{red} are the binding constants for the guest for the oxidised and reduced forms of the host, respectively.

2.95   **2.96**

Anion-binding hosts are often transformed into sensing systems by appending redox-active groups such as ferrocenyl units. For example, the 'Venus' flytrap' anion sensor **2.97** was developed from the chloride binding tripodal host **2.57a**.[21] This receptor also shows high affinity towards chloride, with binding constants of up to $17\,400\,M^{-1}$ in acetonitrile. Chloride is included within a six-fold array of hydrogen bonds from the three NH and three pyridinium CH groups (Figure 2.18). However, the electrochemistry of **2.97** only shows modest shifts upon the addition of halides because of poor communication between the guest species and the redox centers *via* the CH_2 linker group.

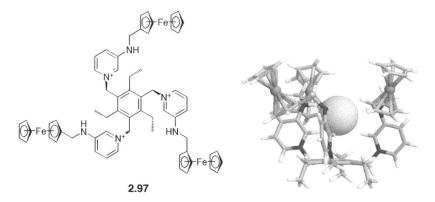

2.97

Figure 2.18 Semiempirical model of chloride binding by receptor **2.97**.

A number of organometallic bowl-shaped anion hosts (Figure 2.19) have been prepared based on Ru(II), Rh(I), Rh(III) and Ir(III) π-complexes of bowl-shaped macrocycles such as the calix[*n*]arenes (*n* = 4 and 5, Section 2.3.6) and cyclotriveratrylene (CTV) in which the redox reporter group is much more rigidly attached to the receptor. CTV derivatives are very versatile hosts for many neutral and anionic species and CTV itself forms solid-phase clathrates (see Chapter 4, Section 4.3). The CTV molecule is bowl-shaped with a shallow cavity and is highly preorganised as there is a significant conformational preference for all aryl rings to point in the same direction. Appending two or more organometallic moieties around the exterior of the bowl produces a highly electron-deficient cavity. The result is strong, co-operative electrostatic host–anion interactions, giving rise to very short host···anion contacts in the solid state. Compound **2.98** binds large tetrahedral anions, such as $^{99}TcO_4^-$. Technetium-99 is a major uranium fission product and a long-lived β-emitter. Cyclic voltammetry shows a modest shift on ReO_4^- binding but larger shifts for HSO_4^- and $H_2PO_4^-$, suggesting that **2.98** is selective for tetrahedral anions. Interestingly, the cyclic voltammograms of the mono-, di- and triruthenium hosts show, respectively, one, two and three reduction waves, so suggesting inter-metallic communication. Such communication is rarely observed for more flexible systems.

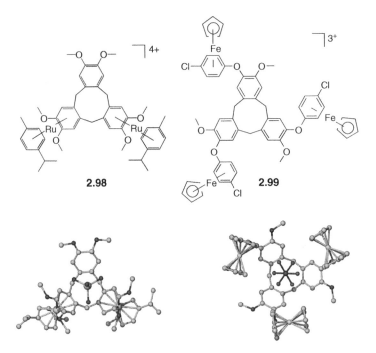

Figure 2.19 X-ray crystal structures of **2.98** · $(ReO_4)(CF_3SO_3)_3$, showing the included ReO_4^- ion, and **2.99** · $3PF_6$, showing the inclusion of a PF_6^- ion.

CTV may also be used as a platform for a deeper host, such as **2.99**, which exhibits a deep cavity surrounded by three cationic cyclopentadienyl arene Fe(II) arms that interact electrostatically with anion guests (Figure 2.19). Upon titration with Br$^-$, the ^{1}H NMR spectrum of the material indicates a slowing of the rotation of the [CpFe(arene)]$^+$ groups.

2.6 Simultaneous cation and anion receptors

Gale, P. A., 'Anion and ion-pair receptor chemistry: highlights from 2000 and 2001', *Coord. Chem. Rev.*, 2003, **240**, 191–221.

In the preceding sections, we looked at the design requirements to bind cations and anions as two separate entities. In both cation- and anion-binding, the guest in question is accompanied by a counter-ion that will influence the binding affinity. To overcome this problem, the chemist deliberately biases the system such that the counter-ion is 'non-competing', for example, tetraalkylammonium cations or tetraarylborate anions are often used. Therefore, the binding constant for a charged host (which acts as that counter-ion for its guest) really represents an empirical selectivity factor for the displacement of one less competing ion for one which is bound more strongly. In the case of neutral receptors, the host is always accompanied by the target guest and its counter-ion and the measured binding constant is for the ion pair, not the target anion or cation. Moreover, in real-world applications, non-competitive counter-ions, such as those mentioned above, are not generally encountered and hence inter-ion competition can be significant. Another disadvantage is that salts rarely, if ever, exist as separate ions unless the medium is highly solvating. In fact, salts usually exist as either solvent-separated ion pairs (Figure 2.20(a)), contact ion pairs (Figure 2.20(b)) or aggregated contact ion pairs (Figure 2.20(c)) (Figure 2.20). In light of these issues, the most effective ionic recognition strategy is to design a receptor capable of explicitly recognising both the anion and cation of an ion pair (either as a contact ion pair or in different regions of the receptor) within the same molecular host. This particular area has become very popular over the last few years, as many such receptors have recently been used to simultaneously transport anions and cations across cell membranes, a process known as *symport*.

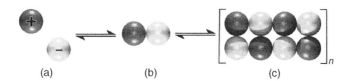

(a)   (b)   (c)

Figure 2.20 Schematic illustrating the various states that ions are generally found in solvents, *i.e.* (a) solvated ion pairs, (b) contact ion pairs and (c) aggregated contacted pairs.

A common approach in the design of such receptors is the combination of common cation-binding groups, such as crown ethers and calixarenes, together with a common anion-binding functionality, for example, a Lewis-acid centre separated by linking appendages or spacers. Broadly speaking, there are three types of receptors (or strategies) that are used for simultaneous ion binding: cascade receptors which function by binding one type of guest (usually the cation) and which use this as a new binding site for the counter-ion, for example, **2.100** (Figure 2.21(a)), ditopic receptors with two distinct binding sites in the host, for example, **2.101** (Figure 2.21(b)) and zwitterion receptors for guests that have both positive- and negative-charged regions in the same molecule as in **2.102** (Figure 2.21(c)) (Figure 2.21).

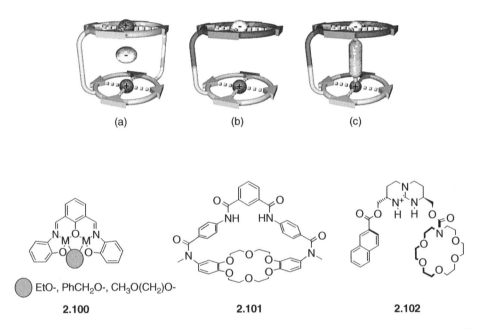

Figure 2.21 Schematics of three different types of simultaneous receptors, *i.e.* (a) cascade receptor, *e.g.* **2.100**, (b) ditoptic receptor, *e.g.* **2.101**, and (c) zwitterion receptor, *e.g.* **2.102**.

2.6.1 Cascade receptors

Cascade receptors represent the earliest simultaneous receptors and some examples date back to the 1970s and 1980s.[33] Typically, more than one metal ion (cation) coordinates to a particular ligand (often a Schiff base or macrocyclic heteroalkane) in a well-defined geometry and the anionic species then coordinates to the metal centre (Figure 2.21(a)) – this complex is known as a *casacade complex.* For example, the macrocycle **2.103** is able to coordinate two Cu(II) ions within the cavity. The X-ray crystal structure shows that the two azide anions bridge across the metal ions (Scheme 2.14). Cascade receptors often find applications as models for enzyme-active sites.[15]

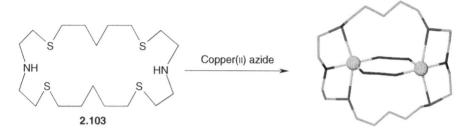

2.103

Scheme 2.14 The macrocyclic receptor (**2.103**) is able to coordinate two Cu(II) ions, with two azide anions bridging across the two metal centres.

2.6.2 Ditopic receptors

Ditopic ion-pair receptors involve the binding of ion pairs, either as contact or separated ion pairs with a separate binding site for the cation and another site for the anion (Figure 2.21(b)). An interesting aspect of binding both cations and anions is that they often exhibit co-operative behaviour (see Chapter 1, Section 1.2.3). A particularly striking example of a simultaneous ion receptor that uses both induced-fit and polarisation effects is the tripodal tris(amido-benzo[15]crown-5) receptor **2.104**. Co-operative binding is achieved with Cl^-, I^- and ReO_4^- anions, in conjunction with a crown ether unit that complexes Na^+ ions. Na^+ binding improves the receptor's affinity for ReO_4^- by a factor of 20 in comparison to Cl^- and I^- anions under conditions that mimic aqueous waste streams of environmental significance in the nuclear fuel industry where the binding of the highly labile pertechnetate anion ($^{99}TcO_4^-$) is of concern. Perrhenate represents a commonly accepted non-radioactive model for pertechnetate.

The ditopic receptor **2.105** is another good example of co-operative ion-pair binding. The host contains ether functionalities at the bottom rim of the calix[4]arene for the coordination of cations, as well as two anion-binding urea moieties. A 'pinched-cone' conformation persists in $CHCl_3$ solution as the calix[4]arene is stabilised due to the hydrogen bonding interactions between the *trans*-urea groups. The pinched conformation prevents anion binding. Upon the addition of a cation such as Na^+, the conformation of the calix[4]arene changes, so forcing the hydrogen bonding interactions to break,

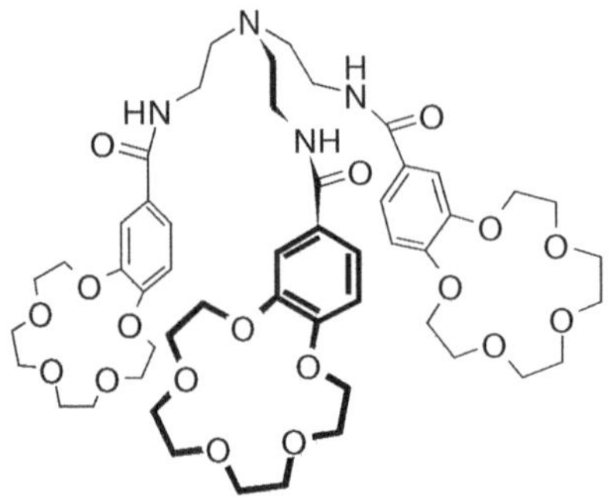

2.104

forming a cavity for anions such as Cl⁻ and Br⁻ to sit within the upper rim (Scheme 2.15).

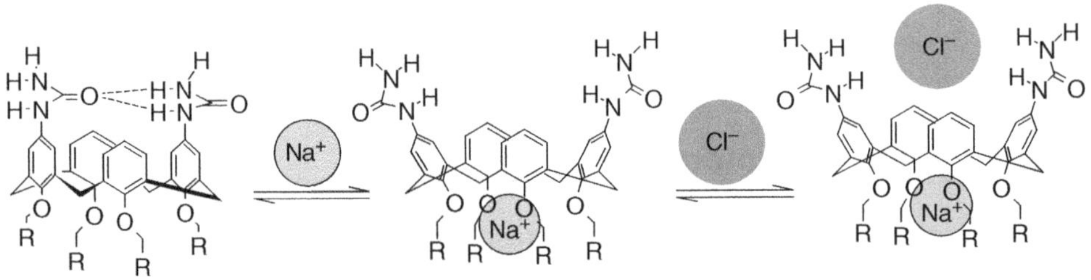

2.105

Scheme 2.15 The ditopic receptor (**2.105**) produces a slight conformational change (induced-fit) upon addition of Na⁺ which influences the affinity to Cl⁻ (R=CO₂Et).

Much work in the area of supramolecular chemistry has tried to mimic biological systems and biological ion transport is particularly topical. There has been a lot of work carried out on cation transport by hosts such as valinomycin (**2.106**) and monesin (**2.107**), biological examples of carrier molecules that transport cations across biological membranes (Figure 2.22). These types of metal ion carriers are called *cationophores*. In contrast to cation binding, there are very few synthetic examples of anion- or salt-binding synthetic receptors for transport of salts across cell membranes. One such example is the very elegant ditopic receptor **2.108** (Figure 2.23).[34] Compound **2.108** binds both sodium chloride and potassium chloride as contact ion pairs (Figure 2.20) and transports them across a vesicle membrane (see Chapter 5, Section 5.4.1). The Cl⁻ efflux through the membrane can be monitored by using a chloride ion-selective electrode (Figure 2.24). Chloride is a particularly interesting target as misregulated Cl⁻ transport is a key factor in the genetic disease cystic fibrosis. Chloride transport is a challenging process for the supramolecular chemist, however, as there is competition between

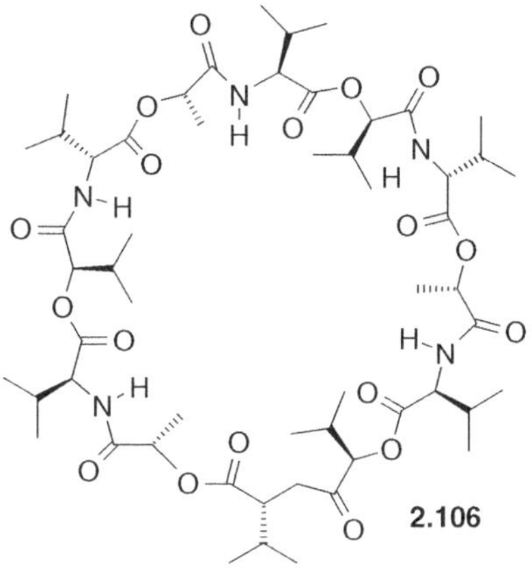

2.106

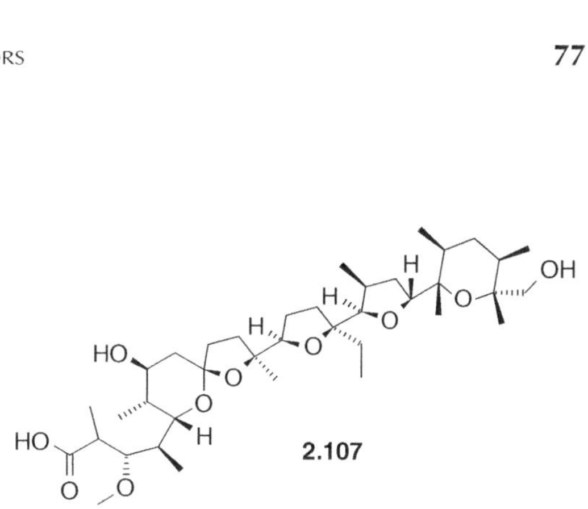

2.107

Figure 2.22 The biological receptors valinomycin (**2.106**) and monesin (**2.107**) which are able to transport cations across cell membranes.

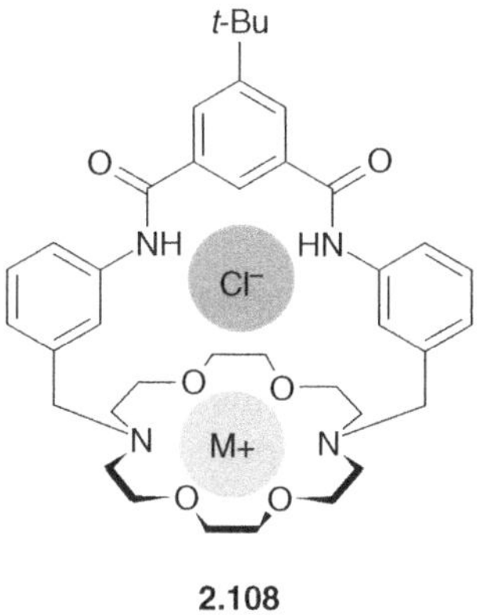

2.108

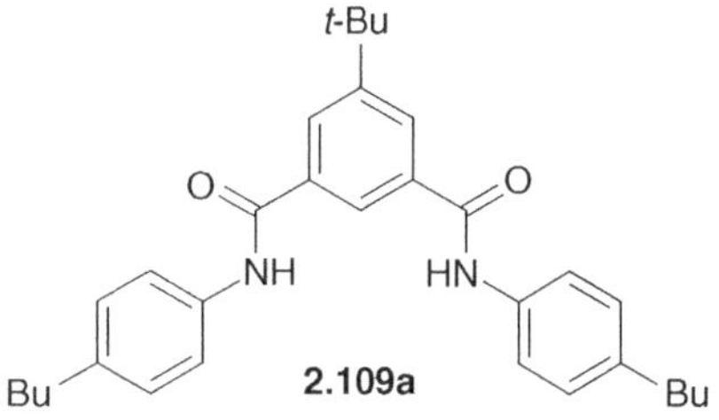

2.109a

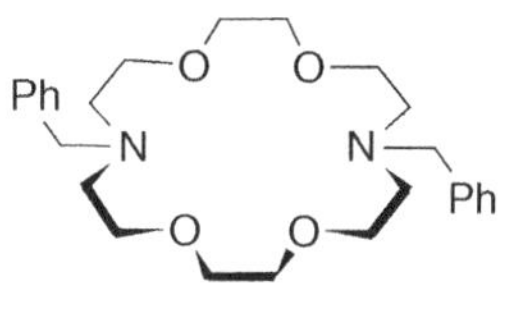

2.109b

Figure 2.23 The ditopic receptor (**2.108**) and its components (**2.109a** and **2.109b**). The anion is hydrogen bonded within the molecular cleft while the cation sits within the azo-crown.

the anion being bound at the surface of the membrane and the high concentration of the anionic phosphodiester residues, and as a consequence the process is very slow. It is known that a net positive charge exists inside an unperturbed bilayer membrane. Therefore, if the molecule is able to pass through the surface of the membrane the remaining transport should occur with relative ease. The macrobicycle **2.108** is a very effective transporter and even in a 2500:1 phospholipid to **2.108** ratio, still releases half of the vesicle chloride content within 300 s. The importance of this receptor is highlighted by carrying out a control experiment with **2.109a** and **2.109b** (Figure 2.23), *i.e.* the two separate components

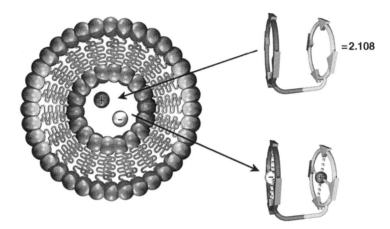

Figure 2.24 Alkali-metal chloride transport (*symport*) through a membrane by the receptor **2.108**.

of the macrobicycle **2.108**. The result is very little chloride efflux, even at very high concentrations of the two separate components.

2.6.3 Zwitterion receptors

Zwitterionic receptors differ from the two previous examples because the charges are both on the same molecule, hence rendering the receptor's overall charge neutral. Two different types of sub-units are required in the receptor design. These sub-units are arranged in the correct geometry and space for the different charged ends of the receptor to bind to the complementary groups on the zwitterion guest. However, there are two disadvantages of these receptors. First, many of the guest targets are highly solvated, so resulting in strong competition between solvent and receptor. Secondly, the receptors may aggregate into dimers or form oligomers because they can be self-complementary. This factor has been employed to engineer supramolecular catalysis in a self-recognition process between a guanidinium group and carboxylate functionality to act as a template for a Schiff-base bond formation, to produce a *duplex e.g.* **2.110** (Scheme 2.16).

Zwitterionic receptors can be split into three classes: (i) those that incorporate charge complementarity between the receptor and the zwitterions; (ii) those in which the receptor binds one of the ionic parts of the substrate, which in turn causes a structural change (induced-fit) such that the other ion group then binds and (iii) betain receptors. Betains, for example, dioctanoyl-L-α-phosphaditylcholine (**2.111**), are nutrients that play a pivotal role in the cardiovascular system. Together with other nutrients, betains reduces the levels of toxic homocysteine (Hcy) that may cause heart disease. Betains contain a tertiary trimethylammonium

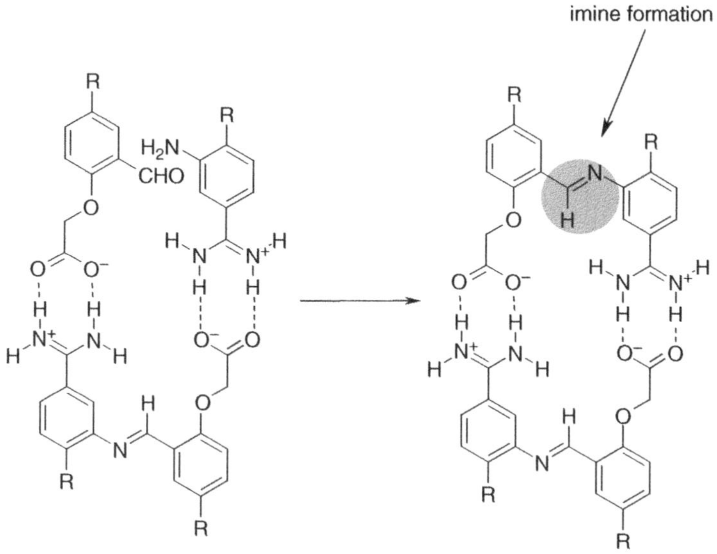

imine formation

2.110

Scheme 2.16 A zwitterion duplex (**2.110**) involved in templation of a condensation reaction between complementary components to give rudimentary self-replication.

end group that cannot form hydrogen bonding interactions with anionic residues. The inspiration for the design of a receptor for the betaine guest **2.111** came from a crystal structure of the Fab domain of the McPC 603 antibody complexed with phosphorylcholine. The betain–antibody complex incorporates an anionic phosphate was bound *via* a salt bridge (cation–anion contact ion pair) with an argenine residue. The ammonium end is bound *via* a cation–π interaction with the aromatic residues of a number of amino acids (Figure 2.25). This series of observations have been incorporated into the elegant design of a lariat calix[6]arene receptor (**2.112**) (Figure 2.25) containing a guanidinium pendant arm. The betain **2.111** forms a strong complex with **2.112** in a mixture of CDCl$_3$ and CD$_2$Cl$_2$. Molecular modelling suggests that the calix[6]arene adopts a cone conformation and the guanidinium side arm comes over the top of the molecule, enveloping the betain which is bound within the cavity, *via* cation–π interactions. The energy-minimised structure indicates that the guanidinium protons are hydrogen bonded to the phosphate group and the choline trimethylammonium head sits deeply within the cavity of the calix[6]arene.

The most straightforward design of a zwitterion receptor is to incorporate complementarity to each ion or ionic portion of the guest by use of opposite charges between the zwitterion receptor and the guest molecule. This type of interaction rarely exists on its own in biological systems and the host–guest complex is enhanced with other non-covalent interactions, such as hydrogen bonding. The chiral macrocylic alkaloid (+)-turocurarine (**2.113**) at pH 9 is in a

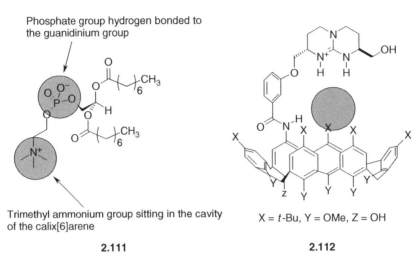

Phosphate group hydrogen bonded to
the guanidinium group

Trimethyl ammonium group sitting in the cavity
of the calix[6]arene

X = *t*-Bu, Y = OMe, Z = OH

2.111   **2.112**

Figure 2.25   A lariat calix[6]arene binding betain **2.111** within the cavity.

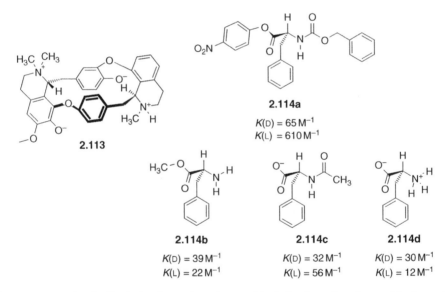

2.113

2.114a

$K(\mathrm{D}) = 65\,\mathrm{M}^{-1}$
$K(\mathrm{L}) = 610\,\mathrm{M}^{-1}$

2.114b

$K(\mathrm{D}) = 39\,\mathrm{M}^{-1}$
$K(\mathrm{L}) = 22\,\mathrm{M}^{-1}$

2.114c

$K(\mathrm{D}) = 32\,\mathrm{M}^{-1}$
$K(\mathrm{L}) = 56\,\mathrm{M}^{-1}$

2.114d

$K(\mathrm{D}) = 30\,\mathrm{M}^{-1}$
$K(\mathrm{L}) = 12\,\mathrm{M}^{-1}$

Figure 2.26 Chiral discrimination of amino acid derivatives by the zwitterions (+)-
tubocurarine (**2.113**).

zwitterionic form (Figure 2.26), and binds various chiral amino acid derivatives
with significant enantioselectivity. The selectivity of **2.113** is attributed to a
balance in charge–charge interactions and the hydrophobic effect. Simulation
of the complex between **2.113** and **2.114c** suggests that the macrocycle forms a

concave surface; however, the cavity is too small for the guest to sit inside and it perches slightly above the hollow.

Crown ethers are able to bind ammonium ions *via* hydrogen bonding interactions within their cavity (Section 2.3.3). Amino acids, under certain conditions, have an ammonium functionality and consequently there has been a plethora of receptors that incorporate crown ethers, with a hydrogen bonding group within the same receptor to bind amino acids, for example, compound **2.115**. The receptor **2.115** extracts amino acids from aqueous solution to dichloromethane. There is selectivity for guests that contain aromatic side chains, for example, L-tryptophan (Scheme 2.17). The ditopic receptor binds to tryptophan *via* two sets of hydrogen bonding interactions, those between the carboxylate group and the guanidinium functionalities, and those between the ammonium ion and the crown ether. The complex is stabilised by π–π interactions between the aromatic side chains.[35]

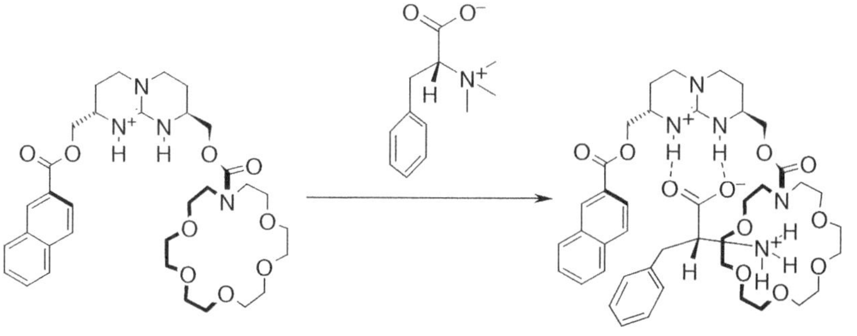

2.115

Scheme 2.17 The ditopic receptor (**2.115**). The carboxylate group is hydrogen bonded to the guaindinium moieties, the ammonium ion is hydrogen bonded within the cavity and the complex is further stabilised by π–π interactions between the aromatic ring systems.

2.6.4 Cation and neutral simultaneous receptors

Simultaneous receptors exist that can bind neutral and ionic guests. These types of receptors are large macrocycles, such as calixarenes, in which the ion is bound to the functional groups at the lower rim of the calixarene and the neutral (often solvent) molecule is encapsulated within the cavity of the bowl, *e.g.* tetramethoxy-*p*-*t*-butylcalix[4]arene (**2.116**). In the solid state, an Na$^+$ion is coordinated to the oxygen atoms of the lower rim of the calix[4]arene and the complex contains one molecule of toluene within the hydrophobic cavity (Figure 2.27).

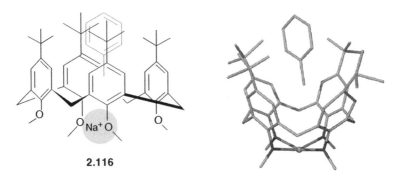

2.116

Figure 2.27 The X-ray structure and chemical diagram of the receptor **2.116**, binding Na$^+$ and a single molecule of toluene within the cavity.

2.7 Neutral-molecule binding

 Cram, D. J., and Cram, J. M., *Container Molecules and their Guests*, The Royal Society of Chemistry, Cambridge, UK, 1994.

In general, interactions to neutral guests are weaker than those to ions. In order to achieve neutral-molecule recognition, the host needs to either self-assemble around the neutral guest (Chapter 3) or it must be highly preorganised. Generally, such preorganised hosts have permanent curvature to form an intrinsic cavity that persists in solution. This type of receptor is generically termed a *cavitand*. The inclusion of the guest within the cavity or cleft of the cavitand yields a *caviplex*. There are a large number of cavitands reported in the literature, many based on the intrinsically curved calixarenes (Section 2.3.6), for example. For convenience, these are broadly classed into different chemical families, which will be discussed in the following sections.

Cavitand: A molecular host that has an enforced concave surface, producing a cavity.

Caviplex: The term given to the complex between a cavitand and a guest molecule.

2.7.1 Cyclophane hosts

The term *cyclophane* includes all molecules that contain a bridged aromatic ring. Many types of bowl-shaped molecules fall into this category – these include carcerands and hemicarcerands (Section 2.6.2), cryptophanes and hemicryptophanes (Section 2.6.3) and calixarenes, resorcinarenes and rigidified recorcinarenes (Section 2.6.4). One of the first examples of a cyclophane molecule was [2.2]metacyclophane (**2.117a**), which was synthesised in 1899 by Pellegrin (Figure 2.28). Cram and Steinberg introduced the modern concept of a cyclophane in 1951 by re-synthesising cyclophane **2.117a**, along with **2.117b**, termed *meta-* and *paracyclophanes*, respectively, in which two aromatic rings are held together rigidly by ethylene bridging groups. Vögtle and Neumann broadened the definition of cyclophanes to include every molecule that contains one aromatic ring and at least one *n*-membered bridged group, with $n > 0$. In 2003, IUPAC produced other rules for the definition of highly complex molecular ring systems, as follows:

1. A cyclophane is a compound that has a MANCUDE ring system, that is to say, a **MA**ximium number of **N**on**CU**mulative **D**oubl**E** bonds, or assemblies of MANCUDE ring systems.

2. Cyclophanes contain atoms and/or saturated or unsaturated chains as alternate components of larger ring systems (*cf.* compounds **2.117a** and **2.117b**).

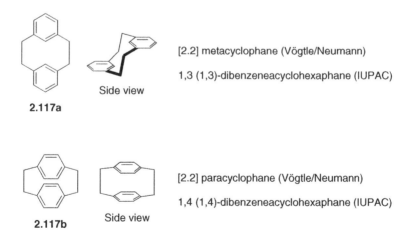

[2.2] metacyclophane (Vögtle/Neumann)

1,3 (1,3)-dibenzeneacyclohexaphane (IUPAC)

Side view

2.117a

[2.2] paracyclophane (Vögtle/Neumann)

1,4 (1,4)-dibenzeneacyclohexaphane (IUPAC)

2.117b   Side view

Figure 2.28 The cyclophanes, [2,2]metacyclophane (**2.117a**) and [2,2]paracyclophane (**2.117b**), showing the original (Vögtle and Neumann) and IUPAC nomenclatures.

The synthesis of cyclophanes requires the closure of a ring system to avoid polymerisation or oligermerisation. As cyclophanes cover many different complex macrocycles, it is not possible to cover all of the types of synthetic procedures. Broadly speaking, the preparation of cyclophanes can be classified under three different methodologies, as follows:

1. The molecular building-block technique, which involves a wide range of organic reactions. For example, nucleophilic substitutions, in particular, those involving sulfur, Schiff-base condensations, Mitsonubu reactions, carbon-carbon bond formation (*e.g.* the Suzuki reaction), amide-bond formation, alkyne – coupling reactions, *etc.*

2. The dilution principle (Section 2.2.1).

3. The template effect (Section 2.2.2).

The use of alkyne and aromatic spacing groups maintains the rigidity of the hosts and presents an easily accessible, preorganised three-dimensional space within their confines. The use of aromatic spacers also promotes their usefulness as hosts for aromatic guests by virtue of the substantial π–π interactions. For example, the binding constant for *p*-nitrophenol by the rigid cyclophane **2.118** is an impressive 9.6×10^4 M^{-1} in CH$_2$Cl$_2$ (Figure 2.29(a)).

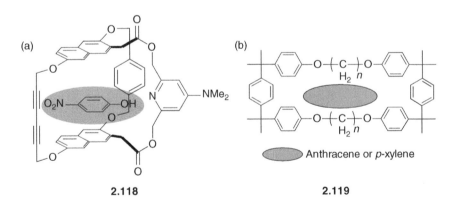

2.118   2.119

Figure 2.29 (a) *p*-Nitrophenol bound within the cavity of the cyclophane (**2.118**) *via* π–π interactions and (b) aromatic neutral guests bound *via* π–π interactions with the cyclophane (**2.119**).

Cyclophane hosts may be classified by the nature of the cavity that they possess. There are those that have an acidic cavity, called *endoacidic cyclophanes* – such cyclophanes are able to bind cationic species. In contrast, *endobasic cyclophanes* are able to act as electron donors and hence are capable of binding metal ions. *Endolipophilic cyclophanes* are able to bind neutral guest molecules through π–π interactions and hydrophobic effects. A series of simple cyclophanes, able to bind anthracene or *p*-xylene, have been synthesised by coupling a bisphenol group and an analogous bis(bromoalkyl) unit, forming a highly electron-rich aromatic system containing a hydrophobic cavity (**2.119**). This type of cyclophane has been called a 'corral' (Figure 2.29(b)). Unfortunately, the binding ability of **2.119** in solution is limited because of solubility constraints. The X-ray crystal structures of **2.119** with *p*-xylene and anthracene confirm that the cavity is highly complementary to these aromatic guests which are locked into place by edge-to-face π-interactions.

2.7.2   Calixarenes and resorcarenes

Calixarenes, for example, **2.120**, have already been discussed as hosts for cations and both cation and neutral molecules simultaneously in Sections 2.3.6 and 2.5.5, respectively. *Resorcarenes* (*e.g.* **2.121**) are very closely related to calixarenes and are sometimes referred to as *calixresorcarenes*. They are prepared from the acid-catalysed condensation of resorcinol (1,3-dihydroxybenzene) with aldehydes, a reaction first carried out by Baeyer in 1872. This macrocycle is arguably one of the most important molecules in the field, as resorcarenes serve as the basic building blocks for many cavitands and (hemi)carcerands. Both calixarenes and resorcarenes posses a shallow bowl-shaped conformation in their most stable forms. However, [4]resorcarenes (the [4] prefix refers to the number of resorcinol units that make up the molecule, just as in calix[*n*]arenes) tend to produce a

Calix[4]arene	Calix[4]resorcarene	A bridged resorcarene
X = OH		
R = H, CH₃, *t*-Bu	R = alkyl	
2.120	**2.121**	**2.122**

more shallow bowl compared to calix[4]arenes. This is because intramolecular hydrogen bonding interactions between the peripheral hydroxyl groups stabilise the molecule in a more splayed conformation.

Both calixarenes and resorcarenes bind a wide range of aromatic guest molecules, predominantly in a 1:1 stoichiometry in the solid state as clathrates (see Chapter 4, Section 4.3). The complexes are generally stabilised by a range of weak interactions, such as C−H···π hydrogen bonds. However, there is very little evidence for significant binding of neutral molecules in organic solvents. Water-solubilising groups, such as sulfonates, can be appended to either rim of the bowl cavity, thereby increasing the affinity for neutral organic guests in water as a consequence of hydrophobic interactions.

Resorcarenes, such as **2.121**, have themselves been used as hosts for cations in basic media. In addition, the hydroxyl groups from adjacent resorcinol-derived units can be covalently linked to form bridged resorcarenes, such as **2.122**, resulting in highly rigid bowl-shaped molecular hosts. For instance, receptor **2.122** is able to form complexes with neutral guests, such as SO_2, CS_2 and CH_2Cl_2.

2.7.3   Carcerands and hemicarands

When two cavitands are covalently linked, the result is a capsular molecule that completely encloses space. These classes of compounds are termed *carcerands*, derived from the Latin word *carcer*, meaning *imprison*. Once the guest species is incarcerated within the host forming a *carceplex* (a carcerand and guest complex), it cannot escape, for example, **2.123** (Figure 2.30). The gaps within the molecular framework are too small for any encapsulated guest of significant size to escape and the guest molecule can only be freed by the breaking of covalent bonds. *Hemicarands* are partial containers or capsules that allow measurable guest exchange to occur, *e.g.* **2.124** and **2.125**. These hosts are capable of neutral-guest complexation by steric imprisonment, presenting a significant kinetic barrier to guest exchange.

> **Carcerand:** Two cavitands covalently linked together, resulting in a completely enclosed cavity. Guests are unable to escape.

One of the earliest examples of a carcerand was prepared by joining two hemi-spherical bowl-shaped components together (*cavitands*, such as calix[4]arenes or the closely related [4]resorcarenes (Section 2.7.2). This was achieved by reacting a thiol-appended resorcarene with an analogous alkyl halide to form a thioether-bridged equatorial seam (**2.123**). This led to a challenging analysis as carcerand

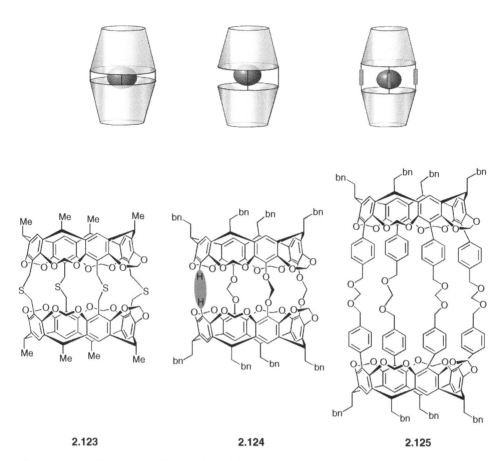

2.123   **2.124**   **2.125**

Figure 2.30 Schematics and examples of the carcarand **2.123**, hemicarcarand **2.124** (one of the bridging links missing–highlighted; bn-benzyl) and hemicarcarand **2.135** (large aperture).

2.123 is extremely insoluble. Characterisation was performed solely by mass spectrometry and elemental analyses. It transpires that on synthesis of **2.123**, a number of guest species are encapsulated, depending on the conditions employed during the reaction and the time of cavity closure. This step of the reaction is called the *guest determining step* (GDS). Various solvent molecules, templating ions (Cs^+ and Cl^-) and even argon (the reaction is carried out under an inert argon atmosphere) are present in various amounts within the carceplexes.[36]

The carcerands were the first examples of single molecules entrapped within single molecules and have opened the door to many different types of host–guest inclusion complexes. The complete lack of guest exchange in the carcerands, however, makes them of only limited interest. Later work has focused on the closely related hemicarcerands. In contrast to the carcerands, hemicarcerands are able to reversibly bind guests at ambient or elevated temperatures, forming *hemicarceplexes*. Guest exchange is frequently slow, despite the lack of significant host–guest interactions. This effect is known as *constrictive binding*. Guest exchange is achieved in one of two ways, either *via* a 'portal' made by omitting one of the four bridging groups between the hemispheres (**2.124**) or by making the bridging groups long enough to provide large holes in the side of the cage (**2.125**) (Figure 2.30). As with the carcerands, the hemicarcerands are produced *via* a templated reaction and there is no empty hemicarcerand formed. The guest-exchange kinetics of the hemicarceplexes with a single portal has been monitored by ^{1}H NMR spectroscopy. The temperature at which decomplexation occurs is related to the size and shape-fit of the guest, compared to the portal through which the exchange takes place. Small molecules, such as diatomic gases and water, can enter and leave relatively easily, whereas larger solvent molecules, such as chloroform, are sterically prevented from doing so. At 140°C, solvent molecules display first-order exchange behaviour with a long exchange half-life. For host **2.125**, half-lives of 14 and 34 h are observed for $(CH_3)_2NCHO$ and $(CH_3)_2NCOCH_3$ guests, respectively, using 1,2,4-trichlorobenzene as a solvent (a solvent that is not included within the host on steric grounds). Smaller solvent molecules, such as acetonitrile, show guest-exchange behaviour at ambient temperatures (see Chapter 4, Section 4.2). For example, the half-life of the 1:1 acetonitrile hemicarceplex of **2.125** for exchange of the acetonitrile guest is 13 h at 22 °C in dichloromethane. Larger guests, such as benzene, can be incorporated but only at very high temperatures. The selectivity of **2.125** in terms of size exclusion is good but it is not possible to discriminate between molecules of a similar size very easily. In a study using O_2, N_2 and H_2O guest, the ^{1}H NMR spectra showed a slow exchange between all three hemicarceplexes, as well as the free host. Formation of the free host is possible, because with one lone portal it is necessary for one single guest to exit the cavity before the next can enter.

> **Hemicarcerand:** Cavitands covalently linked together that either have a portal made by omitting one of the four bridging groups, or by making the bridging groups long enough to provide large holes between the hemispheres for guest exchange.

2.7.4 Cryptophanes and hemicryptophanes

Cryptophanes are related to hemicarcerands – they are composed of smaller and shallower cavitand bowls, such as cyclotriveratrylene (CTV) molecules that have been used for both anion and neutral molecule recognition (Section 2.5.2). The first cryptophane was synthesised in 1981 *via* a 'self-directed synthesis'.[37] This small cage is named, chronologically, cryptophane-A (**2.126**) (Figure 2.31). Space-filling models of cryptophane-A show a closed surface container, with

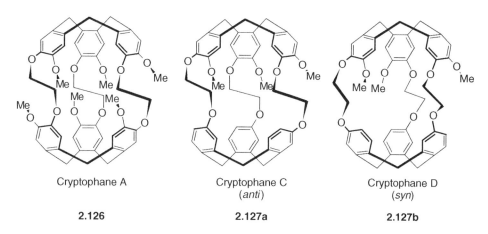

Cryptophane A   Cryptophane C   Cryptophane D
                      (*anti*)   (*syn*)

2.126   **2.127a**   **2.127b**

Figure 2.31 Closely related cyclophanes, A, B and C.

no room for small molecules to pass through the 'methoxy-guarded' apertures. Initial complexation studies were carried out with CH_2Cl_2 as a potential guest, but only very weak binding was observed. The conclusion was that CH_2Cl_2 was too large to fit between the gaps of the cryptophane. To solve this problem, the apertures needs to be larger and this was achieved by omitting OCH_3 groups on one of the CTV bowls to give the related cryptophane-C and cryptophane-D – *anti* (**2.127a**) and *syn* (**2.127b**) – isomers (Figure 2.31). Binding studies carried out by ^{1}H NMR spectroscopy showed that the two cryptophanes, C and D, bind CH_2Cl_2 more effectively than cryptophane-A, even though the binding constants are disappointingly small ($K_a = 1.8$ M^{-1}!). The *anti*-diastereomer (cryptophane-C) binds guest molecules more effectively than the *syn*-diastereomer (cryptophane-D). This is demonstrated by the recognition between optically pure cryptophane-C and one of the simplest chiral molecules, bromochlorofluoromethane (CHFClBr). The association constant between cryptophane-C and CHFClBr in CH_2Cl_2 is 0.22 M^{-1}, whereas for cyclophane-D it is 0.30 M^{-1}. This tiny difference is sufficient to effect the separation of the two resulting diastereoisomers and a weakly resolved sample of this simplest chiral guest can be obtained ($[\alpha]_D^{25} = 1.75$).

The very low binding affinities observed in the early experiments is attributed to solvent competition and a larger non-competitive solvent, 1,1,2,2 tetra-chloroethane, was used for further studies. In this medium, the affinity of cryptophane-A for CH_2Cl_2 and $CHCl_3$ is 475 and 230 M^{-1}, respectively. This highlights the pivotal role of solvent in all binding-constant determinations in supramolecular chemistry. Binding constants must always be interpreted in relation to the solvent system and the particular set of experimental conditions employed – the solvent can drastically influence the binding affinity and indeed can be said to be the dominant factor to a first approximation.

Very recent work has resulted in an anion-binding derivative of crypto-phane E (analogous to cryptophane A, but with propylene bridges instead of ethylene), in which each aromatic ring is capped by a $C_5Me_5Ru^+$ moiety (Section 2.5.2). Because of the resulting 6+ change on the molecule, an anion, $CF_3SO_3^-$, is included within the cavity. As with the neutral hemicarceplexes, the guest exchange is very slow, taking of the order of three weeks to reach equilibrium under ambient conditions in nitromethane, with a complexation activation energy of $75\,kJ\,mol^{-1}$.[38]

Hemicryptophanes also exist that are analogous to cryptophanes but only have one bowl-shaped (CTV) fragment (analogous to hemispherands, such as **2.127**, but note the contrast to hemicarcerands where two bowl-shaped units are present). The hemicryptophane **2.130** is prepared by reacting the CTV derivative **2.128** over several steps to produce the derivatised CTV (**2.129**), which was then capped to produce the hemicryptophane **2.130** (Scheme 2.18). The hemicryptophane **2.130** can exist as two different steroisomers, depending on the orientation of the P=S bond, *i.e.* whether this functional group points into or out of the cavity. Only one of the isomers can be isolated, shown by X-ray crystallography to be the 'outward isomer' (Scheme 2.18). A single toluene molecule was encapsulated within the compound. In addition to being a neutral guest (toluene) receptor, this molecule also possesses basic imine functionalities and hence can also act as a ditopic receptor (Section 2.5) and simultaneously binds cations.

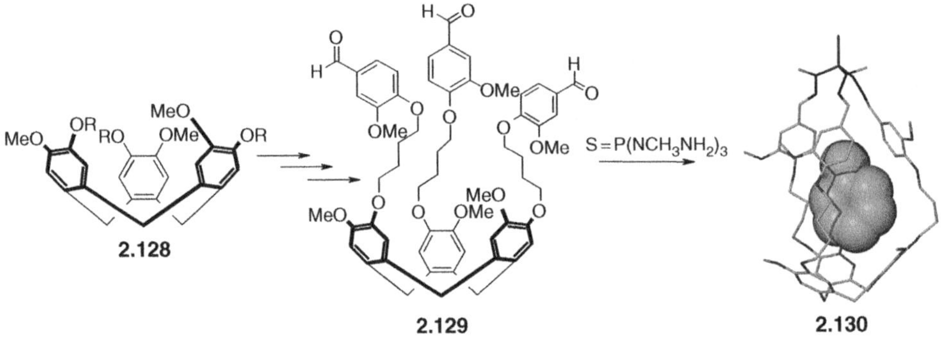

2.128 → **2.129** $S=P(NCH_3NH_2)_3$ → **2.130**

Scheme 2.18   Scheme 2.18 Synthesis of the hemicryptophane **2.130**. The X-ray crystal structure shows a toluene solvent molecule encapsulated within the host.

2.7.5   Cyclodextrins

 Del Valle, E. M. M., 'Cyclodextrins and their uses: a review', *Proc. Biochem.*, 2004, **39**, 1033–1046.

Cyclodextrins are a class of chiral, cyclic oligosaccharides that have molecule-sized cavities. They are preorganised and have a defined bowl shape that is held together by an intramolecular hydrogen bonding network. Cyclodextrins are the most widely used receptors in host–guest inclusion chemistry, with a broad range of applications and industrial production of over a thousand tons per annum. They are used in the food and cosmetics industries and the pharmaceutical sector as stabilising agents, and for the slow release of drugs. Their existence has been known for a long time, but they were merely scientific curiosities until the latter part of the 20th Century. The explosion in their use in recent years is due to a number of factors. Cyclodextrins are semi-natural compounds – they are synthesised from starch *via* a simple enzymatic conversion. As a consequence, the technology employed to synthesise cyclodextrins on a multi-ton scale is both cheap and environmentally friendly. Cyclodextrins are effective complexing agents for a wide range of molecular guests but have negligible toxicity which can be eliminated by selecting the appropriate derivative for a particular application. The importance of cyclodextrins is such that a magazine, *Cyclodextrin News*, is dedicated to all areas of research on cyclodextrins. The total number of cyclodextrin-related papers is over 30 000 as of July 2005 – averaging 4.4 publications per day between 2004 and 2005 alone!

Cyclodextrins commonly comprise between six and eight D-glucopyranoside units that are linked together by a 1,4-glycosidic link. There are three important crystalline, homogeneous and non-hydroscopic cyclodextrins, *i.e.* α-, β- and γ-cyclodextrin, consisting of six (**2.131**), seven (**2.132**) and eight (**2.133**) glucopyranose units, respectively (Figure 2.32). Smaller and larger ring systems are also known, for example, molecules with five, nine and ten units are referred to as pre-α-cyclodextrin, δ-cyclodextrin and ε-cyclodextrin, respectively. This nomenclature is historical, and is generally accepted in industry.

The shapes of cyclodextrins are generally represented as a cylindrical funnel (Figure 2.33) by analogy to the calixarene family. Cyclodextrins have an upper (wide) and lower (narrow) rim. The upper rim consists of the secondary hydroxyl groups and the lower of primary hydroxyl groups. The large number of hydrophilic hydroxyl groups around the rims, plus the hydrophobic nature of the cavity, gives these molecules their unique ability to form inclusion complexes in water.

Cyclodextrins can easily be derivatised by modification of the hydroxyl groups. There are a large number of cyclodextrin derivatives in the literature, containing groups such as alkyl, hydroxyalkyl, carboxyalkyl, amino, thiol and tosyl, which often contain ether or ester linkages. Derivatisation improves the hosts' solubility, increases their affinity for a particular guest target, allows the attachment of specific catalytic groups or allows grafting onto a polymer support for use in chromatographic separation technologies.

Cyclodextrins usually form 1:1 host–guest complexes, although 1:2 and 2:2 complexes are also possible. In aqueous solution, the hydrophobic (or at best 'semi-polar') cavity is filled with water. This water is relatively unstable, due to unfavourable polar–apolar interactions, and the water molecules can be easily replaced by another guest molecule less polar than water. For example *p*-xylene forms a 1:1 complex with β-cyclodextrin (Figure 2.34).

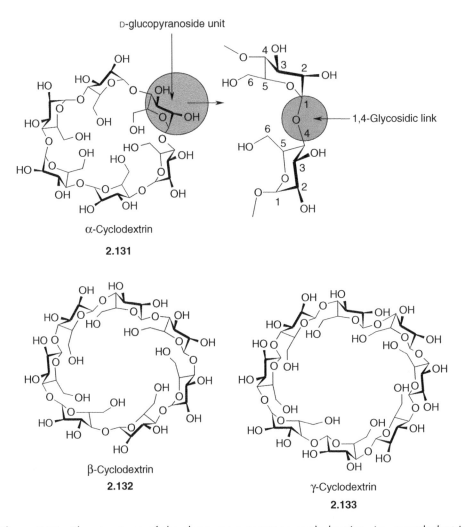

Figure 2.32 The structures of the three most common cyclodextrins, *i.e.* α-cyclodextrin (**2.131**), β-cyclodextrin (**2.132**) and γ-cyclodextrin (**2.133**).

α-CD   174 Å
β-CD   262 Å
γ-CD   427 Å

Figure 2.33 Schematic of the common cyclodextrins (CDs), highlighting the cavity volume.

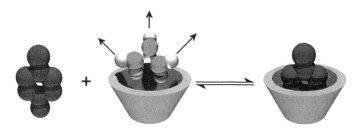

Figure 2.34 Schematic of the cyclodextrin inclusion complex with *p*-xylene in water.

To use a more sophisticated example, we can look to the products of the neem tree (*Azadirzchta indica*), a tropical plant that is known for its pesticidal properties. The seed of this tree is abundant with limonoids and simple terpenoids that are responsible for its biological activity. One particular limonoid found in the seed is Azadirachtin (**2.134**). The bioactivity of Azadirachtin potentially leads to a wide range of applications in herbal medicine and healthcare products for the treatment of malaria and tuberculosis and in anti-worm, clotting, and blood-detoxification preparations. These uses of Azadirachtin as a biopesticide or herbal medicine is limited due to solubility constraints in water and its instability as a result of its propensity to undergo complicated, irreversible rearrangements under acidic, basic and photolytic conditions. Consequently, there has been much research in the structural modification of Azadirachtin to overcome its solubility constraints to increase stability. This process normally involves many protection and deprotection synthetic steps and chromatographic separations.

2.134

To overcome the solubility constraints, a water-soluble complex with β-CD and Azadirachtin has been reported. The binding constants measured by NMR titration for Azadirachtin with α-, β-, and γ-CDs are, *ca.* 0, 238 and $73\,M^{-1}$, respectively. The mutual affinity of Azadirachtin and β-CD shows that both size and shape play a critical role. The NMR studies also revealed the presence of both 1:1 and 2:1 cyclodextrin–guest complexes in solution.

Cyclodextrins: industrial applications

Cyclodextrins are used extensively in the pharmaceutical industry. One major problem often encountered is a drug's poor solubility, as in the case of Azadirachtin, which affects the bioavailability of the drug and the time it takes to reach the blood stream. Another problem is the stability of the drug with respect to photochemical degradation or moisture absorption upon contact with air, so causing formulation problems.[39] Pharmaceutical companies prefer to administer drugs in an orally delivered solid form, such as a powder or tablet. Cyclodextins have often been used in order to address these issues as they have been found to improve bioavailability and to enhance the shelf-life. Moreover, often the cyclodextrin–drug complex is a freely flowing powder and therefore easier to handle. Even though there are numerous advantages of using cyclodextrins in the pharmaceutical world, there are also a number of disadvantages. Not all drugs form stable inclusion complexes, in particular, inorganic-derived drugs (in their salt form). In addition, bulky molecules, such as proteins, aromatic amino acids and polypeptides, are not bound in the cavity. This problem is overcome by binding a particular part of the drug, for example, a side chain in prostaglandin E_1 (Figure 2.35).

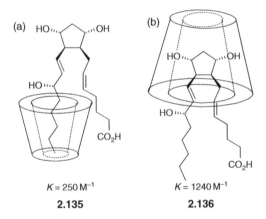

Figure 2.35   The binding of prostaglandin E_1 by (a) α-cyclodextrin and (b) β-cyclodextrin.

Cyclodextrins are also widely applied as food additives. Much of the food we eat contains flavourings. These substances are generally volatile and decompose readily, limiting shelf-life and increasing costs. Many foodstuffs have various quantities of cyclodextrins, in particular, β-cyclodextrin, incorporated into the food to form stable complexes with flavouring agents, so significantly increasing the shelf-life and obviating the need for other preservatives. For example, a lemon-peel oil and β-cyclodextrin complex is mixed with powdered sugar to make pastries. Many of the spicy flavours in canned meat are often complexed with cyclodextrins. However, cyclodextrins are not flavourless – they tend to have a sweet taste, which can be undesirable. Another use for cyclodextrins is in the preparation of oil and water emulsions, such as in mayonnaise and salad dressings. One of the most elegant examples of the use of cyclodextrins is the removal of cholesterol from butter, a separation technique commonly used in Belgium. In this process, β-cyclodextrin is added to a melted butter mix forming a stable complex with cholesterol. This complex is easily removed to produce butter greater than 90 % cholesterol-free. Note that this is a separation technique, and no β-cyclodextrin is found in the final product.

The use of cyclodextrins in the cosmetic industry is very similarly widespread. Most vitamins form stable complexes with cyclodextrins. These complexes have been found to increase the bioavailability of compounds such as vitamin A and retinol. The added advantage is the protection of vitamin A and retinol against oxidation. Again, as with the food industry, emulsions are easily prepared. Cyclodextrins have also been incorporated into many sun-block lotions. Complexation increases the photostability of the cream, increases its protective properties, reduces odour and decreases the staining of fabrics.

Cyclodextrins (and their derivatives) have been used widely in analytical chemistry, in particular, in separation technology. They have been incorporated into chromatographic applications, such as thin layer chromatography, gas chromatography, capillary electrophoresis and high performance liquid chromatography (HPLC), for the separation of similar chemical substances and even enantiomers (cyclodextrins are chiral). Conventional chiral HPLC columns are costly. A normal column containing an immobilised chiral cyclodextrin, such that one enantiomer forms a more stable complex over the other, can drastically reduce the cost.

2.7.6 Clefts and tweezers

Molecular 'clefts' or 'tweezers' can be described as a special type of podand system that are preorganised so that the binding sites are separated by a spacer and the guest is bound between the two pendant side arms in a 'pincer' style. Many of these systems contain aryl rings, as these represent a rigid scaffold for preorganising the host into a specific shape. The advantage of preorganisation is that the side arms can be arranged so that a required minimum distance between the host and guest is enforced. Early work on molecular clefts utilised Tröger's base (**2.137**) and Kemp's triacid (**2.138**) as core building blocks, which possess well-defined geometries. This rigidity allows the controlled assembly of molecular-cleft complexes in which the host contains functional groups that are complementary to the recognition sites of a target guest. Rebek Jr and co-workers have synthesised a series of preorganised molecular clefts incorporating Kemp's tri-acid derivatives and various diamines, to form different-sized molecular clefts (**2.139–2.141**) (Scheme 2.19).[40] The spacers between the two Kemp acid units have methyl substituents on the aryl cores. This functionality is a key factor in the preorganisation of the molecular cleft. It hinders rotation around the N–C_{aryl} bonds, forcing the molecule to have a fixed orientation.

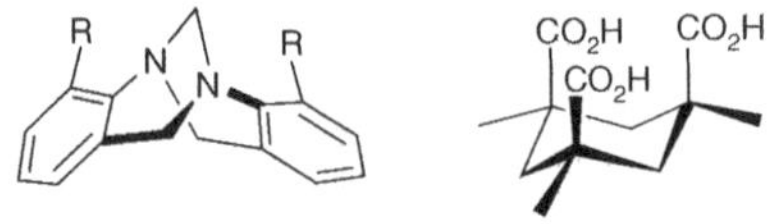

Tröger's base	Kemp's triacid
2.137	**2.138**

Tröger's base (**2.137**) can be resolved into two enantiometric forms which have also been utilised as scaffolds for molecular clefts due to the curvature of the backbone and the chiral nature of the molecule (Scheme 2.20). All of the resultant systems have been found to bend sharply to produce the cleft-like architecture. Depending on the substituent that is appended to the aryl rings, the curvature ranges from 92 to 104°. For example, host **2.142** binds 2-aminopyrimidine as a 1:1 host–guest complex. The binding constant is $2.4 \times 10^4\,M^{-1}$ in rigorously dry dichloromethane.[41] This compound was used as a test case in a fascinating

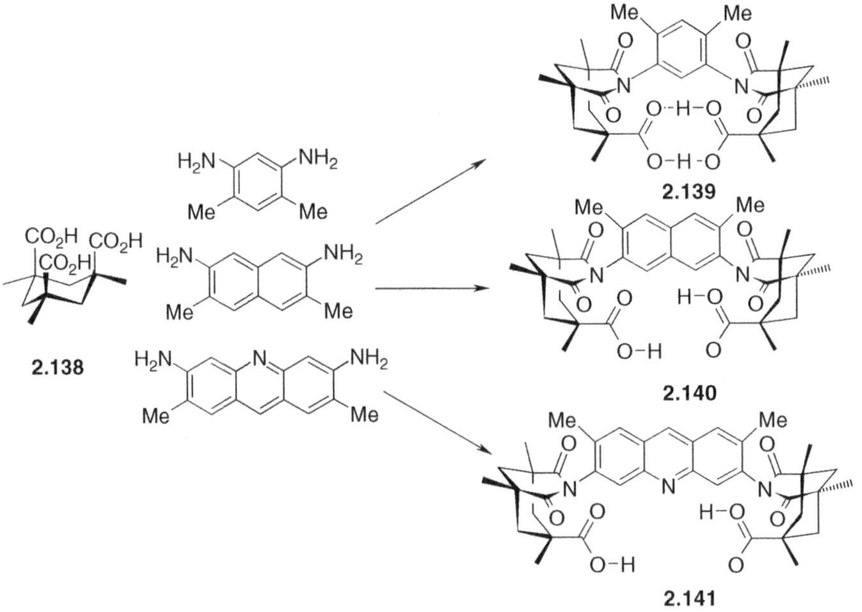

Scheme 2.19 Different rigid amine spacers, using Kemp's triacid, to form various-sized molecular clefts.

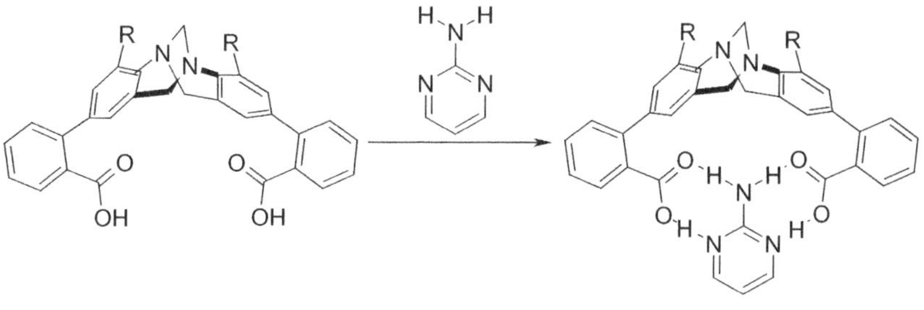

2.142   **2.142a**

Scheme 2.20   The use of Tröger's base to form a molecular cleft.

exploration of the effect of solvation on affinity. Rigorous drying of solvents is difficult and many binding constant determinations in the literature are carried out in solvents containing varying amounts of adventitious water. Remarkably, the binding constant calculated for **2.142** and 2-aminopyrimidine does not change significantly when a dichloromethane-saturated water solution is used. This observation is not as reassuring as it seems, however, because it arises from the accidental cancelling of large changes in enthalpic and entropic contributions in the two different media. The water competes strongly for the guest binding sites, so reducing binding enthalpy, but the entropic contribution associated with release of bound water is more favourable.

2.8   Supramolecular catalysis and enzyme mimics

Maksimov, A. L., Sakharov, D. A., Filippova, T. Y., Zhuchkova, A. Y. and Karakhanov, E. A., 'Supramolecular catalysts on the basis of molecules – receptors', *Ind. Eng. Chem. Rev.*, 2005, **44**, 8644–8653.

2.8.1   Zinc complexes as catalysts

We have seen throughout this chapter that the supramolecular chemist has derived inspiration from nature. This is especially true when it comes to mimicking enzymes – the biological catalysts. Before we delve into examples of enzyme mimics, we must first have an understanding of how enzymes themselves catalyse reactions. One of the most important catalytic biological reactions is the hydrolysis of the amide linkage in a protein. For this to occur in nature, three things are required to happen. The first is an enhancement in the rate of nucleophilic attack on the substrate, the second is the stabilisation of the intermediate and the third is enhancement of the rate of leaving-group departure.

One of the most studied classes of enzymes that cleave the peptide bond is the *peptidases*, of which *chymotrypsin* is one of the best known. The nucleophile in the reaction catalysed by this enzyme is from the alcohol group on the side chain of serine-195. The reaction is base-catalysed from the adjacent side chain, histidine-57. A stable intermediate is formed due to the hydrogen bonding occurring in the cavity, known as the 'oxy-anion packet'. The leaving-group departure occurs from an acid-catalysed reaction (Scheme 2.21).

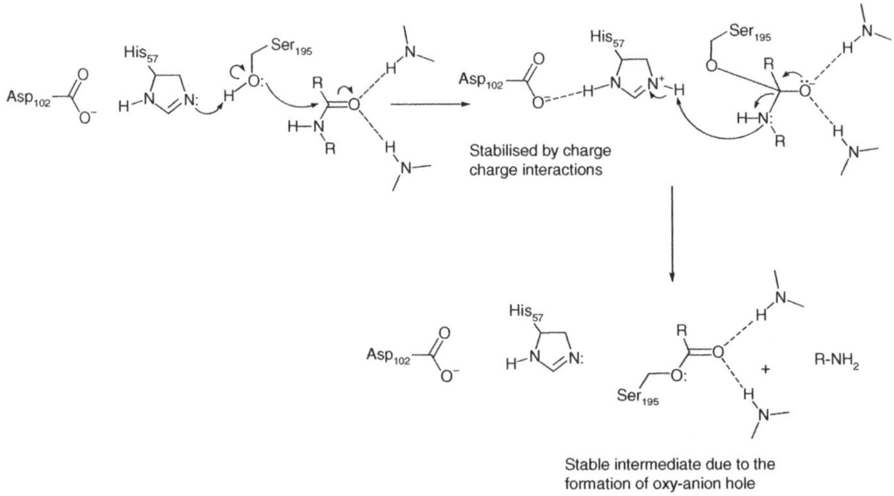

Scheme 2.21   Peptide-bond cleavage by chymotrypsin.

There have been many examples of synthetic supramolecular compounds that have been designed to mimic the type of reaction shown in Scheme 2.21. For example, a class of compounds containing Zn(II) triazacyclododecane and cyclam units such as **2.143**, have been used extensively as enzyme mimics due to the strong interaction between the azacrown and the metal centre and the Lewis acidity of Zn(II). The metal also has one or more labile coordination sites, often bound to water. Complex **2.143** has a single water molecule coordinating to the Zn(II) centre. The bound water molecule's acidity is dramatically enhanced by zinc coordination (pK_a of 7.30, compared to a value of 14 for uncoordinated water) and hence the water is readily deprotonated (Scheme 2.22). Other anions can also bind to the metal centre and their strength of binding correlates to the pK_a of the conjugate acid of the anion. The more basic anions have a higher affinity to the Lewis-acidic Zn(II) centre. This anion recognition-property makes these receptors good enzyme mimics.

The Zn(II) complex (**2.144**) mimics the *esterase* reaction, as shown in Scheme 2.23. The complex uses lariat azacrown ethers for the hydrolysis of ester groups. This is of significance interest because of the hydrolysis of phosphate

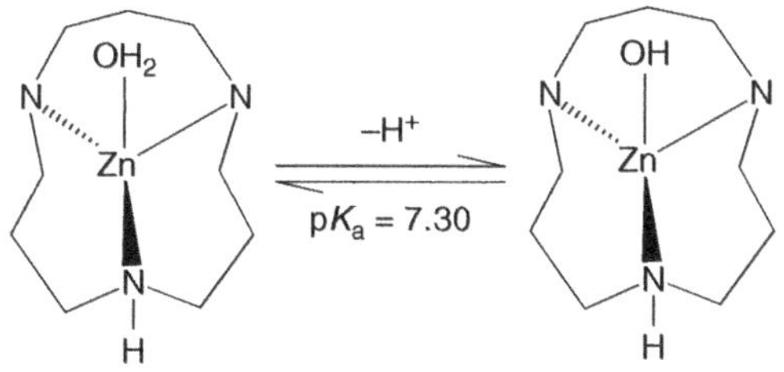

2.143

Scheme 2.22 Deprotonation of the complex **2.143**.

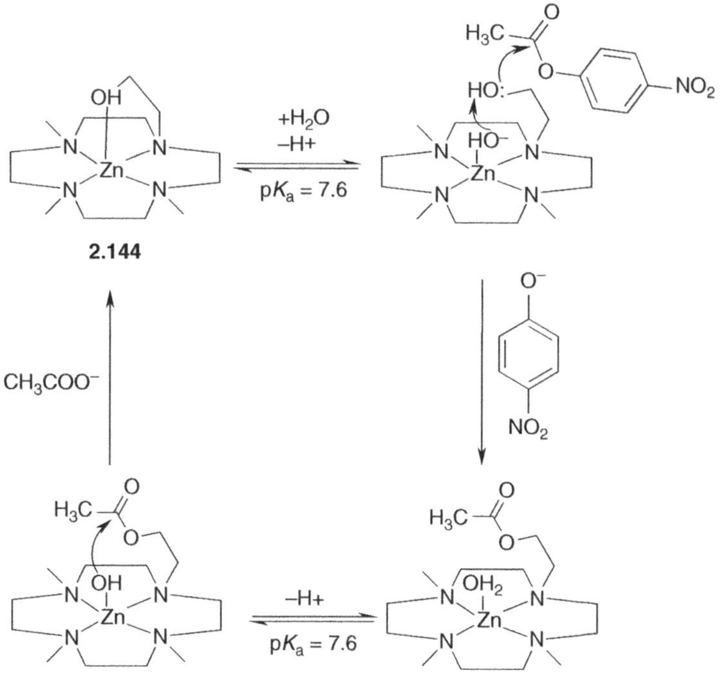

Scheme 2.23 The catalytic cycle for the hydrolysis of 4-nitrophenyl acetate.

esters that are found in DNA and RNA. The hydrolysis of 4-nitrophenyl acetate has been measured in aqueous conditions at various pH values (6.4–9.4). The OH group of the lariat arm in azacrown **2.144** becomes activated as a hydroxide ion coordinates to the metal centre, which is able to extract the proton from the OH group on the lariat arm which becomes highly nucleophilic and attacks the carbonyl group of 4-nitrophenyl acetate. The kinetics are second-order, with a rate constant of $0.46\,M^{-1}\,s^{-1}$. This reaction is ten times faster than the analogous reaction with no alcohol group on the lariat ether side arm. In alkaline solution, the acyl intermediate is deprotonated and the acetate is hydrolysed to give the starting materials and CH_3COO^-.

2.8.2 Calix[4]arenes and cyclodextrins as catalysts

We have previously seen that calix[4]arenes and cyclodextrins have been used as host molecules for the recognition of various guest molecules (Sections 2.3–2.6) As a consequence of their hydrophobic cavity in conjunction with reactive functional groups, both calix[4]arenes and cyclodextrins, in particular, β-cyclodextrin, have been used in many catalytic applications, for example, ester hydrolysis, oxidation reactions, hydroformylation, hydrogenation and cross-coupling reactions.

To take just one example, an important industrial process is the *Wacker process*, *i.e.* the oxidation of ethene to ethanal, which traditionally uses a Pd(II) heterogeneous catalyst or a Pd(II) salt, together with a phase-transfer catalyst. Oxidation of higher alkenes using the traditional conditions is problematic, as longer-chain alkenes generally form isomerised ketones and therefore the selectivity is low. To overcome this problem, both derivatised cyclodextrins and calixarenes, for example, **2.145** and **2.146**, have been prepared. The nitrile functionalities coordinate to the Pd(II) in the catalytic cycle. The growth of activity is caused by a co-operative binding (see Chapter 1, Section 1.2.3) of the substrate in the cavity of the cyclodextrin **2.145** or the calixarene **2.146**, with simultaneous coordination of the nitrile group with the Pd(II) reaction centre.

2.145   NC   **2.146**

Derivatised β-cyclodextrin   Derivatised calix[4]arene

The use of supramolecular catalysts has achieved the conversion of 1-octene to its corresponding ketone (2-octanone), using β-cyclodextrin **2.145** in 2–4 h, whereas the traditional surfactant phase-transfer process takes significantly longer and requires more Pd(II) in the cycle.[42]

2.8.3 Dendimers as nano-reactors for catalysis

Dendrimers are branch-like polymers (see Chapter 5, Section 5.8) and have been used as *nano-reactors*.[43] These branched molecules have large voids in their structure that can incorporate a catalytic species into the inner nanoscale environment of the dendrimer. The dendrimer poly(propylene imine) (PPI) has been used as a nano-reactor for *Heck reactions* and alkyl aminations. 4-Diphenylphosphinobenzoic acid is fixed into the voids of the dendrimer *via* ionic contacts between the positive charges on protonated amine functional groups and the deprotonated benzoic acid and used to anchor the palladium catalyst (Figure 2.36).

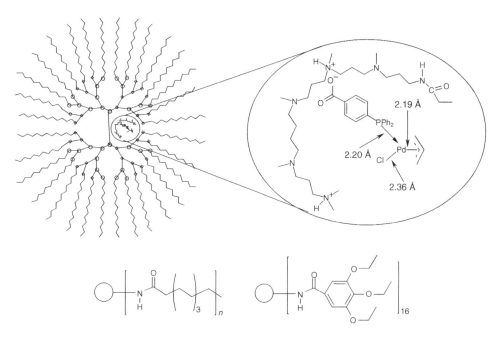

Figure 2.36 Proposed structures of a series of PPI dendrimers incorporating a Pd catalyst. The chemical structures of the dendrimers are shown at the bottom ($n = 16, 32$ or 64).

Higher generations of dendrimer (larger and more branched) lead to high catalytic activity and stability of the encapsulated Pd complex. For example, the 5th-generation dendrimer–Pd complex used for the Heck coupling between an alkyl iodide and styrene yielded the Heck product **2.147** in 99 % yield in 12 h (Scheme 2.24).[44]

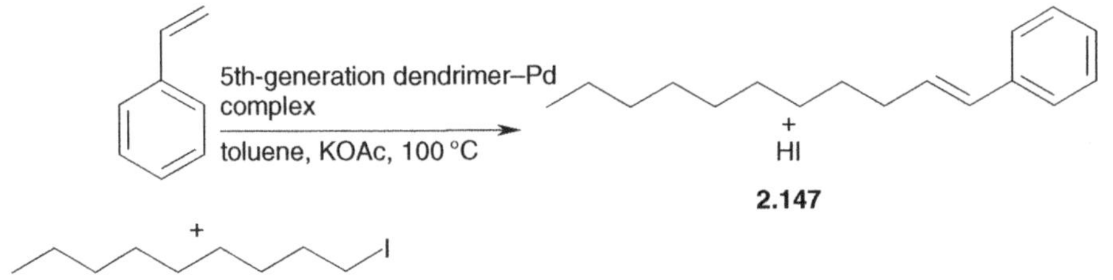

Scheme 2.24   Heck reaction between an alkyl halide and styrene with a Pd(0) catalyst supported in a 5th-generation dendrimer.

2.8.4   Enzyme mimics

We have previously described how versatile β-cyclodextrins are as host molecules (Section 2.6.6). If a single hydroxyl group is deprotonated, a nucleophilic centre is generated in close proximity to the hydrophobic cavity, hence resulting in enzyme-like activity.[45]

The 3′, 5′-cyclic monophosphate of adenosine (cAMP) (**2.148**) is an important secondary messenger for intercellular communication in biochemistry. When the cell is stimulated by the first messenger, compound **2.148** is formed from adenosine triphosphate (ATP) (Scheme 2.25). This reaction is catalysed by an adenosine cyclase enzyme. The cAMP then goes on to activate other intracellular enzymes, so producing a cell response. The response is terminated by the hydrolysis of cAMP by phosphodiesterase (a phosphate-ester-hydrolysis enzyme). The action of adenylate cyclase has been mimicked successfully with a β-cyclodextrin complex of Pr(III) and other lanthanide(III) metals, under physiological conditions. The

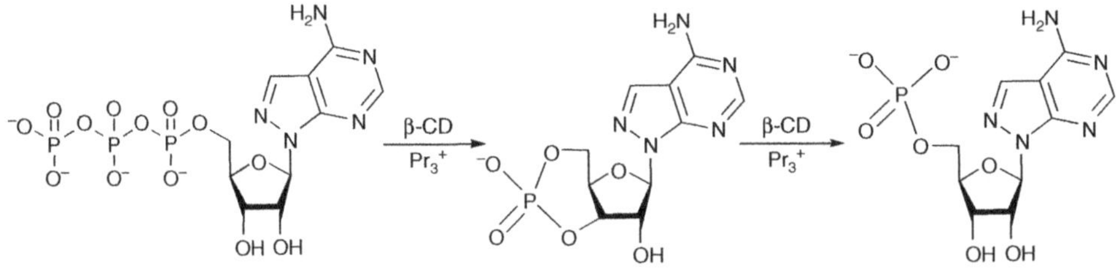

Scheme 2.25   Adenylate cyclase/phosphodiesterase activity by cyclodextrin complexes.

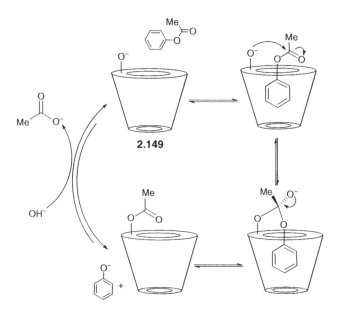

Scheme 2.26 Catalytic cycle of ester hydrolysis by β-cyclodextrin.

half-life for the catalysed conversion into adenosine monophosphate (AMP) is 6 s
(Scheme 2.26), whereas the half-life for the uncatalysed reaction is 3 000 000 years!
The acceleration factor of 10^{13} is attributed to the strong interaction between the
ligand and the hard, oxophilic metal ion.

2.8.5   Corands as ATPase mimics

Adenosine triphosphate (ATP) is important in biological systems as an energy
storage molecule. The energy that comes from ATP is produced by ATPases, which
hydrolyse the terminal phosphate residue of the triphosphate tail of ATP to produce
adenosine diphosphate (ADP) and inorganic phosphate ($H_2PO_4^-$). We have already
seen that protonated azacrowns (*cf.* Section 2.3) can be effective anion-binding hosts
due to the $N^+-H\cdots X^-$ hydrogen bonding interaction. The protonated corand **2.150**
has a cleft-like structure in the solid state and molecular modelling supports the
hypothesis that this structure also exists in solution. The corand has a remarkably
strong binding affinity for ATP (log $K = 11$) when fully protonated. It accelerates
the rate of ATP hydrolysis by a factor of 100 over a range of pH values. The hydrolysis has been followed by ^{31}P NMR spectroscopy which has revealed the production
of phosphorylated intermediates, and led to the proposal of the mechanism shown
in Scheme 2.27.[46]

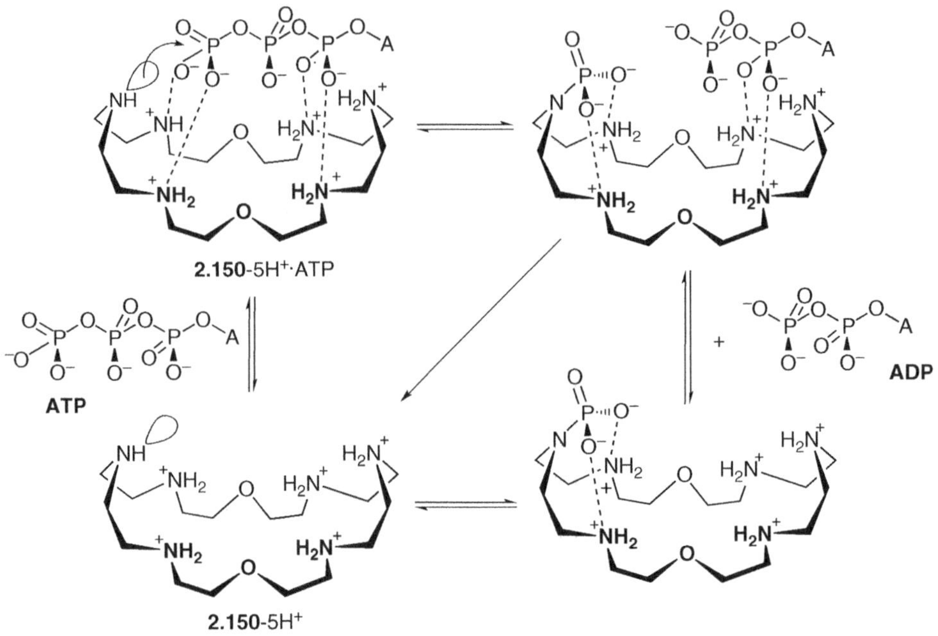

Scheme 2.27 Mechanism of ATP hydrolysis by the protonated corand **2.150** (A = adenosine).

References

1. Nieto, S., Pérez, J., Riera, V., Miguel, D. and Alvarez, C., 'Cationic fac-tris(pyrazole) complexes as anion receptors', *Chem. Commun.*, 2005, 546–548.

2. Zhu, L. and Anslyn, E. V., 'Signal amplification by allosteric catalysis', *Angew. Chem., Int. Ed. Engl.*, 2006, **45**, 1190–1196.

3. Richman, J. E. and Atkins, T. J., 'Nitrogen analogs of crown ethers', *J. Am. Chem. Soc.*, 1974, **96**, 2268–2270.

4. Pedersen, C. J., 'Cyclic polyethers and their complexes with metal salts', *J. Am. Chem. Soc*, 1967, **89**, 7017–7036.

5. Leznoff, C. C. and Lever, A. B. P. (Eds), *The Phthalocyanines*, Vols 1–4, John Wiley & Sons, Inc., New York, NY, USA, 1986–1993.

6. Pedersen, C. J., 'The discovery of crown ethers (Nobel Lecture)', *Angew. Chem., Int. Ed. Engl.*, 1988, **27**, 1021–1027.

7. Lehn, J.-M., 'Supramolecular chemistry – scope and perspectives, molecules, super-molecules and molecular devices', *Angew. Chem., Int. Ed. Engl.*, 1988, **27**, 89–112.

8. Simmons, H. E. and Park, C. H., 'Macrobicyclic amines. I. Out–in isomerism of 1, (K + 2)-Diazabicyclo K.L.M alkanes', *J. Am. Chem. Soc.*, 1968, **90**, 2428–2429.

9. Cram, D. J., 'The design of molecular hosts, guests and their complexes (Nobel Lecture)', *Angew. Chem., Int. Ed. Engl.*, 1988, **27**, 1009–1020.

10. Steed, J. W., 'First- and second-sphere coordination chemistry of alkali metal crown ether complexes', *Coord. Chem. Rev.*, 2001, **215**, 171–221.

11. Pearson, R. G. (Ed.), *Chemical Hardness*, John Wiley & Sons, Inc., New York, NY, USA, 1997.

12. Gokel, G. (Ed.), *Crown Ethers and Cryptands*, The Royal Society of Chemistry, Cambridge, UK, 1991.

13. Glenny, M. W., van de Water, L. G. A., Vere, J. M., Blake, A. J., Wilson, C., Driessen, W. L., Reedijk, J. and Schroder, M., 'Improved synthetic methods to mixed-donor thiacrown ethers', *Polyhedron*, 2006, **25**, 599–612.

14. Gokel, G. W., Barbour, L. J., Ferdani, R. and Hu, J., 'Lariat ether receptor systems show experimental evidence for alkali metal cation – pi interactions', *Acc. Chem. Res.*, 2002, **35**, 878–886.

15. Fenton, D. E., 'Metallobiosites and their synthetic analogues – a belief in synergism – 1997–1998 Tilden Lecture', *Chem. Soc. Rev.*, 1999, **28**, 159–168.

16. Doyle, D. A., Cabral, J. M., Pfuetzner, R. A., Kuo, A., Gulbis, J. M., Cohen, S. L., Chait, B. T. and MacKinnon, R., 'The structure of the potassium channel: molecular basis of K^+ conduction and selectivity', *Science*, 1998, **280**, 69–77.

17. Matthews, S. E., Schmitt, P., Felix, V., Drew, M. G. B. and Beer, P. D., 'Calix[4]tubes: A new class of potassium-selective ionophore', *J. Am. Chem. Soc.*, 2002, **124**, 1341–1353.

18. Miller, M. J. and Malouin, F., 'Microbial iron chelators as drug delivery agents – the rational design and synthesis of siderophore drug conjugates', *Acc. Chem. Res.*, 1993, **26**, 241–249.

19. Kiggen, W. and Vogtle, F., 'Functionalized, oligocyclic large cavities – a novel siderophore', *Angew. Chem., Int. Ed. Engl.*, 1984, **23**, 714–715.

20. Metzger, A., Lynch, V. M. and Anslyn, E. V., 'A synthetic receptor selective for citrate', *Angew. Chem., Int. Ed. Engl.*, 1997, **36**, 862–865.

21. Wallace, K. J., Belcher, W. J., Turner, D. R., Syed, K. F. and Steed, J. W., 'Slow anion exchange, conformational equilibria and fluorescent sensing in Venus' flytrap aminopyridinium-based anion hosts', *J. Am. Chem. Soc.*, 2003, **125**, 9699–9715.

22. Hartshorn, C. M. and Steel, P. J., 'Coelenterands: a new class of metal-encapsulating ligands', *Angew. Chem., Int. Ed. Engl.*, 1996, **35**, 2655–2657.

23. Davis, A. P. and Joos, J. B., 'Steroids as organising elements in anion receptors', *Coord. Chem. Rev.*, 2003, **240**, 143–156.

24. Kyba, E. P., Helgeson, R. C., Madan, K., Gokel, G. W., Tarnowski, T. L., Moore, S. S. and Cram, D. J., 'Host–guest complexation. 1. Concept and illustration', *J. Am. Chem. Soc.*, 1977, **99**, 2564–2571.

25. Gale, P. A., Anzenbacher, Jr, P. and Sessler, J. L., 'Calixpyrroles II', *Coord. Chem. Rev.*, 2001, **222**, 57–102.

26. Wedge, T. J. and Hawthorne, M. F., 'Multidentate carborane-containing Lewis acids and their chemistry: mercuracarborands', *Coord. Chem. Rev.*, 2003, **240**, 111–128.

27. Hamilton, A. D. and Linton, B., 'Formation of artificial receptors by metal-templated self-assembly', *Chem. Rev.*, 1997, **97**, 1669–1680.

28. Turner, D. R., Smith, B., Spencer, E. C., Goeta, A. E., Evans, I. R., Tocher, D. A., Howard, J. A. K. and Steed, J. W., 'Anion binding by Ag(I) complexes of urea-substituted pyridyl ligands', *New J. Chem.*, 2005, **29**, 90–98.

29. Filby, M. H. and Steed, J. W., 'A modular approach to organic, coordination complex and polymer-based podand hosts for anions', *Coord. Chem. Rev.*, 2006, **250**, 3200–3218.

30. Harding, L. P., Jeffery, J. C., Riis-Johannessen, T., Rice, C. R. and Zeng, Z. T., 'Anion control of the formation of geometric isomers in a triple helical array', *J. Chem. Soc., Dalton Trans.*, 2004, 2396–2397.

31. Martínez-Máñez, R. and Sancenón, F., 'Fluorogenic and chromogenic chemosensors and reagents for anions', *Chem. Rev.*, 2003, **103**, 4419–4479.

32. Beer, P. D. and Bayly, S. R., 'Anion sensing by metal-based receptors', *Top. Curr. Chem.*, 2005, **255**, 125–162.

33. Motekaitis, R. J., Martell, A. E. and Murase, I., 'Cascade halide binding by multiprotonated bistren and copper(II) bistren cryptates', *Inorg. Chem.*, 1986, **25**, 938–944.

34. Koulov, A. V., Mahoney, J. M. and Smith, B. D., 'Facilitated transport of sodium or potassium chloride across vesicle membranes using a ditopic salt-binding macrobicycle', *Org. Biomol. Chem.*, 2003, **1**, 27–29.

35. Metzger, A., Gloe, K., Stephan, H. and Schmidtchen, F. P., 'Molecular recognition and phase transfer of underivatized amino acids by a foldable artificial host', *J. Org. Chem.*, 1996, **61**, 2051–2055.

36. Cram, D. J., Karbach, S., Kim, Y. H., Baczynskyj, L., Marti, K., Sampson, R. M. and Kalleymeyn, G. W., 'Host guest complexation. 47. Carcerands and carcaplexes, the 1st closed molecular container compounds', *J. Am. Chem. Soc.*, 1988, **110**, 2554–2560.

37. Collet, A., Dutasta, J.-P., Lozach, B. and Canceill, J., 'Cycletriveratrylenes and cryptophanes: their synthesis and applications to host–guest chemistry and to the design of new materials', *Top. Curr. Chem*, 1993, **165**, 103–129.

38. Fairchild, R. M. and Holman, K. T., 'Selective anion encapsulation by a metalated cryptophane with a pi-acidic interior', *J. Am. Chem. Soc.*, 2005, **127**, 16364–16365.

39. Hilfiker, R (Ed.), *Polymorphism in the Pharmaceutical Industry*, Wiley-VCH, Weinheim, Germany, 2006.

40. Rebek, J., Askew, B., Killoran, M., Nemeth, D. and Lin, F. T., 'Convergent functional groups. 3. A molecular cleft recognizes substrates of complementary size, shape and functionality', *J. Am. Chem. Soc.*, 1987, **109**, 2426–2431.

41. Adrian, J. C. and Wilcox, C. S., 'Chemistry of synthetic receptors and functional-group arrays. 15. Effects of added water on thermodynamic aspects of hydrogen bond-based molecular recognition in chloroform', *J. Am. Chem. Soc.*, 1991, **113**, 678–680.

42. Karakhanov, E. E., Maksimov, A. L., Runova, E. A., Kardasheva, Y. S., Terenina, M. V., Buchneva, T. S. and Guchkova, A. Y., 'Supramolecular catalytic systems based on calixarenes and cyclodextrins', *Macromol. Symp.*, 2003, **204**, 159–173.

43. Crooks, R. M., Lemon, B. I., Sun, L., Yeung, L. K. and Zhao, M. Q., 'Dendrimer-encapsulated metals and semiconductors: synthesis, characterization and applications', *Top. Curr. Chem.*, 2001, **212**, 81–135.

44. Ooe, M., Murata, M., Mizugaki, T., Ebitani, K. and Kaneda, K., 'Supramolecular catalysts by encapsulating palladium complexes within dendrimers', *J. Am. Chem. Soc.*, 2004, **126**, 1604–1605.

45. Breslow, R. and Dong, S. D., 'Biomimetic reactions catalysed by cyclodextrins and their derivatives', *Chem. Rev.*, 1999, **98**, 1997–2011.

46. Hosseini, M. W., 'Supramolecular catalysis of phosphoryl anion transfer processes', in *Supramolecular Chemistry of Anions*, Bianchi, A., Bowman-James, K. and Garcia-España, E. (Eds), John Wiley & Sons, Inc., New York, NY, USA, 1997, pp. 421–448.

3
Self-assembly

3.1 Introduction

3.1.1 Self-assembly

Self-assembly, in its most general sense, means the spontaneous association of two or more molecules or ions to create a larger, aggregate species through the formation of reversible (generally supramolecular) interactions. This generic definition has significantly broadened over time to incorporate many aspects of biochemistry and nanotechnology which employ the basic principles of strict chemical self-assembly, albeit with subtle variations. These various disciplines have bled into each other, leading to a sharing of ideas and terminologies and developing a wide scope of work that now comes under the umbrella term of 'self-assembly'.

Relatively simple molecules with complementary functionalities may, under certain conditions, interact with one another to form significantly more complex supramolecular species, held together only by virtue of non-covalent interactions. Examples of such self-assembly processes are repeatedly found in nature. The DNA double-helix requires two complementary strands to become entwined *via* hydrogen bonds and π–π stacking in a self-assembly process (see Chapter 1, Section 1.3). The strands recognise each other and join together to form the most thermodynamically stable assembly product. Protein folding and viral assembly operate in a broadly similar manner.

> **Self-Assembly:** The spontaneous and reversible association of molecular species to form larger, more complex supramolecular entities according to the intrinsic information contained in the components.

Core Concepts in Supramolecular Chemistry and Nanochemistry Jonathan W. Steed, David R. Turner and Karl J. Wallace
© 2007 John Wiley & Sons, Ltd ISBN: 978-0-470-85866-0 (Hardback); 978-0-470-85867-7 (Paperback)

Synthetic self-assembling systems rely upon the ability of the chemist to design molecules containing complementary functionalities. The synthetic chemist has no direct control over the assembly process – otherwise it would not be termed *self*-assembly. Ultimately, the chemical system itself rearranges to the most thermodynamically stable product. It is possible, however, to make intelligent and informed decisions as to what compounds may interact well together by studying existing chemistry, a task for which databases such as the Cambridge Structural Database (CSD) are invaluable (see Chapter 4, Section 4.5.2).

Among the most widely used interactions in synthetic self-assembly processes are metal–ligand interactions, primarily due to their lability and high degree of directionality as a result of predictable metal-ion coordination environments.[1] Metals with well-defined coordination preferences may be coupled with rigid ligands to provide routes towards predictable self-assemblies. The highly directional nature of hydrogen bonds also leads to many examples of self-assembled hydrogen bonded materials in synthetic systems, particularly in cases where multiple hydrogen bonding interactions reinforce one another, as in nucleobase pairing in DNA.

It is possible to employ a retrosynthetic methodology towards the design of synthetic self-assembling systems. Say, for example, that the desired assembly is a molecular square (Figure 3.1). A square can be disassembled into simple geometrical components, *i.e.* four corners with 90° angles and four linear edges.

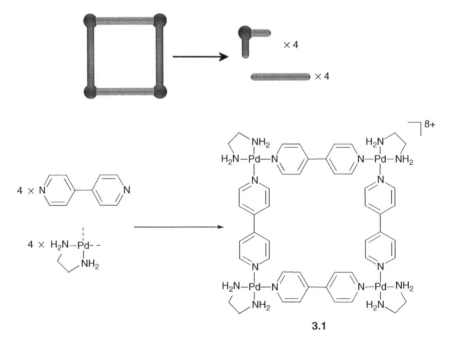

Figure 3.1 A retrosynthetic approach towards a self-assembling square (**3.1**).[2]

If suitable molecular analogues for these components can be found, then the spontaneous formation of a square complex can be designed. A 90° angle is frequently created from a square-planar metal ion, such as Pd^{2+} or Pt^{2+}, which has two *cis* coordination sites blocked off. The linear sides require a ligand that has metal binding sites directly opposite each other, such as 4,4'-bipyridine.

3.1.2 Definitions and basic concepts of self-assembly

Swiegers, G. F., 'Self-assembly: Terminology', in *Encyclopedia of Supramolecular Chemistry*, Steed J. W. and Atwood, J. L. (Eds), Marcel Dekker, New York, NY, USA, 2004, pp. 1263–1269.

In its strictest sense, the term self-assembly can be applied only to systems in which the assembly process is kinetically rapid and both completely reversible and replicable, *i.e.* if taken apart (*e.g.* by change of conditions) the supramolecular aggregate will be reformed under a particular set of conditions so as to be indistinguishable from the original. The reversible nature of the assembly process gives rise to an important feature of self-assembling systems, namely their ability to correct 'mistakes' during assembly and gradually work their way towards the most thermodynamically stable product. When several molecules are able to join together, it is highly likely that there is more than one possible combination. One product of the assembly, however, will be predominant due to greater thermodynamic stability over the others under the reaction conditions. Self-assembled systems can therefore be said to be thermodynamically selective in the product that is formed.

Statistically, it is highly unlikely that a stable aggregate species composed of many molecules will be formed in a single, concerted step. Instead, smaller aggregates, which will go on to act as the building blocks of the final assembly, assemble first (Figure 3.2). For example, in a system containing a metal and ligands, it can be expected that the metal–ligand complex will assemble initially before forming

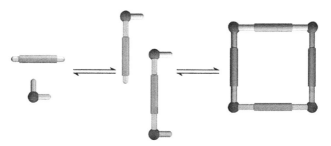

Figure 3.2 A simplistic cartoon of how a thermodynamically stable assembly, a molecular square, may assemble from simple molecular building blocks (Figure 3.1). Small aggregates are initially formed that then go on to form larger assemblies which are thermodynamically stable and may also be relatively kinetically inert.

a supramolecular assembly with other complexes. These smaller assemblies are continually forming and disassociating within the reaction mixture, since they do not represent the thermodynamic minimum of the system. However, once the most stable product is formed it will predominate until the conditions are changed in such a way as to lower its stability or equilibrium concentration. Essentially, any system that is capable of forming products by self-assembly can be thought of as a *dynamic combinatorial library* – many products are capable of forming and their formation and destruction is an ongoing process, tending towards the most thermodynamically stable product. The positions of the equilibria of these systems can be influenced by changes in concentrations and temperature. In this way, one desired product among several potential products that are close in terms of their stability, may be obtained.

It is common within self-assembling systems for there to be more than one type of interaction present. For example, assemblies are often formed using coordination interactions that are supported by hydrogen bonding. In such cases, the system acts to optimise the product by forming those structures that have the lowest overall free energy. This process is dominated by the most stabilising interactions in a descending hierarchy. The stepwise formation of the assembly therefore occurs in a manner in which the interactions that result in the greatest stabilisation of the product are optimised first, followed by the weaker interactions. For instance, the formation of metal–ligand coordination interactions will take higher priority than weaker interactions such as π-stacking. The use of different interactions with differing strengths can be thought of as supplying the system with a set of instructions for the order of the assembly, with the interactions offering the most thermodynamic stability taking precedence in the assembly process. For example, discrete metal–ligand complexes may form initially which then combine into a larger architecture *via* hydrogen bonding. This is an example of *hierarchical assembly*, a self-assembly process that can be broken down into distinct steps that cannot proceed until the preceding step is complete (Section 3.2.2). Such processes are often referred to as a *linear hierarchy*, as the separate processes must occur in a successive fashion with no other possible routes.

It is well-known that certain molecules, metals or functional groups have preferences for binding in certain ways. These set geometries and binding preferences are often considered to dictate the manner in which the assembly process proceeds – they act as 'assembly instructions'. An analogy with computing has been made here, suggesting that these molecular preferences act as an 'interactional algorithm' that programs the self-assembly system. Taking this analogy a step further, it can be said that there exists a linear algorithm taking the system from a high-energy state towards the lowest-energy situation.

Self-assembled systems can be classified according to the diversity of interactions that they contain. *Single-interaction self-assembly* (sometimes referred to as *single code self-assembly*) refers to systems in which only one specific interaction is present (Figure 3.3(a)). For example, the palladium/bipyridyl square (**3.1**) in

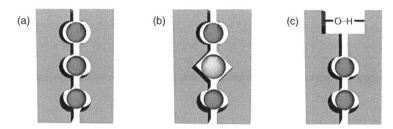

Figure 3.3 (a) A single-interaction assembly, using one specific type of metal–ligand interaction. (b) A unimediated multiple-interaction assembly, using two different metal–ligand interactions. (c) A multimediated multiple-interaction assembly, using both metal–ligand interactions and hydrogen bonding.

Figure 3.1 is referred to as single-interaction self-assembly as it is assembled solely from Pd–N interactions that are all equivalent. The self-assembly of carboxylic acids in the solid state is also deemed to be single-interaction self-assembly as only OH$\cdots$O hydrogen bonds are present.

Self-assembly processes in which more than one interaction of whatever type is present are referred to as being *multiple-interaction self-assembly* (or *multiple-code self-assembly*). Multiple-interaction assemblies can be sub-divided into *unimediated* and *multimediated* assemblies, depending on whether there is more than one category of interaction present. If the self-assembled system contains different categories of interaction (*e.g.* both metal–ligand bonds and hydrogen bonds), then the term *multimediated assembly* is used (Figure 3.3(c)). However, if an assembly contains two different metal coordination environments then we refer to it as a *unimediated assembly* because even though there are two distinct interactions, they are both metal–ligand interactions (figure 3.3(b)). The same terminology applies to systems that contain two or more distinct examples of hydrogen bonding interactions to hold the assembly together. The terms 'multiple interaction assembly' and 'multimediated assembly' are not interchangeable and they should *not* be confused. A further term that is frequently encountered to describe the complexity of a self-assembled species is '*n*-component self-assembly', whereby n is simply the number of separate parts that combine to make the final product. The palladium/bipyridyl square is therefore an example of an 8-component assembly (four Pd^{2+} corners and four bipyridyl edges).

Hierarchical Assembly: A self-assembled system that comprises several levels of complexity which are built up over several, successive self-assembly processes. One level cannot exist without the preceding one being in place.

Single/Multiple Interaction Self-Assembly: Descriptive terms pertaining to the number of different kinds of interaction that are present within the assembled structure.

Uni/multimediated Assembly: A description of how many distinguishable examples of one type of interaction are present in multiple interaction assemblies.

3.1.3 Enthalpic and entropic considerations

Self-assembling systems selectively produce the most thermodynamically stable products and therefore both the enthalpic and entropic contributions towards the final species must be considered. The formation of a self-assembled product, by definition, necessitates the formation of new, favourable interactions, *i.e.* the process is enthalpically favourable. However, the formation of aggregate species occurs at an 'entropic cost' as many degrees of freedom in the system are lost. The entropic penalty is offset somewhat by the release of solvent molecules that were previously interacting with the binding areas of the assembly components, in a *solvophobic effect* (see Chapter 1, Section 1.3.5).

Entropic arguments can also be used to show that there is a preference for discrete oligomeric structures to form rather than polymeric networks, assuming that the connectivity of the components allows for this possibility. Comparing closed-cyclic systems to open-chain oligomers and polymers, cyclic structures form slightly more interactions per structural unit, as they have no loose ends and therefore they are more enthalpically favourable. However, this effect becomes much less pronounced as the number of structural units increases. The formation of discrete structures is also less entropically unfavourable than the formation of polymeric chains, an effect which increases with the number of units in the oligomer. Although polymeric chains allow for more conformational freedom than rigid, closed structures, there is a significant loss in the number of degrees of translational freedom (*i.e.* the number of species present) which outweighs the entropic gain obtained from the flexible species.

3.1.4 Self-assembly with modification

Whilst *strict* self-assembly must be fully reversible and occur spontaneously between two or more components, there are further sub-classes in the field of self-assembly that do not carry such rigid rules. In these cases, the assembly process may require help, such as *templation*, in order to proceed, or contain one or more non-reversible steps:

- **Directed self-assembly**. The common usage of this term has altered somewhat from its original definition. When first introduced, directed self-assembly referred to a templated process in which the template was not included in the final product. The definition has been diluted somewhat and now refers to a templated synthesis, regardless of whether the templating species is part of the product or not. Many of the synthetic examples of self-assembled systems are in fact examples of directed self-assembly, whereby templating agents are deliberately employed.

- **Self-assembly with post-modification**. The most common form of modification that is utilised in synthetic systems is the alteration of a system after the self-assembly step has taken place. Many common systems, such as rotaxanes and catenanes (Section 3.4), are formed from a self-assembled structure, held loosely together by non-covalent interactions, which are locked in place once the self-assembly process is complete by the formation of covalent bonds, much like tying off the ends of a knot.

- **Irreversible self-assembly**. The final product of the self-assembly process cannot revert into its component pieces without covalent bonds being broken. The last stage in the assembly process is therefore usually the formation of a covalent bond in a process that is under kinetic control.

- **Assisted self-assembly**. In many biological systems, the self-assembly process can be aided by species that are not themselves incorporated into the final assembly (*cf.* directed self-assembly). This process can almost be thought of as catalysed self-assembly, for the comparison with chemical catalysis is easy to make as the assisting species work by lowering the kinetic barriers to the formation of the products.

- **Precursor modification followed by self-assembly**. Self-assembly, being a spontaneous process, will happen whenever it can. Sometimes, in biology, this is not advantageous and the self-assembly process may be prevented by subtly altering the components of the assembly process. At the right time, these precursors can be changed back into a reactive state through the action of enzymes and the self-assembly process can occur.

3.1.5 Occurrences and uses

This chapter aims to detail some of the more common areas in which self-assembly is utilised in both the natural world and synthetically. It is important to note that while this chapter deals with discrete species, self-assembly is also involved in the formation of infinite networks and solids which are dealt with separately in Chapter 4. In addition, self-assembly plays a major role in the 'bottom-up' approach towards nano-scale machines and devices (Chapter 5).

The self-assembly of discrete species is commonly seen in nature, although the discrete molecular species in question are often of tens of nanometres in size. The ubiquity of self-assembly in nature has seen the method become common-place in laboratories. It is possible to generate quite complex geometries by simply letting the components assemble by themselves. Threaded rings, knots, grids and capsules are just a few of the examples that have been studied to date and a rich diversity of intriguing systems has been established.

3.2   Biological self-assembly

Hamilton, T. D. and MacGillivray, L. R., 'Self-assembly in biochemistry', in Encylopedia of Supramolecular Chemistry, Steed, J. W. and Atwood, J. L. (Eds), Marcel Dekker, New York, NY, USA, 2004, pp. 1257–1262.

3.2.1   Biological self-assembly

Processes that rely upon self-assembly are ubiquitous in nature. In fact, life could not exist without them. Vital biological processes, such as the replication of DNA and the folding of proteins, are governed by countless weak interactions that cause highly complex architectures to form reliably and spontaneously. There are three main non-covalent interactions that are frequently found in biological systems – hydrogen bonding, ion-pair interactions and hydrophobic effects. In particular, metal-coordination interactions in metalloenzymes' are also crucial, for example, coordination of Zn^{2+} in carbonic anhydrase.

It is relatively easy to comprehend why self-assembling systems play such important biological and biochemical roles. If every biological process required strong chemical bonds to be formed or broken, then biological systems would consume enormous amounts of energy and the synthesis of biological molecules would be extremely complex and time-consuming. For example, everyday processes such as DNA replication, would involve many thousands of covalent bonds being broken and reformed for a duplicate of just a single strand to be produced. Rather than synthesise large and complex molecules from scratch, nature uses smaller building blocks that are coded with the right information to form the larger species by themselves. By 'coded', we mean that the building blocks possess all of the correct functionalities in the correct geometrical arrangements to interact successfully with other species. It must be remembered, however, that the products of a self-assembled system represent the thermody-namic minimum of the system and mistakes can be corrected during the assembly process. This means that there is a very selective reproducibility associated with self-assembled systems which is vital from a biological viewpoint – any errors or unwanted byproducts could have disastrous consequences. For example, many serious diseases, such as BSE, diabetes and Alzheimer's Disease, are caused by the misfolding of proteins into toxic, aggregated β-sheet fibrils. These represent cases where the biological self-assembly process is perturbed from its natural sequence or equilibrium in some way.

3.2.2 Proteins

 Okabayashi, H.-F., Ishida, M. and O'Connor, C. J. 'The self-assembly of oligopeptides', in *Self-Assembly*, Robinson, B. H. (Ed.), IOS Press, Amsterdam, The Netherlands, 2003, pp. 331–338.

Proteins are responsible for numerous functions throughout the body, acting as transporting agents, as the scaffold of biological catalysts (enzymes), as hormone receptors and in many more roles, including structural ones. Although proteins form a part of most undergraduate courses, it is well worth stressing the point that supramolecular chemistry, and in particular self-assembly, plays a major part in protein structure and behaviour.

When viewed at their most simplistic, proteins are essentially linear molecules, composed of typically 200–300 amino acid residues held together through peptide bonds. There are 20 naturally occurring amino acids with the generic formula $NH_2-C^*HR-CO_2H$, with C^* being a chiral carbon centre (except in glycine where $R = H$). The side group that is attached to the chiral carbon atom is the only factor to vary between amino acids and it is the order in which the amino acid residues are joined that is responsible for the overall behaviour of the protein. This cova-lent sequence is referred to as the 'primary protein structure' (abbreviated to the 1° structure) (Figure 3.4). However, proteins do not exist in the body as simple linear covalent strands – rather they are folded upon themselves (secondary and tertiary structures, 2° and 3° respectively) and joined with other proteins in larger aggregates (quaternary structure). All of the assembly beyond the primary struc-ture is supramolecular in nature and aggregation is by self-assembly. Proteins represent biological examples of hierarchical assemblies – the tertiary structure cannot form until the secondary structure has been assembled and similarly the quaternary structure cannot assemble until the tertiary folding is complete. This stepwise assembly demonstrates the need for assembly on a local level before aggregation to form larger, more complex structures. Proteins are able to fold into their most stable forms within a few minutes *via* this hierarchical mechanism, as opposed to the estimated 10^{27} years it should take if the assembly proceeded through a random search for the lowest-energy conformer (this value is for a relatively

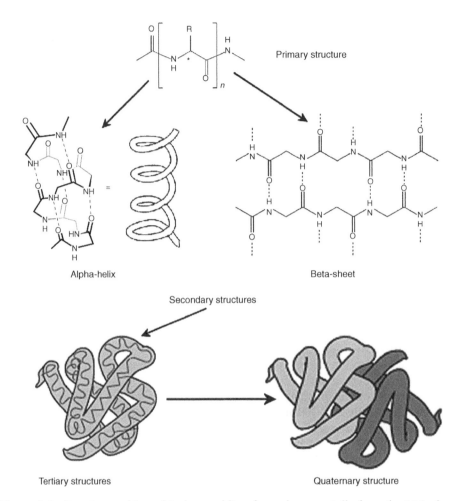

Figure 3.4 Proteins are hierarchical assemblies, formed sequentially from the 1° (polypeptide), 2°, 3° and 4° structures.

small protein with only 100 amino acid residues). Clearly, folding at this speed is of no use for any organism, being 10^{18} or so times longer than the age of the universe!

When proteins fold, there are regions that take on well-defined shapes as they adopt one of two common folding motifs. This is referred to as the secondary (2°) structure. The two predominant secondary structures of proteins are the α-helix and the β-pleated sheet (Figure 3.4). Both of these structures are formed by the folding of the polypeptide chain upon itself in such a way as to ensure that the hydrophobic side chains are packed within the core of the resultant structure with a hydrophilic surface left exposed. This folding is achieved primarily through the formation of $NH \cdots O{=}C$ hydrogen bonds. The interactions between the amine and carbonyl groups also allow for these relatively hydrophilic groups to be incorporated within the hydrophobic protein interior as they effectively 'neutralise' one another. The order of the residues in the polypeptide chain dictates exactly how this folding will occur, therefore the same structure is obtained every time for a given primary sequence. There is the risk, however, that the proteins may associate with other molecules rather than with themselves and therefore most biological systems utilise 'molecular chaperones' that protect the protein until it is folded. This is an example of *assisted self-assembly*. Any errors in the polypeptide chain, as a result of miscoding from genetically impaired DNA, will generally assert themselves at this stage, or the tertiary stage, of the protein's assembly. If the protein cannot fold properly due to incorrect sequencing, then it is effectively useless. This is termed *protein misfolding*. The effects of many genetic diseases arise through such dysfunctions.

The tertiary (3°) structure of a protein arises from the folding of the secondary regions into the final three-dimensional structure. However, the interactions at this stage are between the various side groups along the peptide chain, rather than between the groups of the backbone and hence the already-folded regions close in upon each other. A number of different types of interactions play roles in determining the tertiary structure, *e.g.* hydrogen bonds, hydrophobic influences, ionic attractions and disulfide bridges (although these are covalent in nature). It is at this stage of the assembly process that the functional regions of the protein become assembled, such as catalytic sites and binding surfaces.

The quaternary protein structure is the manner in which individual proteins come together, through non-covalent interactions, to form larger assemblies comprised of several separate polypeptide chains. For example, haemoglobin is made up from four separate myoglobin sub-units.

A similar terminology is occasionally applied to synthetic structures. The primary and secondary structures represent individual molecules and discrete supramolecular complexes, respectively. Tertiary structure can be observed between supramolecules and quaternary structure may exist in some macroscale systems.

3.2.3 Viruses

 Harrison, S. C., 'Principles of virus structure', in *Fundamental Virology*, 4th Edn, Knipe, D. N. and Howley, P. M. (Eds in Chief), Lippincott Williams and Wilkins, Philadelphia, PA, USA, 2001, Ch. 3, pp. 53–86.

Viruses are biologically active entities that act by carrying a small piece of genome into foreign cells and replicating in a parasitic manner. They are essentially nanoscale host molecules with an RNA strand acting as the guest species within a self-assembled outer casing comprised of protein sub-units that are virally encoded. This outer casing is called the *capsid* and it is held together solely through non-covalent interactions between protein sub-units. Individual viruses are constructed from very few different types of protein (in some cases only one) and form highly symmetrical geometries. There are two main geometrical forms in which viruses assemble – icosahedral and helical structures.

Icosahedral viruses

The structure of icosahedral viruses was first proposed by Crick and Watson (most famous for their role in the discovery of the double-helical structure of DNA). An icosahedron consists of 20 equilateral triangular faces and is the largest closed Platonic solid structure that can be assembled from a repeating component with identical interactions between the sub-units (Figure 3.5(a)). Many common viruses adopt an icosahedral-based geometry, including the polio virus, herpes virus and the human immunodeficiency virus (HIV). Some viruses are contained within a further lipid bilayer referred to as an envelope (Figure 3.5(b)). The icosahedral symmetry requires a certain number of *structure units* in order for the virus to assemble. These structure units, or *capsomers*, are symmetrical although they themselves may be assembled from asymmetrical components. The majority of viruses are actually based around a quasi-icosahedral structure.

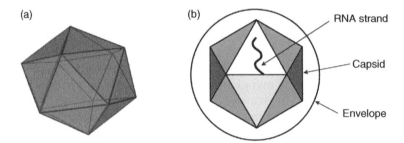

Figure 3.5   (a) A regular icosahedron, the geometrical basis of many viruses. (b) Many of these, such as the herpes virus, are surrounded by a bilayer envelope.

The self-assembly of protein sub-units into hollow entities is not unique to virus formation. Beneficial and indeed essential proteins within the human body also self-assemble in a similar manner. An example of such a protein is *ferritin*, which is responsible for iron storage. The correct storage and release of iron is crucial to maintaining life – there must be enough present to provide the blood with a healthy concentration of haemoglobin, but not enough for toxic effects to be felt. Ferritin is composed of 24 identical sub-units that combine to form a hollow shell with an internal diameter of 80 Å. Inside the protein, iron is stored as a mineral core with the formula $[FeO(OH)]_8[FeO(H_2PO_4)]$. The release of iron occurs as $[Fe(H_2O)_6]^{2+}$ through polar channels at a point where the sub-units meet.

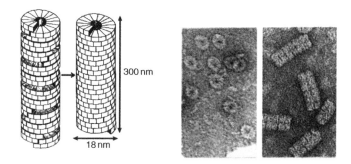

Figure 3.6 The tobacco mosaic virus (TMV) consists of identical protein sub-units arranged around an RNA strand. The units involved are large enough to be imaged by electron microscopy.[3] Reproduced with permission from P. J. G. Butler, 'Self-assembly of tobacco mosaic virus: the role of an intermediate aggregate in generating both specificity and speed', *Phil. Trans. R. Soc., London, B*, **354**, 537–550 © 1999 The Royal Society.

Helical viruses

Rather than form *pseudo*-spherical capsules, some viruses form helical capsids. The helix is made up of identical protein sub-units spiralled around the central nucleic acid. The size of the capsid is dependant upon the enclosed RNA strand. The most common example of a helical virus is the *tobacco mosaic virus* (TMV) (Figure 3.6). The size and complexity of self-assembled capsids can readily be seen in the structure of TMV. Each capsid consists of 2130 individual protein sub-units, with 158 amino acids in each (a molecular weight of approximately 18 000 Da per protein). The fully assembled capsid measures 18 nm in diameter and is 300 nm long. The virus forms by the threading of the nucleic acid strand through two stacked protein discs, templating the beginning of the helix in the process. Subsequent helical turns are then threaded onto the RNA until it is totally encased by the protein sheath. The truly remarkable aspect of the TMV virus is that if it is broken down into its component parts and they are then re-mixed under physiological conditions, the virus reassembles itself into a fully functional replica indistinguishable from the original. This process represents true self-assembly on a large scale, with over 2000 units forming a fully replicable structure.

3.2.4 DNA

Calladine, C. R. and Drew, H. R., *Understanding DNA: The Molecule and How it Works*, 3rd Edn, Academic Press, New York, NY, USA, 2004.

Deoxyribonucleic acid (DNA) is the molecule that is responsible for encoding the genetic information that makes life possible. The structure of DNA itself arises from supramolecular interactions. Just one strand of DNA contains all of the necessary information for an entire organism, an incredible feat of biological data compression! Single strands of DNA are comprised of polyphosphate/sugar backbones to which are appended nucleobase residues of four types: adenine (A), cytosine (c), guanine (G) and thymine (T). The sequence of the nucleobases is how the genetic information is held. Triplets of these nucleobases correspond to one of twenty amino acids that assemble to form proteins (Section 3.2.2).

The structure of DNA is the famous *double helix*. Two complementary strands are wound around each other in a (usually) right-handed helix in such as way as to minimise the volume occupied by the molecule. Complementary pairs of nucleobases (A–T and C–G) form strong hydrogen bonds that tightly hold the strands together, supported by π–π stacking interactions. These two pairs are the only stable combinations and the ability of the nucleobases to recognise each other is responsible for the effectiveness of DNA replication. The base pairs are held together by medium strength $NH \cdots N/O$ hydrogen bonds (the $N \cdots O$ and $N \cdots N$ distances are in the range 2.8–2.9 Å) although there is extra strength imparted through the use of multiple interactions, two in the A–T pair and three in the more stable G–C coupling (Figure 3.7). The overall lengths of each pair, from backbone to backbone, are equal at 10.85 Å. The manner in which the DNA helix is assembled is such that all of the hydrophobic groups, *i.e.* the nucleobase side chains, are located on the inside of the structure away from the aqueous cell nucleus interior. The two single DNA strands are able to come together spontaneously in a *strict self-assembly process*, partly assisted by a hydrophobic driving force. The self-assembly of the helix occurs through a cascade process, rather like the sealing of a zipper. As successive pairs come together, the entropic penalty that arises from

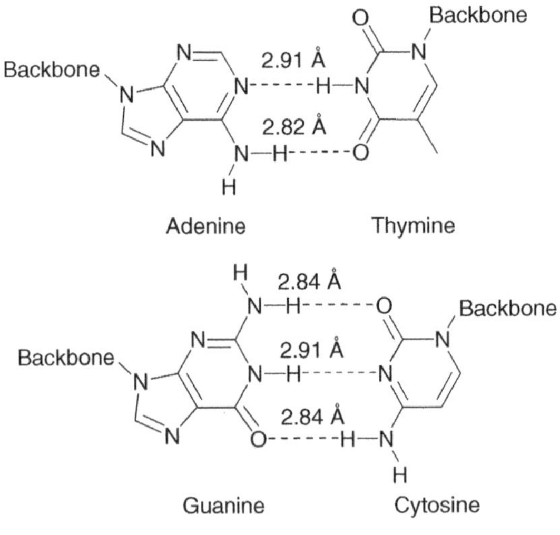

Figure 3.7 The two pairs of nucleobases that are responsible for holding the DNA double helix together through hydrogen bonding. Distances quoted represent the $D \cdots A$ separations.

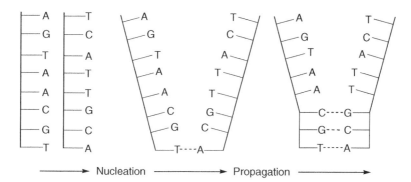

Nucleation ⟶ Propagation ⟶

Figure 3.8 DNA self-assembly occurs *via* two distinct processes, *i.e.* nucleation followed by propagation.

a loss of degrees of freedom, is gradually overcome, with the effect becoming less with each pair that binds. There are in fact two distinguishable phases in DNA self-assembly – nucleation and propagation (Figure 3.8). Two strands are initially brought together in the nucleation stage. It is this part of the process where the high entropic cost is paid (the formation of several hydrogen bonds is a negligible enthalpic contribution at this stage). After the nucleation event, the chain begins to 'zip-up' during the propagation phase, with each new pair that forms contributing a favourable enthalpy, with little further change in entropy. The large number of interactions that are ultimately formed counteracts the entropic loss of bringing together the two long chains.

Self-assembly processes are also at the heart of DNA replication and protein encoding. The intracellular DNA is able to unwind and template the formation of t-RNA (transfer ribonucleic acid). The RNA backbone contains an extra oxygen atom at each of the sugar groups which allows for it to pass out of the cell nucleus. It is this RNA strand that codes the production of proteins within the ribosome of the cell. It has recently been found that DNA helices can also coil into several more complex shapes, such as knots (Section 3.4.6).

3.3   Ladders, polygons and helices

3.3.1   Self-assembly using metal templates

 Fujita, M. 'Metal directed self-assembly of two- and three-dimensional synthetic receptors', *Chem. Soc. Rev.*, 1998, **27**, 417–426.

Perhaps the simplest way to begin understanding some synthetic self-assembled systems is to look at assemblies that contain regular arrays of templating metal ions. Metal-templated systems have the potential to be more predictable than their hydrogen bonded counterparts due to the strict coordination environments that transition metals in particular possess. A well-defined coordination geometry is especially important in the synthesis of structures of regular geometry, such as grids or polygons.

One of the over-riding concerns when developing the requisite building blocks for metal-templated synthesis is the choice of *convergent* or *divergent* units (Figure 3.9). Convergent components are those that focus binding sites towards a central point, whereas divergent components promote binding in multiple directions. Different combinations of convergent and divergent binding sites will give rise to different products, either discrete complexes or polymeric networks. Using a ligand that possesses convergent binding sites combined with a naked metal ion, which is divergent in nature, should result in a discrete species in which ligands surround the metal (Figure 3.9(a)). The opposite mixture of building blocks, a 'semi-protected' metal (a metal with part of its coordination sphere occupied by a spectator chelate ligand, such as ethylene diamine) and a divergent ligand, also generally forms discrete species (Figure 3.9(b)). Both of these methodologies

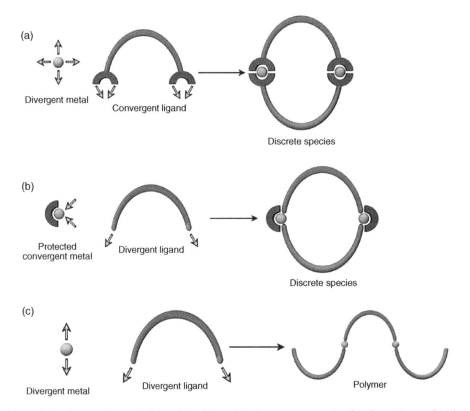

Figure 3.9 Appropriate choice of building blocks can promote the formation of either discrete or polymeric species.

are frequently employed in the synthesis of self-assembled systems, due to the high degree relatively of predictability. If both the metal and the ligand are divergent, then a polymeric structure is usually obtained as there is very little directional control (Figure 3.9(c)). In general, when constructing discrete species, protected metals (*i.e.* convergent metals) are used as the vertices of polygonal or polyhedral structures, whereas naked metals with convergent ligands are more commonly used to join together ligands midway along their length.

It is also important to establish whether *monotopic* or *polytopic* ligands are required for the formation of a particular architecture. Monotopic ligands are those that contain only one binding site, therefore assisting in the formation of discrete species by acting as terminal ligands. They may also act as protecting groups on metals that are part of larger assemblies. Polytopic ligands contain multiple binding sites, enabling the formation of complexes containing more than one metal centre. It is important to note that the term *n*-topic refers to the number of distinct binding sites, not to individual atoms capable of coordination.

> **Mono/Polytopic Ligands:** A descriptive term for how many individual binding sites a ligand contains.

3.3.2 Racks, ladders and grids

Racks, ladders and *grids* are closely related structures that consist of a regular array of metal ions joined by rigid, linear, multidentate ligands (Figure 3.10). Such systems are exclusively the domain of metal–ligand complexes, due to the rigid coordination geometries provided by most metals. Correctly designed ligands, or mixtures of ligands, can therefore produce fascinating, well-defined and controllable structures.

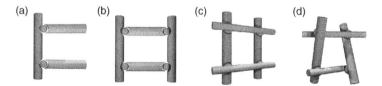

Figure 3.10 Cartoon representations of (a) a [2]-rack, (b) a [2.2]-ladder, (c) a [2 × 2]-square grid and (d) a chiral [2 × 2]-grid.

The simplest of these three architectures is the *rack*, which consists of a single, linear polytopic ligand and several monotopic ligands (Figure 3.10(a)). Naked metal ions act as linkers between the two types of ligand so that the resulting structure contains the monotopic ligands pointing away from the backbone perpendicularly, somewhat like spokes. Racks are named according to how many spokes they contain, for instance, the example in Figure 3.10(a) is called a *[2]-rack*, as there are two monotopic ligands attached to a ditopic ligand. The ligands used must obviously contain binding sites that are capable of binding to the same metal. For example, if the backbone ligand has sites capable of chelating a tetrahedral metal (*e.g.* a pair of suitably arranged donor atoms) then it is no use if the monotopic ligands are complementary to an octahedral centre (*e.g.* contain three donor atoms in a meridonal arrangement). Both tetrahedral metal ions (*e.g.* Ag^+ and Cu^+) and typically octahedral metal ions (*e.g.* Co^{2+} and Ru^{2+}) are capable of connecting two ligands in a perpendicular fashion (Figure 3.11) with octahedral metals commonly adopting a distorted coordination geometry in order to satisfy interactions to the two ligands. This same ligand design philosophy can also be applied to ladder and grid structures. An example of a rack, **3.5**, assembled using Ru^{2+} as the templating ion, is shown in Figure 3.12.[4] The partially blocked ruthenium ions (the blocking terpyridyl units become the spokes) are able to sit within the tridentate pockets in the backbone ligand,

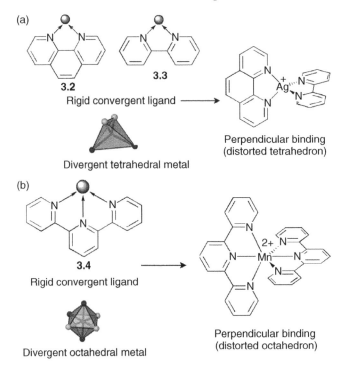

Figure 3.11 Examples of how ligands can be joined at perpendicular angles templated by (a) tetrahedral and (b) octahedral metals, where (a) shows the common bidentate ligands 1,10-phenanthroline (**3.2**) and 2, 2′-bipyridine (**3.3**) and (b) shows the tridentate ligand 2, 2′: 6′, 2″-terpyridine (**3.4**).

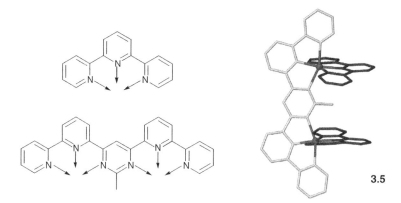

Figure 3.12 The X-ray structure (right) of a [2]-rack (**3.5**) constructed using Ru^{2+} as an octahedral templating ion. The binding pockets of the free ligands are indicated by arrows.

fully satisfying the coordination environment of the metal and resulting in the planes of the ligands being at right-angles to one another. It is important to note that these metal/ligand structures are almost exclusively positively charged and require a counter-anion to be present. This anion must be weakly coordinating (*e.g.* BF_4^-, PF_6^-, $CF_3SO_3^-$) so as not to compete with the binding pockets of the ligands for the metal coordination sites.

Ladders are very closely related to racks, with the difference being the addition of a second polytopic ligand. Instead of spokes protruding from a backbone the two polytopic ligands are connected by a number of ditopic rungs (Figure 3.10(b)). The nomenclature of ladders takes the form [2.n]-ladder, whereby n is the number of rungs and the number '2' refers to the two sides which the rungs connect (the prefix '2' is similar to the naming of grids – see below). The cartoon example shown in Figure 3.10(b) is therefore a [2.2]-ladder. An example of a [2.3]-ladder (**3.6**) is shown in Figure 3.13(a)). The sides of the ladder each contain three bidentate binding sites, suitable for tetrahedral ions. The rungs are simple 2, 2'-bipyrimidine-based ligands which contain complementary binding sites to complete the laddered structure.

An extension to work on ladders is the construction of 'multi-compartmental capsules'. Rather than using a linear ditopic ligand bridging between two sides, more disc-like tritopic ligands can be employed to act as triple bridges between three parallel strands (**3.7**). The resultant structures appear to have 'floors' held in place one above the other effectively separating cavities that are potentially able to hold anions and solvent as guest species (Figure 3.13(b)). Using the standard ladder nomenclature, this could be referred to as a [3.3]-ladder.

Grid-like structures can be formed from several linear polytopic ligands (either the same or different) assembled around metal ions (Figure 3.10(c)). It is most common for one set of ligands (*e.g.* those that run horizontally) to be stacked on the opposite side of the metal ions to the other set. Grids are named according

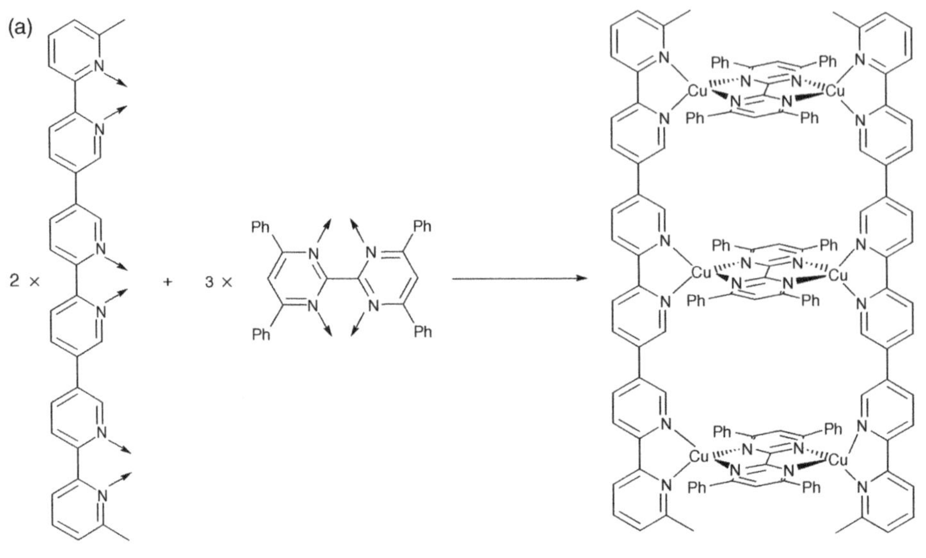

(a)

2 ×   +   3 ×

3.6

(b)

3 ×   +   3 ×

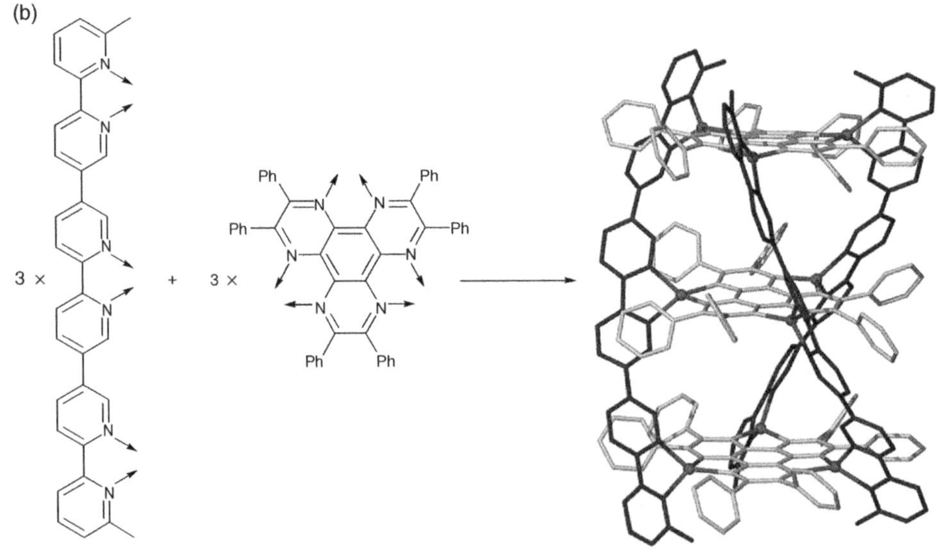

3.7

Figure 3.13 Examples of (a) a [2.3]-ladder and (b) a multi-compartmental assembly, both using Cu$^+$ as a tetrahedral templating ion. The convergent binding pockets of the ligands are indicated by arrows.[5]

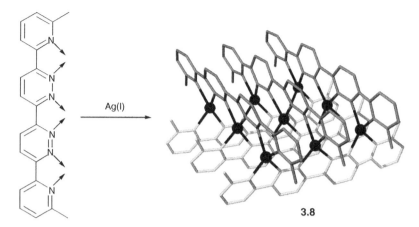

Figure 3.14 An example of a $[3 \times 3]$-square grid made from six tritopic ligands and nine Ag^+ cations. The tetrahedral binding pockets of the ligand are indicated by arrows.[6]

to the number of ligands that occur in the columns and the rows (or by the *n*-topic nature of the ligands), $[m \times n]$-grids. Grids are referred to as square grids when *m* and *n* are equal and as rectangular grids when they are not. It is also possible to construct a chiral square grid, whereby the four ligands bind both over and under the metal ions at opposite ends (Figure 3.10(d)). A silver(I)-templated $[3 \times 3]$-square grid is shown in Figure 3.14. The ligand used is tritopic and hence a $[3 \times 3]$-grid is formed using nine (3^3) metal atoms and six ligands, making a total of fifteen components to assemble the final structure. The complex is readily characterised by ^{109}Ag NMR spectroscopy which shows three resonances in the ratio 4:4:1, corresponding to the Ag^+ ions at the corners of the grid, in the middle of each edge and in the centre.

3.3.3 Helicates

Helicates are coordination complexes that adopt a helical geometry. Although much of the fascination of helical structures lies in their aesthetically pleasing shape and the sheer challenge of producing them, they are also useful as precursors to more complex, interlocked molecules – molecular knots (Section 3.4.6).

Helical structures can be thought of as a thread wound around a central axis, similar to a common household screw. The screw thread can run in either of two directions which are not superimposable, thereby imparting *helical chirality*. This is a fascinating property of helicates – the generation of chirality from achiral components. The distance between one turn and the next along the screw is called the *pitch* (analogous to wavelength, *i.e.* the distance between two adjacent peaks in a wave). Helicates with two threads (analogous to DNA) or three threads are termed double- and triple-stranded helicates, respectively. The specific chirality of helicates is expressed by using the Cahn–Ingold–Prelog notation: *M* (left-handed) and *P* (right-handed), standing for minus and plus, respectively (Figure 3.15).

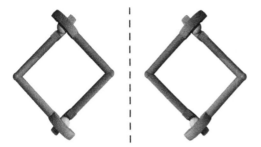

Figure 3.15 Schematics of a left-handed (*M*) double-stranded helicate (left) with its right-handed (*P*) enantiomer (right).

Helicates: Metal complexes that possess a helical and therefore chiral structure geometry akin to a screw thread. The ligands are called **helicands**.

Helicates bear a passing resemblance to ladders as they are usually constructed from threads that are joined by naked templating metal ions (rather than using bridging ligands). The differences arise due to the use of more flexible threads when making helicates as the ligands must twist around in order to form the helical structure. Molecular threads that are used in the assembly of helicates are termed *helicands*.

It is possible to form helicates using either tetrahedral or octahedral metal ions, by applying similar geometrical arguments as for racks, ladders and grids

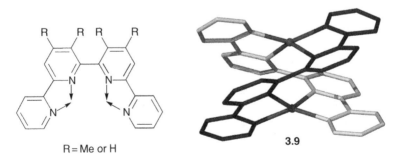

Figure 3.16 A simple example of a double-stranded helicate templated by tetrahedral Cu$^+$ ions (R = H). The bidentate binding sites of the quaterpyridine ligand are shown by arrows.

(Figure 3.11), although tetrahedral metals combined with ligands containing bidentate binding sites are more common. A simple helicand (**3.9**) is shown in Figure 3.16. The ligand $(2, 2' : 6', 2'' : 6'', 2'''$-quaterpyridine) is capable of forming a double-stranded helicate using tetrahedral Cu(I) ions as templates.[7] The importance of the metal ion is demonstrated by the platinum-containing structure in which the ligand adopts a non-helical planar geometry to fulfil the square planar coordination geometry of the Pt^{2+} ion. An analogous helicate in which the ligand has hydrogen atoms in place of the methyl groups has also been synthesised and the pitches of the two dicopper(I) complexes are 3.90 and 3.17 Å, respectively. Thus, substituents can have a significant effect upon structure. The methyl groups exert a steric effect that serves to elongate the pitch of the helicate. Replacing Cu^+ with Ag^+ further reduces the pitch as the ligands are pushed out sideways to accommodate the larger metal ion with a subsequent shortening of the length of the helicate.

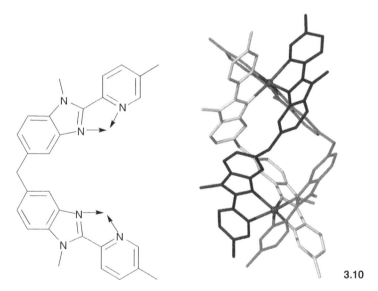

3.10

Figure 3.17 Ligands containing bidentate binding sites can form double helices (**3.10**) with tetrahedral metal ions (Cu^+) or triple helices with octahedral metals, such as Co^{2+} (as pictured).

The Cu^+ helicate shown in Figure 3.16 is an example of a $[4+4]$ helicate. The nomenclature $[m+n]$ helicate refers to the coordination numbers of the two metals that are present, with four being the coordination number for a tetrahedral Cu^+ ion. Octahedral metal ions therefore form $[6+6]$ helicates. In some cases, it is possible to use the same helicands with both tetrahedral and octahedral metals to produce double- and triple-stranded helicates, respectively (Figure 3.17).[8] The well-defined coordination preferences of metal ions are of over-riding importance in the assembly and are the most important 'instructions' coded into the system. It is more common for octahedral metal ions to be used in combination with helicands that contain tridentate binding sites which are more complementary to the coordination geometry of the metal.

A double-stranded [6 + 6] helicate (3.11) can be formed by using a ligand such as 2, 2' : 6', 2'' : 6'', 2''' : 6''', 2'''' : 6'''', 2'''''-sexipyridine, shown in Figure 3.18.[9] Sexipyridine can be thought of as containing either two tridentate binding sites or three bidentate sites. When combined with octahedral metal ions (*e.g.* Fe^{2+}, Co^{2+} and Cd^{2+}), a [6 + 6] double-stranded helicate is formed, with two tridentate binding sites. However, when combined with tetrahedral metal ions (*e.g.* Cu^+), a [4 + 4 + 4] double-stranded helix is formed (the nomenclature is [$m + n + o$] as there are now three metals present). The [4 + 4 + 4] Cu^+ helicate is also formed on electrochemical reduction and disproportionation of the [6 + 6] Cu^{2+} helicate.

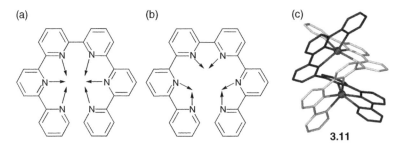

Figure 3.18 A sexipyridine helicand that can form helicates using (a) two tridentate or (b) three bidentate binding sites, as indicated by arrows, when combined with octahedral and tetrahedral metals, respectively. (c) The X-ray structure of the Cd^{2+}-based [6 + 6] helicate.

Although synthetic helicates synthetic are almost exclusively formed by using metal–ligand interactions, it is worth noting that there are examples of helical structures that are brought about by hydrogen bonding, with the DNA double helix being the most obvious example (Section 3.2.4). Hydrogen bonds are, like coordination interactions, highly directional in nature, although they are less strong. The α-helical secondary structure of proteins forms single-stranded helices and in wool fibres three α-helices wrap around each other to form triple helices – effectively, a 'helix of helices'. Figure 3.19 shows another natural system, a chlorophyll derivative, which assembles into a double-stranded helical structure. The two strands are held together solely by six $C = O \cdots H - N$ hydrogen bonds. More simple organic molecules also form helical architectures in the solid state, such as the solid state structure of urea (see Chapter 4, Section 4.3.1).

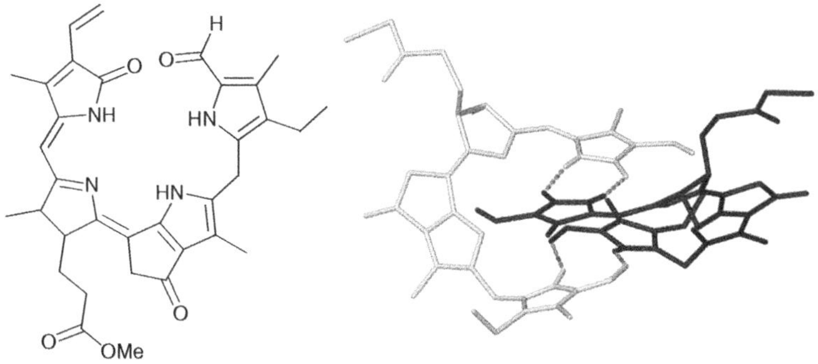

Figure 3.19 A methyl ester chlorophyll-a derivative and the helical structure it adopts in the solid state.

3.3.4 Molecular polygons

The rigid coordination geometries of metal ions can be exploited to create regular polygons when combined with appropriate ligands. Since polygons have, by definition, fixed angles and straight sides it is relatively simple to deconstruct a shape into its constituent parts, vertices and edges, as we have already seen (Figure 3.1). The vertices are constructed from semi-protected metal ions, forming a convergent binding site that the 'edge' ligands may attach to. The metal is protected in such a way that the vacant coordination sites are the correct geometry to form the angles of the polygon.

If semi-flexible ligands are used, then the geometry around the metal ion may be of lesser importance in the assembly and polygons that do not reflect the angle at the metal can be formed. An example of this phenomenon is based around a molecular-square system similar to that seen at the beginning of this chapter (Figure 3.1). Rather than use a 4, 4'-bipyridine ligand, the longer *trans*-bis(4-pyridyl)ethylene ligand is reacted with semi-protected platinum groups (with PMe_3 as the protecting ligand) (Figure 3.20(a)).[10] The square that is assembled is exactly what would be expected from semi-protected metals with two vacant sites perpendicular to each other and rigid, linear ligands. However, the system is in equilibrium with a molecular triangle, containing the same components in the same 1:1 ratio, (Figure 3.20(b)), an unusual result considering that square planar Pt^{2+} is being used and one that reflects the flexibility of the ligand. The different complexes can be isolated as crystals, depending on which solvents and anions are used.

Unlike helicates, molecular polygons are hollow, containing a central cavity similar to simple macrocyclic host molecules (see Chapter 2, Section 2.2). The assembly processes can be templated around a guest in a novel variant of the

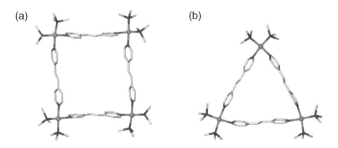

Figure 3.20 The formation of both (a) an M_4L_4 molecular square and (b) an M_3L_3 molecular triangle using semi-protected Pt^{2+} vertices and a flexible linker ligand.

thermodynamic template effect (see Chapter 2, Section 2.3.3). Metal–ligand polygons are almost always positively charged complexes and therefore templation is usually brought about by an anionic guest or neutral molecule. The presence of different anions during the assembly can exert a significant influence on the product that is formed. For example, an M_4L_4 molecular square can be created using octahedral Ni^{2+} ions that are occupied at two *cis* positions by acetonitrile molecules. The bridging ligands have two binding sites, thereby allowing two ligands to connect perpendicularly to the octahedral metals (Figure 3.21(a)). The molecular square is templated by either a BF_4^- or ClO_4^- anion which resides within the cavity. If the counter-anion is the larger SbF_6^-, then a molecular pentagon is formed instead, which is able to accommodate the larger templating anion within its confines (figure 3.21(b)).[11]

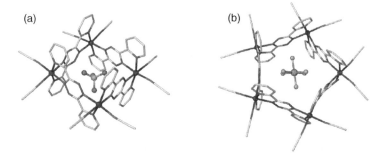

Figure 3.21 X-ray crystal structures of (a) a BF_4^- templated molecular square and (b) a SbF_6^- templated molecular pentagon, both containing the same ligand and same semi-protected metal vertices.

The use of multi-bridging ligands or of metals that have three or more vacant coordination sites can bring about the assembly of larger structures that are polyhedral in nature, rather than polygonal. These structures are discussed in more detail in Section 3.5.

3.4 Rotaxanes, catenanes and knots

3.4.1 Topological connectivity

Rouvray, D. H. and King, R. B. (Eds), *Topology in Chemistry – Discrete Mathematics of Molecules*, Horwood Publishing, Chichester, UK, 2002. Suavage, J.-P. and Dietrich-Buchecker, C. (Eds), *Molecular Catenanes, Rotaxanes and Knots – A Journey Through the World of Molecular Topology*, Wiley-VCH, Weinheim, Germany, 1999.

Perhaps one of the most fascinating aspects of self-assembly is not the discrete assemblies that can be fashioned but the *mechanical interlocking* of molecules. *'Mechanical interlocking'* means that two or more species become connected to each other, not *via* chemical bonds but rather they are threaded through or around one another to form a single entity. The assemblies are usually supported by supramolecular interactions between the individual, covalent components. Species that are linked in this manner are said to be *topologically connected*. In order for compounds to be truly topologically connected, a chemical bond must be broken in order in extricate the individual components of the assembly from each other or, in the case of knots, from themselves. The interest in such systems stems from two sources. First, and most simply, the complex topologies that are created are aesthetically pleasing and represent a significant synthetic challenge. Secondly, and more important, many interlocking structures have interesting physical properties or the potential for them. Such systems are currently topical as models for future nanoscale *molecular machines* and *molecular electronic devices* (Chapter 5).

> **Topological Connection:** The interlocking of two or more species or the knotting of a single entity without the need for a chemical bond between the components – a mechanical interlocking rather than a chemical one.

Topological connectivity is often represented by using diagrams known as *molecular graphs*. A molecular graph is a simplistic line-drawing of a molecule or species that depicts its connectivity, although not necessarily the actual geometrical shape. Therefore, a macrocycle, at its most simple, can be depicted as a regular circle. Similarly, an unbranched linear molecule, such as an *n*-alkane, can be represented as a straight line. It is impossible to draw topologically connected species without there being crossing points (or *topological nodes*) in the molecular graph (Figure 3.22(a)). A molecular graph with crossing-points is called a *nonplanar graph* and the topology of the corresponding molecule is termed *non-trivial*.

Two unlinked macrocycles are considered to be the *topological isomer* of the same two macrocycles that are interlocked (Figure 3.22(a)). There is no difference in the atomic connectivity between the two isomers (*i.e.* the covalent bonds are identical), only in their topological connectivity, and the two situations cannot be interconverted without chemical bonds being broken. Rotaxanes (a molecule comprising a thread through a ring – Section 3.4.2) are not considered to be topological isomers of the individual components as, with significant deformation, it is possible to separate the parts within a molecular graph, despite the fact that it may not be achievable in reality if the 'stopper' at the end of the thread is sufficiently large. It is also possible for the formation of interlocking species, and of knots in particular, to result in a chiral product starting from non-chiral reagents. Such chirality is known as *topological chirality* (Figure 3.22(b)).

> **Topological Isomers:** Compounds that have the same covalent connectivity but are topologically unique. Inter-conversion cannot occur with any amount of deformation without bonds being broken.

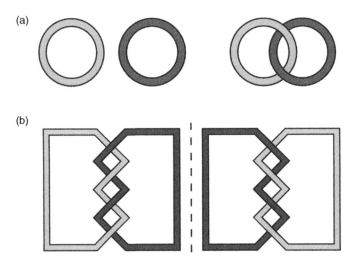

Figure 3.22 Molecular graphs show that (a) two unconnected macrocycles are considered to be the topological isomer of two linked macrocycles as the two situations cannot be interconverted without breaking bonds and (b) topological chirality can be imparted by molecular species becoming entangled.

Figure 3.23 shows how it may be possible for a simple, self-assembling macrocycle based on a coordination compound to form an interlocked aggregate, in this case a *catenane* (mutually interlocked rings – Section 3.4.3). The formation of the macrocycle itself occurs when a linear ditopic ligand has its ends joined by a metal cation in a process which is reversible (Figure 3.23(a)). If, however, the macrocycle

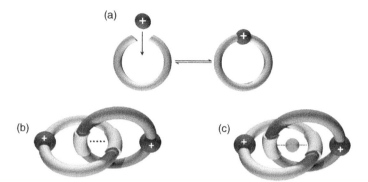

Figure 3.23 A metal-linked macrocycle may (a) form a discrete molecule (a metallomacro-cycle) or, if the conditions are favourable, an interlocked species may be formed by either (b) interlocking directly through non-covalent interactions between parts of the ligands or (c) interlocking around a templating species (sphere). Note that the final products cannot be dismantled without breaking chemical bonds, although the template in (c) can, in principle, be removed after the catenane has formed.

also possesses a second binding site, directly opposite the metal binding site, then the interlocking of the rings may be bought about by favourable supramolecular interactions involving this secondary site in one of two ways. The sites may contain complementary functionalities, *e.g.* a hydrogen bond donor/acceptor pair or groups capable of mutual π-interactions, and therefore be able to interact directly with each other (Figure 3.23(b)), templating the formation of a cate-nane. Alternatively, the formation of an interlocked species may proceed *via* the use of an additional templating species such as a second metal ion or an anion (Figure 3.23(c)). Although these approaches seem quite different, they act to form the interlocking species in the same manner – by bringing the individual compo-nents together in such a way spatially that the macrocyclisation occurs while the two components are held in an interlocked geometry. The nature of such templating effects is discussed in more detail later in this chapter.

In order for topologically connected species to form, it is necessary for the templation to occur before the rings are closed. In the case of catenanes, this may occur in either a stepwise process (one ring forms first and the second is threaded through before sealing), Figures 3.24(a) or in a concerted manner (both rings close at the same time), Figure 3.24(b). In a self-assembly regime, the process is governed by thermodynamic selectivity and hence the lowest-energy product will form selectively and the reversible nature of the interactions allows for 'connectivity mistakes' to be corrected. In addition, there are also many known examples in which the template species acts as a kinetic template with the closing of the rings occurring through the formation of covalent bonds (Figure 3.24(c)). In such cases, the template is referred to as an *auxiliary linkage*. This process is not strict self-assembly, rather it is self-assembly of the intermediate followed by synthetic modification to form the catenane. These kinetically templated species

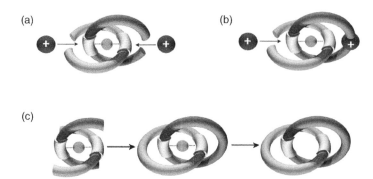

Figure 3.24 The ring-closing steps of two components may occur (a) simultaneously or (b) in a stepwise manner. (c) Kinetic templates may also be used which template an intermediary species, hence allowing the macrocycles to be completed by covalent modification and subsequent demetallation.

can be demetallated to yield a purely organic catenane with no inter-ring templating interactions.

Complex topologies, as their name suggests, are complicated structures to assemble and spatially difficult to interpret in many cases. In fact, even the topologies that appear simple, when viewed as cartoons, are extremely challenging to synthesise. In the following sections, we will examine three main classes of interlocking structures – *rotaxanes*, *catenanes* and *knots*, approximately in order of increasing complexity. A fourth category, *interpenetrating polymeric networks*, will be dealt with in Chapter 4, Section 4.6.5.

3.4.2 Rotaxanes

A *rotaxane* consists of a linear molecule which is threaded through a ring with the ends of the thread, or axel, capped in such a manner that the ring cannot slip off (Figure 3.25(a)). Traditional bulky capping groups have included $-CPh_3$ PPh_3 ligands, although recently there has been a trend towards the exotic, with porphyrins and fullerenes being utilised, for example. Strictly speaking, rotaxanes cannot be described as topologically connected species as, with much deformation of the molecular graph, it is possible to pull the ring off of the thread. However, in practice the macrocyclic ring cannot usually be removed without breaking any bonds and in practical terms the components are inextricably held together. The term 'rotaxane' is derived from the Latin words *rota* and *axis*, meaning wheel and axel, respectively, and is therefore highly descriptive of these systems. Closely related to rotaxanes are *pseudorotaxanes*, which possess no terminal groups at the ends of the thread to prevent dissociation of the ring (Figure 3.25(b)). Pseudorotaxanes are commonly used as intermediates in the synthesis of rotaxanes (and also catenanes, Section 3.4.3). The nomenclature

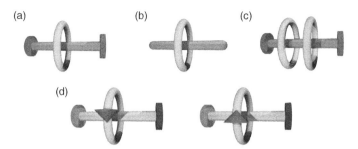

Figure 3.25 Schematic diagrams of (a) a [2]-rotaxane, (b) a [2]-pseudorotaxane and (c) a [3]-rotaxane. (d) An enantiomeric pair of chiral rotaxanes using an asymmetric thread and a macrocycle in which directionality can be defined.

used for rotaxanes is [n]-(pseudo)rotaxane, where n is the number of components that comprise the complex. Hence a [2]-rotaxane consists of only two components, a ring and a thread, whereas a [3]-rotaxane consists of two rings positioned along the same thread. While this terminology is effective for simpler compounds, it can be somewhat ambiguous for more complicated examples and may require further clarifying terms. Topological chirality may be introduced into rotaxanes by using an asymmetric thread (*i.e.* utilising two different terminal groups) with an asymmetric macrocycle (in which directionality can be defined by the covalent ordering of atoms) (Figure 3.25(d)). In such cases, the orientation of the macrocycle with respect to the thread can result in one of two enantiomers that are non-interchangeable without disassembling the rotaxane and rethreading it in the opposite orientation. For [3]-rotaxanes diastereomers may also exist.

> **Rotaxanes:** Interpenetrating compounds in which one or more macrocycles are threaded onto a linear molecule and secured in place by the use of bulky terminal groups.

> **Pseudorotaxanes:** Interpenetrating compounds in which one or more macrocycles are threaded onto a linear molecule but held in place only through weak, non-covalent interaction without mechanical interlocking.

The templated synthesis of rotaxanes can be achieved through one of four methods, as shown in Figure 3.26. These methods are as follows:

- **Threading**. Possibly the most common way in which rotaxanes are be formed is by using a pseudorotaxane as a precursor. The axel is threaded through the wheel, akin to cotton thread through the eye of a needle, and is held in place by one or more templating interactions between the two components. The termini of the thread are appropriately functionalised so that once the ring is secure they can readily be capped in a post-assembly synthetic modification step.

- **Trapping.** If a thread is synthesised in such a way so that it is already blocked at one end, then only the uncapped end needs to be capped once the ring is in place. This approach can be advantageous as there is statistically less chance of the ring slipping off of one end of the thread. Furthermore, this synthetic strategy can be used to create non-symmetrical rotaxanes, with different capping groups, giving rise to the possibility of chirality or directionality in the products.

- **Clipping.** It is possible to form rotaxanes by using a fully assembled axel, complete with bulky terminal groups, as one of the starting components. The wheel can be introduced by closing it around the thread at a suitably templated point. The two ends of the macrocycle may simply be joined together through a covalent reaction (like putting on a necklace) or the wheel may be formed from two separate species that combine to form the complete macrocycle.

- **Slipping.** In some cases, it may be possible for a macrocycle to be forced over one of the blocked ends of a preformed axel at elevated temperatures. This process is referred to as 'thermally induced slippage' or simply 'slipping'.

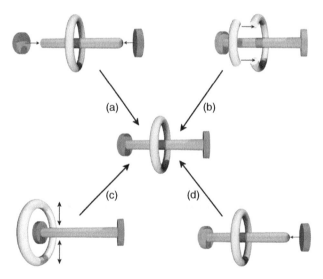

Figure 3.26 The synthesis of rotaxanes can be achieved *via* (a) threading, (b) clipping, (c) slipping or (d) trapping (see text for details).

A simple example of the synthesis of a [2]-rotaxane *via* threading is shown in Figure 3.27(a)).[12] A secondary ammonium group is positioned at the centre of a linear molecule that has appropriate functional groups at both ends. This thread is able to pass through the centre of a dibenzo-[24]-crown-8 macrocycle, forming a pseudorotaxane precursor (**3.11**) that is held together solely by charge-assisted $NH_2^+ \cdots O$ hydrogen bonds. Once in place, the axel is capped by bulky triphenylphosphonium substituents, so preventing the macrocycle from slipping

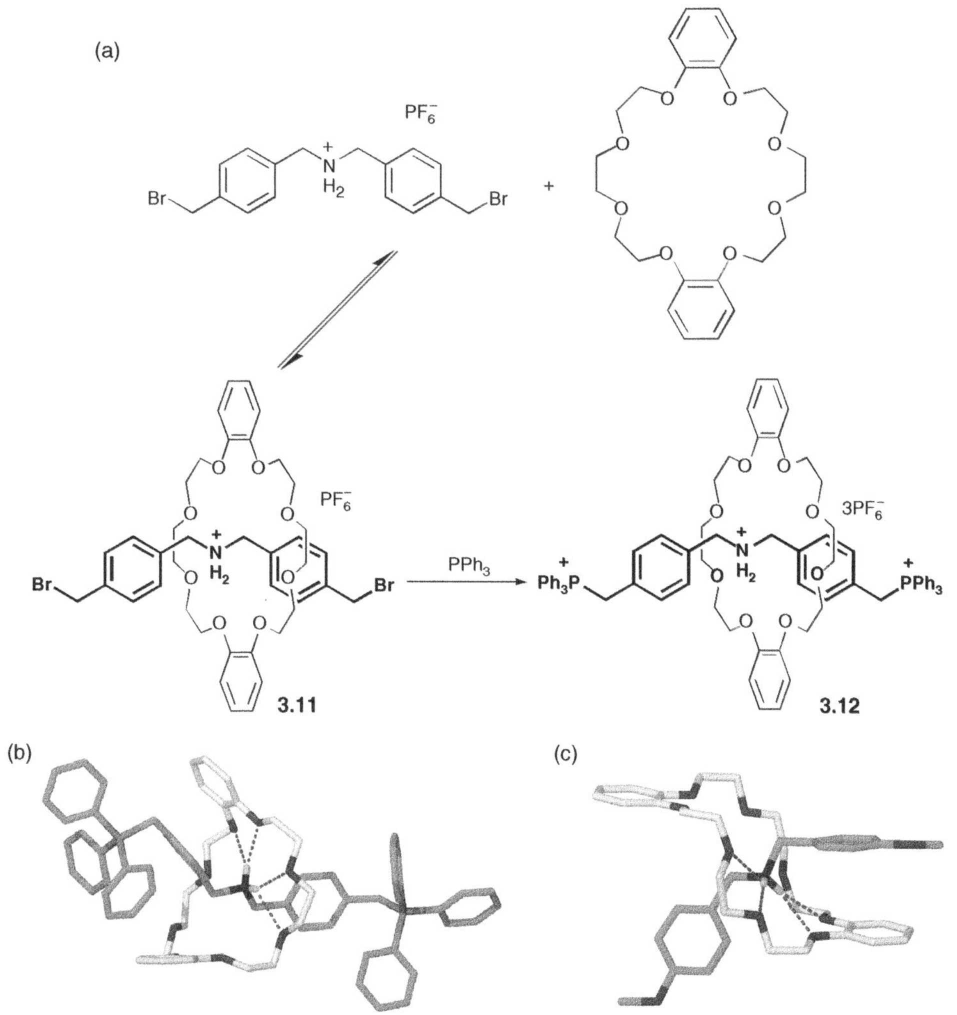

Figure 3.27 (a) The synthesis and (b) the X-ray crystal structure of an ammonium/crown ether-based [2]-rotaxane (**3.12**). (c) A simple [2]-rotaxane in which the $NH_2^+ \cdots O$ interactions are supported by π–π interactions[13] (counter-anions are omitted for clarity).

off and producing the final rotaxane (**3.12**). The X-ray crystal structure of the rotaxane reveals the hydrogen bonding interactions that are responsible for holding the macrocyclic ring in place (Figure 3.27(b)). The relative bulk of the capping groups comparative to the size of the macrocycle is also evident.

The secondary ammonium/crown-ether motif enables the ring to be held in place by several $NH^+ \cdots O$ interactions. In many cases, this relatively strong interaction is further enhanced by $CH \cdots O$ hydrogen bonds from methylene groups adjacent to the ammonium centre. The incorporation of aromatic groups into both the axel and the wheel can also lead to favourable π–π interactions between the components. An example is given in Figure 3.27(c)), which shows the X-ray crystal structure of the cationic complex formed between dibenzo-[24]-crown-8 and di(p-methoxybenzyl)ammonium hexafluorophosphate. The positioning of the aromatic rings, parallel to each other and slightly offset, can be clearly seen.

Crown ethers are commercially available in a variety of sizes and secondary ammonium species are readily synthesised. It is not surprising, therefore, that there are many examples of ammonium/crown ether rotaxane systems, ranging from simple [2]-rotaxanes to much more complicated structures involving multiple rings, multiple threads or both (Figure 3.28). It is unconventional rotaxanes such as these for which the standard nomenclature begins to fail. Figure 3.28(a) shows a [3]-rotaxane, with one thread that passes through two

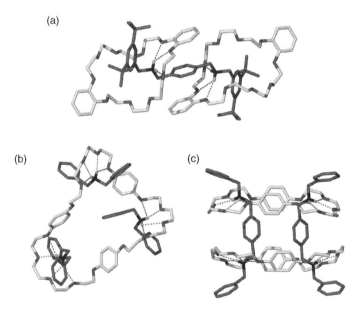

Figure 3.28 Complex ammonium/crown ether (pseudo)rotaxane-type structures with (a) two rings on a thread, (b) three threads within one ring and (c) two threads through two rings.

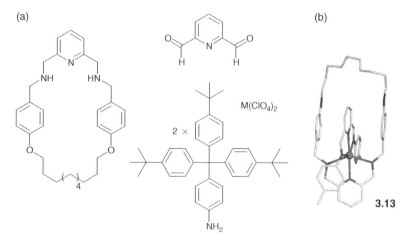

(a)

(b)

M(ClO$_4$)$_2$

2 ×

NH$_2$

3.13

Figure 3.29 (a) The five components used in the self-assembly of a metal-templated [2]-rotaxane (**3.13**) and (b) the X-ray crystal structure of the Cd complex with bulky −CPh$_3$ terminal groups (hydrogen atoms omitted for clarity).

rings. The species shown in Figures 3.28(b) and 3.28(c) are technically pseudorotaxanes, consisting of unblocked threads that pass through rings, despite the fact that they are far from the classical expectations of a rotaxane.

Rotaxanes may also be constructed by utilising a templating species. Traditionally, such a template has been a metal ion, bringing components together in a similar manner to the formation of racks and grids (Section 3.3). An example of a rotaxane that is synthesised by using a cationic metal template is shown in Figure 3.29.[14] The rotaxane **3.13** is formed through a five-component self-assembly reaction, *i.e.* the macrocycle, a small linear molecule, two bulky terminal groups and the metal salt. The thread of the rotaxane is formed reversibly through a condensation reaction to give imine bonds between pyridine-2,6-dicarboxaldehyde and *p*-aminophenyl-tris(*p-tert*-butylphenyl)methane. The rotaxane forms as the sole metal-containing species, indicating its thermodynamic stability, although it is unclear whether such a product would form *via* a threading or trapping method. The metal cation in the final product resides within an array of six nitrogen atoms and adopts a distorted octahedral coordination geometry. Metal templated complexes may sometimes be demetallated, for example by EDTA^{4-} (see Chapter 1, Section 1.2.3), to form a neutral, interlocked organic rotaxane. In the case of the example in Figure 3.29, however, the imine bonds are not stable without the metal present. If the imine bonds are first reduced to yield the amine derivative though, then demetallation yields a stable, organic rotaxane.

Recently there have been examples of pseudorotaxanes that are templated by a chloride anion, although similarities can be observed with their metal-bound counterparts. Similar design principles apply for the two cases, although the binding sites must be adapted for an anionic template. Figure 3.30 shows a

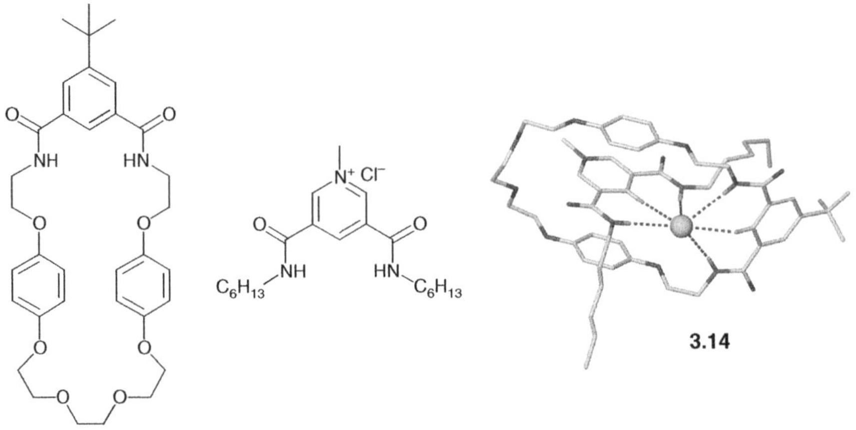

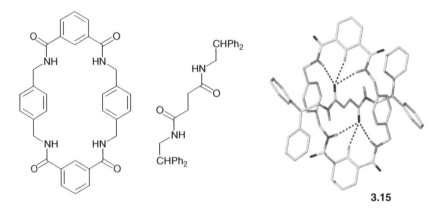

Figure 3.30 A chloride-templated pseudorotaxane is formed between a neutral macrocycle and a cationic thread, both of which contain amide groups.

Figure 3.31 A neutral rotaxane held together by $N-H\cdots O=C$ hydrogen bonds.

pseudorotaxane (**3.14**) that utilises amide groups within both the macrocycle and the thread to coordinate to a templating chloride anion.[15] Macrocyclic polyamides are well-known as anion hosts (see Chapter 2, Section 2.4) and the inclusion of a thread bearing a positive charge, in addition to two amide hydrogen bond donor groups, provides the driving force for the assembly of the pseudorotaxane. The similarity in approach for both metal and anion-templated rotaxanes can be seen, with both templates being firmly held within an array of six interactions.

All of these (pseudo)rotaxanes involve charged species, either as one of the components or as a template. It is also possible, however, to make rotaxanes from neutral species, although they lack the benefits of stronger, charge-assisted interactions. Neutral supramolecular complexes therefore utilise functional groups with intrinsically strong hydrogen bond donor/acceptor properties. Figure 3.31 shows an example of a neutral rotaxane (**3.15**) that makes use of amide groups as both hydrogen bond acceptors (*via* the carbonyl) and donors (*via* the amine).[16] As with the previous examples, the complementarity between the two components is evident from the crystal structure.

3.4.3 Catenanes

One of the magic tricks that everyone remembers from their childhood is when a magician takes seemingly unbreakable rings and inexplicably joins them together. The chemical analogue of this illusion is the formation of *catenanes*, although we all realise that, as with the magician's trick, something has to give in order to unite the rings. The term 'catenane' is derived from the latin word *catena*, meaning chain. Unlike rotaxanes, catenanes represent examples of true topologically connected species. Once joined, the rings cannot be separated, no matter how they are distorted, without at least one chemical bond being broken.

> **Catenanes:** Mechanically interlocked macrocycles or rings, akin to links in a chain.

The most fundamental form of catenane is shown in Figure 3.32(a) in which only two rings are interlocked. The terminology used for catenanes is relatively straightforward, with the general form *m*-crossing-[*n*]-catenane. We use *n* to represent the number of rings that are in the catenane and *m* to stand for the number of times that one ring crosses the other along its length, which can readily be determined by examination of the molecular graph. In the case of the two interlocked rings, we can define the structure as being a 2-crossing-[2]-catenane, as there are two rings that cross each other twice. For 2-crossing catenanes, the most common type, the '2-crossing' prefix is omitted, giving simply [*n*]-catenanes, with *n* defining the number of rings. Catenanes are known with a variety of different numbers of rings, for example, a [5]-catenane termed 'olympiadane' because of its resemblance to the Olympic symbol, although the most common remains the [2]-catenanes. Although topologically catenanes appear relatively straightforward, their synthesis is often very challenging. We saw in Chapter 2 (Section 2.2) that the synthesis of a single macrocycle is difficult to achieve, let alone the interlocking of two macrocycles as they form.

Catenanes are not limited to species involving two macrocycles and several examples of oligomeric and polymeric complexes are known. Polymeric catenanes are termed [*n*]-polycatenanes, whereby *n* is the number of rings in the chain although there is no clear point at which the term [n]-catenane ends and [n]-polycatenane begins (Figure 3.33(a)). A variation on polycatenanes are

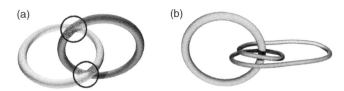

Figure 3.32 (a) A 2-crossing-[2]-catenane (or simply a [2]-catenane) with the two crossing points highlighted and (b) a 4-crossing-[2]-catenane.

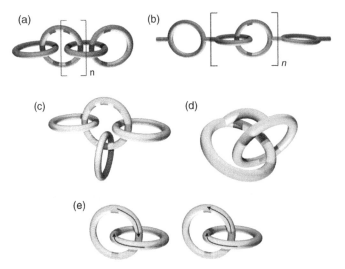

Figure 3.33 More complicated catenanes include (a) [*n*]-polycatenanes and (b) poly-[2]-catenanes. (c) molecular necklaces, (d) Covalent modifications can be made to produce pretzelanes and (e) chirality can be imparted by using macrocycles in which directionality can be defined by the covalent order of atoms in the ring.

poly-[2]-catenanes in which the repeating backbone is composed of [2]-catenanes which are joined together rather than single rings (Figure 3.33(b)). More complicated catenanes are also possible which involve several rings all interlocked with a central ring, in a manner similar to a key-ring. These topologies are referred to as *molecular necklaces* (MNs) (Figure 3.33(c)). These necklace structures are named simply by the total number of rings that they possess, *e.g.* a [4]-MN would have one central ring surrounded by three others, giving a total of four rings. More complex species are also known such as 'pretzelanes' (Figure 3.33(d)), in which the two rings are covalently joined, resulting in a molecular graph that looks like a certain kinds of baked goods! We saw in Figure 3.22(b) that chirality exists in multiple-crossing catenanes. Chirality can also be imparted by using two macrocycles in which directionality can be defined by the order of the atoms in the ring, *i.e.* the macrocycles themselves are unsymmetrical in terms of their covalent assembly (Figure 3.33(e)).

Catenanes can be templated in a variety of ways, *e.g.* by using metal-mediated synthesis, around anionic templates or *via* hydrogen bonding between the component parts. Metal-directed catenanes, as with other examples of self-assembled species, account for a large number of the known examples of catenanes. The ligands and metals that are used must be mutually complementary and be of the appropriate geometry to form closed-cyclic structures. An example that involves two separate ligands forming a cyclic species is shown in Figure 3.34.[17] A 4, 4′-bipyridine ligand and a ligand containing a biphenyl spacer between pyridyl groups form an interlocked catenane when combined with semi-protected palladium. The bipyridyl molecule acts as one side of the rectangular-shaped metallomacrocycle, with the more flexible ligand able to form the remaining three sides. The space within one macrocycle is the correct size and shape for another copy of itself to thread through and fill the void. The catenane persists in solution, as well as in the solid state and also remains intact in mass spectrometry analysis.

A more common use of metals in the synthesis of catenanes is when they template the formation of an intermediary species before the closure of the macrocyclic rings, as depicted in Figure 3.24(c). An example is shown in Figure 3.35, in which a copper(I) ion acts as a tetrahedral auxiliary linker between two substituted phenanthroline units.[18] The terminal hydroxyl groups of the two ligands can be further reacted in a post-assembly synthetic procedure to form polyether macrocycles, closing the structure into a [2]-catenane (**3.17**). A similar synthetic strategy can be employed to make a [3]-catenane (**3.18**) using a diynyl-containing macrocycle as the central ring (Figure 3.35(c)).[19]

Templation is also possible when using an anionic template in a progression from work on their rotaxane counterparts (*cf.* Figure 3.30).[20] A pseudorotaxane precursor is templated by a chloride anion, using NH···Cl⁻ hydrogen bonds and

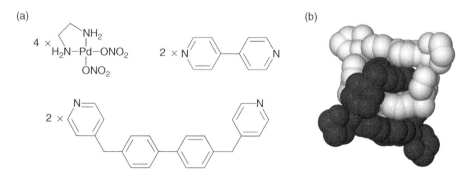

Figure 3.34 (a) The eight components and (b) the crystal structure of a catenane consisting of two metallo-macrocycles. The void with one cyclic species is an ideal size/shape fit for another copy of itself to fill.

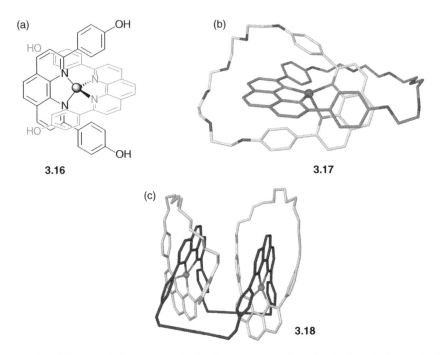

(a)

(b)

3.16

3.17

(c)

3.18

Figure 3.35 (a) The Cu(I) ion acts as a linker between two phenol-substituted phenanthroline units, forming a self-assembled intermediate (**3.16**) which can be reacted to form a polyether catenane (**3.17**), as shown in the crystal structure (b). (c) The synthetic strategy can be extended to produce a [3]-catenane (**3.18**).

assisted by a positively charged pyridinium group (Figure 3.36). By appending ethylene groups at the ends of the rotaxane thread, it is possible for synthetic modification to be performed by using a ruthenium catalyst to join the ends of the thread together through a ring-closing metathesis (RCM) reaction to form the catenane **3.19**.

Catenanes do not necessarily require the presence of an intermediary templating species to form. Organic macrocycles may template themselves into a suitable orientation to favour catenane formation during synthesis by virtue of hydrogen bonding, similar to the formation of ammonium/crown ether rotaxanes. Figure 3.37 shows an example of such a system. The component macrocycles are formed from the reaction between a diamine and isophthaloyl chloride. One of the products of this reaction is the interlocked species **3.20**, with a free macrocycle being the other main species formed. The macrocycles are held together by several NH···O hydrogen bonds, utilising the amide groups as both donors and acceptors.[21]

While the [2]-rotaxanes account for the bulk of work in catenane synthesis, it is worth looking at some more complex examples to see how the basic

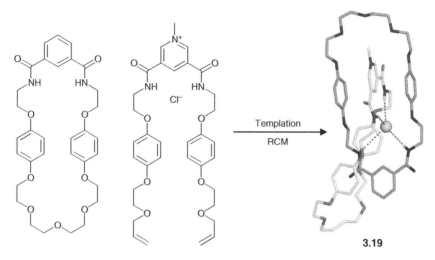

Figure 3.36 An anion-templated [2]-catenane (**3.19**) formed by post-synthetic modification to a pseudorotaxane precursor.

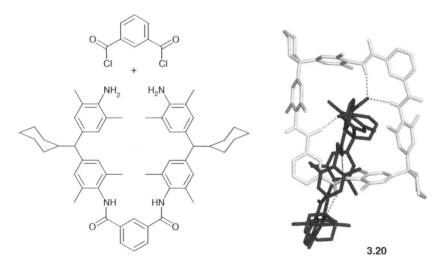

Figure 3.37 A hydrogen bonded [2]-catenane (**3.20**) is assembled from two components, held together by NH····O interactions between amide functionalities.

principles behind the design of catenanes can be extended to larger assemblies. Figure 3.33(c) introduced the concept of molecular necklaces (MNs) in which several macrocycles are joined to a central ring. An elegant example of this is shown in Figure 3.38, in which three ring-shaped cucurbituril molecules (related to glycoluril, see Figure 3.57) are arranged around a metal-templated

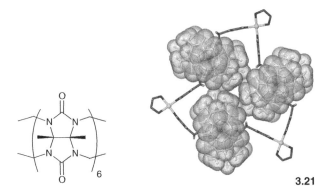

3.21

Figure 3.38 The cyclic cucurbituril molecule and the formation of a [4]-MN around a platinum-templated central ring.

central macrocycle giving the [4]-MN (**3.21**).[22] This work has also been expanded to a square central ring that has four cucurbituril molecules attached to it (a [5]-MN).

3.4.4   Rotaxanes and catenanes as molecular devices

 Raehm, L. and Sauvage, J.-P., 'Molecular machines and motors based on transition metal-containing catenanes and rotaxanes', *Struct. Bonding*, 2001, **99**, 55–78.

Rotaxanes and catenanes are not merely chemical curiosities – they also have the potential to act as 'molecular machines' (see Chapter 5, Section 5.3). For example, it is known that there are naturally occurring rotaxanes that act as processive catalysts, promoting a chemical changes while moving along a linear species.[23] Rotaxane threads can be designed in such a manner that they contain two different binding sites for the macrocycle. The preference for the macrocycle to bind to one site or the other can then be influenced by external factors, either chemically, electrochemically or photochemically. The ring can therefore be moved back-and-forth along the axle and such functional assemblies are known as *molecular shuttles*. Shuttles are driven by processes that involve one binding site becoming more preferable than another. A simple shuttle is shown in Figure 3.39. The thread contains two binding sites (coloured black and white), with the ring initially situated at the white site (left). In order for the ring to move, one of two things must happen – either the black site becomes more favourable or the white site becomes less favourable; either way, the relative affinities are changed in a manner that promotes binding to the black site.

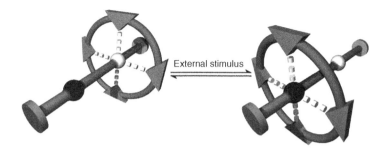

Figure 3.39 A molecular shuttle, based around a rotaxane with two potential binding sites. A change in affinities leads the macrocycle to be shifted to the alternate site.

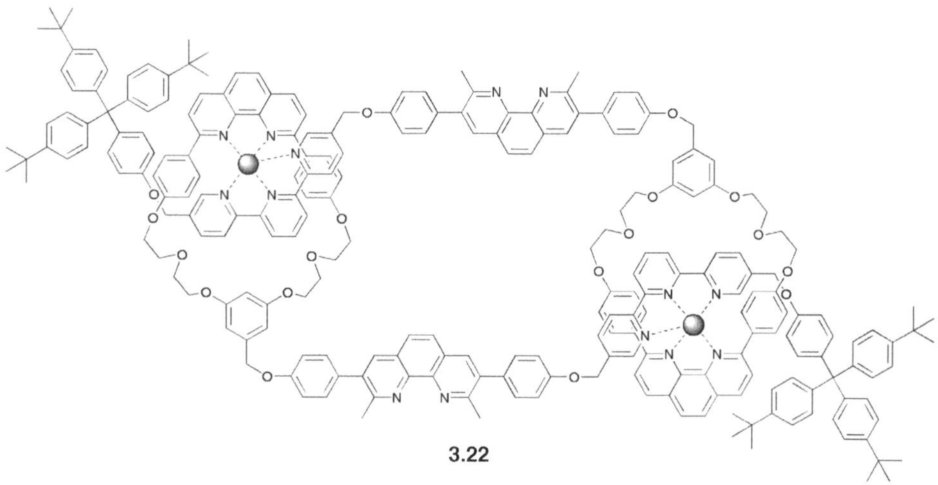

3.22

Figure 3.40 A molecular muscle (**3.22**), in which contraction and expansion is controlled by the coordination of metals with different coordination preferences to determine which binding site on the thread is utilised (this figure illustrates the Zn(ɪɪ) complex).

An example of a pseudorotaxane that alters its conformation in response to chemical changes is shown in Figure 3.40.[24] The compound **3.22** is an example of a *hermaphroditic rotaxane*, in which the ring and the thread are contained in the same molecule. The species is capable of acting as a 'molecular muscle' (see Chapter 5, Section 5.3.4), able to contract and stretch due to the two distinct binding sites within the thread, a bidentate 1,10-Phenanthroline unit and a tridentate terpyridine group. The rotaxane is synthesised from a smaller pseudorotaxane precursor in which the thread terminates after the phenanthroline unit. This pseudorotaxane assembly is templated by two Cu(ɪ) ions which are bound in their preferred tetrahedral geometry between the phenanthroline groups on the thread and in the ring. The terpyridine groups and the bulky stopper are then added in a post-assembly modification step. Once assembled, the complete rotaxane assembly can be demetallated (using KCN). Addition of octahedral Zn(II) induces the formation of a contracted conformation in which the metal is preferentially bound to the macrocycle and to the terpyridine group further along the thread, so reducing the length of the complex by some 20 Å. More details on molecular muscles are given in Chapter 5, Section 5.3.4.

Preferentially, for these molecules to act as molecular devices a change in molecular conformation has to be brought about by a physical, rather than a chemical, event. An example of this is the photoresponsive rotaxane **3.23** shown in Figure 3.41, an extension to the example shown in Figure 3.31.[25] The benzylic amide ring is attached to the thread in a five-component clipping reaction using *p*-xylylene diamine and isophthaloyl dichloride. The initially preferred binding site for the macrocycle is to the left of the thread where it interacts with the two amide functionalities in a succinamide group. The far end of the thread contains a naphthalimide unit which is a much poorer hydrogen bond acceptor. After a laser pulse and photoreduction using an external donor, the naphthalimide site becomes a radical anion and as a result a much stronger hydrogen bond acceptor than the succinamide group. This change causes the macrocycle to shuttle to the radical end of the thread where it binds preferentially. The movement of the macrocycle occurs in a very short timescale, approximately 1 μs. Charge recombination takes approximately 100 μs, after which the macrocycle returns to the succinamide site. The whole process can be repeated 10 000 times per second, generating 10^{-15} W of power per molecule. The system acts as a molecular piston, in which it has a power stroke, in which the work is done, and a recovery stroke in which the system returns to its starting position. The piston process is exactly like that of a macroscale piston in everyday machinery

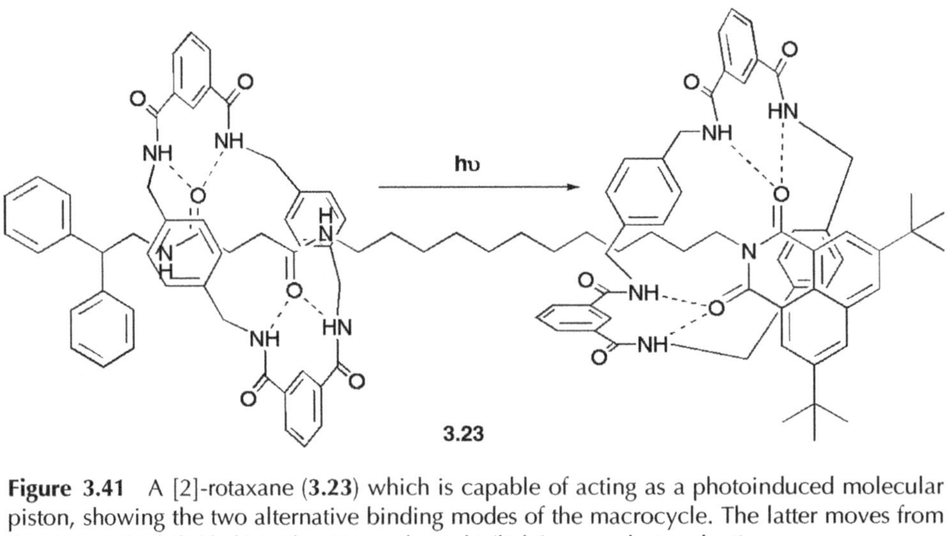

3.23

Figure 3.41 A [2]-rotaxane (**3.23**) which is capable of acting as a photoinduced molecular piston, showing the two alternative binding modes of the macrocycle. The latter moves from the site on the left (dark) to the site on the right (light) upon photoreduction.

and demonstrates how chemical systems can be constructed that have potential to act as nanoscale devices (see Chapter 5, Section 5.3).

Catenanes may also be used as nanoscale devices. Whereas rotaxanes interconvert between conformers through a translation of the macrocyclic component, catenanes are able to act in a pivoting or rotating manner when responding to an external stimulus. The example shown in Figure 3.42 is a model for a molecular device based on a [2]-catenane.[26] One of the macrocycles contains two different binding sites on opposing sides of the ring – terpyridyl (terpy) and diphenylphenanthroline (dpp). The other macrocycle contains only a single dpp binding unit. There are two possibilities for metal binding between the two macrocycles, *i.e.* a four-coordinate dpp/dpp interaction (**3.24a**) and a five-coordinate dpp/terpy interaction (**3.24b**). The former of these two interaction sites favours a tetrahedral metal ion, whereas the latter is suitable for a five-coordinate ion. Cu(I), which preferentially adopts a tetrahedral geometry, causes the formation of the dpp/dpp conformer of the catenane. Upon changing the oxidation state of the copper to Cu(II), either chemically, electrochemically or photochemically, the dpp/terpy conformation is adopted instead.

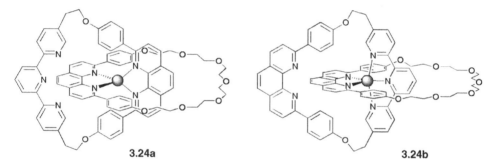

Figure 3.42 A model for a catenane-based molecular device. The two orientations of the catenane are redox-switchable by control of the copper oxidation state. (a) The tetrahedral Cu(I) complex and (b) the square-pyramidal Cu(II) complex.

3.4.5 Borromeates

 Cantrill, S. J., Chichak, K. S., Peters A. J. and Stoddart, J. F., 'Nanoscale Borromean rings', *Acc. Chem. Res.*, 2005, **38**, 1–9.

It is possible for three rings to become mutually interlocked without any catenane-like connections. This is the situation observed within *Borromean rings*. The Borromean rings were used historically by the Italian Borromeo family as a

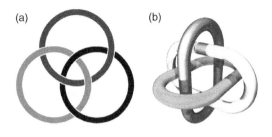

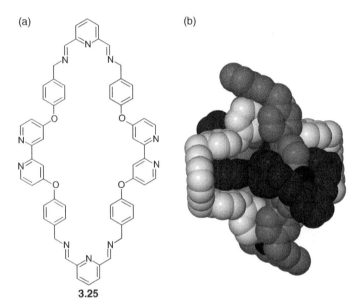

Figure 3.43 Representations of Borromean rings:(a) the molecular graph; (b) three-dimensional view.

3.25

Figure 3.44 (a) The ligand **3.25** is synthesised *in situ* and templates the formation of a discrete Borromean complex. (b) The X-ray crystal structure of the hexa-zinc Borromeate.

symbol of unity – separate parts that are unable to be separated. The molecular graph of Borromean rings is shown in Figure 3.4.5 – no two rings share a link. This situation is impossible to obtain in a two-dimensional system and a three-dimensional representation of the rings makes it clearer as to how such an arrangement may arise.

One example of a discrete Borromean complex has recently been synthesised.[27] The ligand **3.25** contains metal-binding sites on both the exterior and interior of the molecule (Figure 3.44). In this manner, it is possible for three rings to be inserted through one another in a mutually perpendicular fashion by a Zn(II)-templated process. The zinc adopts a distorted octahedral geometry, to suit the tridenate donor group, with a trifluoroacetate ligand occupying the final coordination site. It has also proved possible to demetallate the hexa-zinc complex and preserve the Borromean architecture. Although this remains the only example of a discrete

Borromean complex, infinite two-dimensional sheets are known with a similar interlocking motif (see Chapter 4, Section 4.6.5).

3.4.6 Knots

Dietrich-Buchecker, C., Colasson, B. X. and Sauvage, J. P., 'Molecular Knots', *Top. Curr. Chem.*, 2005, **249**, 261–283.

Knots are occur frequently in everyday life, both in our macroscopic world (ties, shoelaces, *etc.*) and within natural systems (DNA can form knotted structures, for example). In a mathematical and chemical sense, however, a knot is classified as a single, continuous thread that is interlocked with itself and cannot be untied without being broken, *i.e.* without breaking chemical bonds (unlike a tie or shoelace). Knots can therefore be classified as 'self-crossing' macrocycles and are therefore topologically non-trivial. The simplest knot is the *trefoil knot* which can be represented by a molecular graph with three crossing points. To date, no chemical knots have been synthesised with a more complex topology than this, although aggregates of trefoil knots are known.[28]

Knots are considered to be topological isomers of a macrocycle as, although they may have the same connectivity, the two cannot be interconverted with some kind of deformation (Figure 3.45). Perhaps the most important aspect of knots, besides their attractive geometrical form, is that they are intrinsically chiral – they do not require special design routes to make them chiral, unlike catenanes and rotaxanes. This property may ultimately lead to applications of knots as chiral catalysts, components of nanoscale motors and in determining structures of naturally occurring knots. Trefoil knots have been synthesised by two routes – *via* a metal-templated helicate and by hydrogen bond-mediated macrocyclisation.

Figure 3.45 (a) A trefoil knot and (b) its topological enantiomer are topological isomers of (c) a macrocycle.

Metal-templated knots
The first synthetic knot was made from a ligand that incorporates two 1,10-phenanthroline units in an extension of work conducted on the formation of

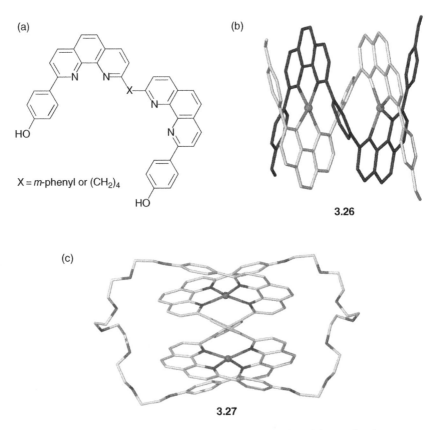

(a)

HO

X = m-phenyl or (CH₂)₄

HO

(b)

3.26

(c)

3.27

Figure 3.46 (a) The helicand precursor to a metal-templated knot, (b) the X-ray crystal structure of its methoxy-derivatised helicate (X = *m*-phenyl) and (c) the X-ray crystal structure of the metal-templated trefoil knot with polyethyleneglycol bridges.

metal-templated [2]-catenanes (Figure 3.35).[29] The two phenanthroline groups are tethered by a butenyl chain or a *meta*-disubstituted aryl ring, resulting in a [4 + 4] double-stranded helicate **3.26** (Section 3.3.3) in essentially quantative yields using Cu⁺ as a templating ion (Figure 3.46). The crystal structure of the methoxy-substituted version of the helicate shows that the functionalised termini of the threads are ideally positioned for further reaction. A post-assembly synthetic modification procedure can then be carried out on the helix to tether the end of one helical strand to the opposite end of the other strand using polyethyleneglycol bridges. The reaction to join together the two strands results in one continuous covalent thread that is interlocked with itself – it has formed a knot, **3.27**. The use of the more rigid *m*-phenylene spacer, rather than the more flexible (CH₂)₄ unit, affords the product in greatly improved yields (27 % versus 3 %). Analogues of this knot have been made using variations in the nature of the bridging group added during the post-assembly synthetic stage. Helicates with terminal ethylene groups have also been used, with the knots closed off using Grubbs' catalyst in a ring-closure metathesis (RCM) procedure. The knots were synthesised as a mixture of enantiomers which proved difficult to separate. Separation was eventually achieved by using a chiral counter-anion, *i.e.* binaphthyl phosphate. Once the knot is assembled, the Cu^+ templates can be removed by treatment with an excess of potassium cyanide, leaving a purely organic knotted molecule.

Hydrogen bonded knots

An alternative method for the synthesis of knots is essentially a macrocyclisation reaction that utilises appropriate hydrogen bond templating within the macrocycle to cause interlocking to occur. The reaction is, at first glance, a standard condensation cyclisation reaction that proceeds by the elimination of HCl to create amide linkages (Figure 3.47). Four isolable products are formed from this reaction as it is relatively non-specific with no rigid templates present (unlike the metal-templated synthesis). The products are macrocycles containing 1:1, 2:2, 3:3 and 4:4 ratios of the starting materials. The major product is the kinetically favoured 1:1 species (30–50 %) with the yield reducing stepwise for each larger macrocycle. The 3:3 product, **3.28**, which forms in yield no greater than 20 %, is the

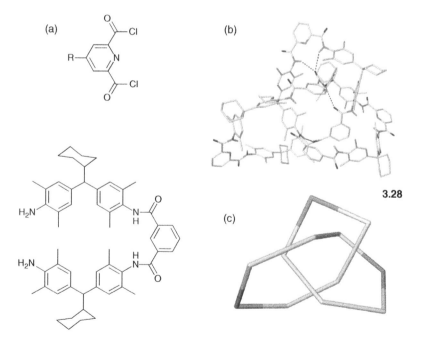

Figure 3.47 (a) The components used to synthesise a molecular trefoil knot *via* a macrocyclisation reaction, (b) its X-ray crystal structure and (c) a simplified schematic of the knot.

only one which does not form a simple macrocycle, but rather folds into a knot *via* hydrogen bonding at some point during its synthesis.[30]

3.5 Self-assembling capsules

3.5.1 Molecular containers

 Turner, D. R., Pastor, A., Alajarín, M. and Steed, J. W., 'Molecular containers: Design approaches and applications', *Struct. Bonding*, 2004, **108**, 97–168.

Molecular containers or *capsules* are three-dimensional cage-like compounds with an internal cavity that is capable of holding one or more guest species. Capsules are of great interest as they not only act as effective host species, able to completely surround their guest, but also due to their promising catalytic properties as nanoscale reaction vessels. Carcerands and hemicarcerands are examples of covalently synthesised capsules, formed by joining together two hemispherical macrocycles (see Chapter 2, Section 2.7.3). However, carcerands are of limited practical use due to the fact that guest species become permanently trapped within them and cannot be released without breaking one or more chemical bonds. In recent years, there has been an emergence of capsules that are constructed using self-assembly. Because these capsules are formed under equilibrium conditions, they can dissociate into their constituent parts and reform, releasing and trapping guests as they do so. In many cases, these processes occur on timescales that are suitable for practical applications, such as sensing and catalysis, neither holding the guest indefinitely (*cf.* carcerands) nor releasing it too quickly.

> **Self-Assembled Molecular Containers/Capsules:** Discrete, hollow self-assembled structures that are capable of holding one or more guests.

The rapid binding and release of guests makes many self-assembled containers comparable to single-molecule hosts that are able to bind guests in solution (Chapter 2) with the advantage that the guest is surrounded on all sides and completely isolated from the bulk environment, potentially allowing for more host – guest interactions and reduced solvent competition for binding sites. The effective removal of a guest species from the bulk environment also aids the catalytic properties of so-called nanoscale reaction vessels. The presence of only one or two molecules within a cavity creates a very high effective concentration at a formally elevated pressure. This situation can allow reactions to occur under much milder conditions than those that would usually be required or, in some cases, allow reactions to proceed that would not normally occur at all. The confined space within capsules has also been used to promote stereoselective reactions, as one isomer of the product can fit within the cavity while the other cannot.

As with many self-assembled systems, the synthesis of molecular containers can be divided into two broad approaches, *i.e.* the use of metal–ligand interactions and the use of hydrogen bonds. The most commonly utilised method is metal-directed self-assembly, using either one-dimensional or two-dimensional ligands. This approach generally creates more robust cages that are more persistent in solution. The use of hydrogen bonds is, however, becoming more common and has been used to great effect in generating capsules with large numbers of components.

3.5.2 Metal-directed capsules

The most common role of metal ions in capsule synthesis is to act as the vertices of a geometrical structure, connected by organic ligands to form the requisite shape. The correct choice of ligands and metal ions is vital to ensure a closed structure, rather than a polymeric one. In general, it is possible to classify ligands into two broad classes – one-dimensional and two-dimensional. One-dimensional ligands bridge only between two metal centres. They can be thought of as topologically equivalent to straight lines and form the edges of the structure. Such an assembly process is termed *molecular scaffolding* (Figure 3.48(a)). Ligands that bridge between more than two metal atoms can be considered two-dimensional and form the faces of the final structure. The assembly of polyhedral capsules using two-dimensional ligands has been termed *molecular panelling* (Figure 3.48(b)).

Molecular scaffolding
Molecular scaffolding involves one-dimensional ligands acting as polyhedral edges and hence it is not unusual for the faces of the polyhedral capsules to contain significant voids through which guest species may squeeze in and out of the cavity. In the vast majority of cases, however, it appears to be the presence of a guest that templates the formation of the capsule. In the example shown in Figure 3.48(a), the M_2L_4 capsule carries a positive charge from the two Cu^{2+} ions and is templated by the presence of a perchlorate anion (ClO_4^-). The guest interacts with the capping copper ions through $Cu^{2+} \cdots O$ interactions. Systems such as this can be totally reliant upon the presence of the correct anion to form. For example, the ligand shown in Figure 3.49, amidinothiourea (**3.29**) (which exists in two tautomeric forms), can be added to nickel salts of nitrate, acetate and perchlorate without any significant association occurring; however, once NiX_2 ($X = Cl$, Br) is added to these solutions, a distorted square-prismatic cage compound is formed that is templated around the halide anion through eight $NH \cdots X^-$ interactions.[31]

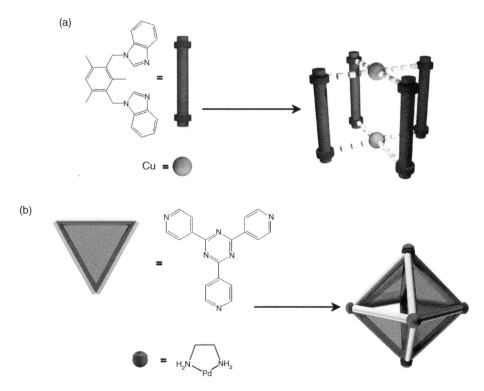

Figure 3.48 Examples of cages made by (a) molecular scaffolding and (b) molecular panelling.

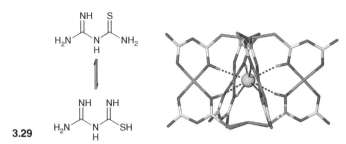

Figure 3.49 A spherical, anionic template (Cl^-/Br^-) is required to template this cationic capsule using amidinothiourea ligands. The guest is held by eight $NH \cdots X^-$ hydrogen bonds.

Although we refer to the ligands used in scaffolding assembly as forming the edges of a polyhedron, they do not have to be rigid, linear molecules to achieve this. The example given in Figure 3.50 shows the formation of a tetrahedral cavity using the deprotonated forms of four relatively flexible ligands (**3.30**) to join together four gallium ions.[32] The framework of the Ga_4 capsule carries a high negative charge (due to quadruple deprotonation of the ligands) and can therefore

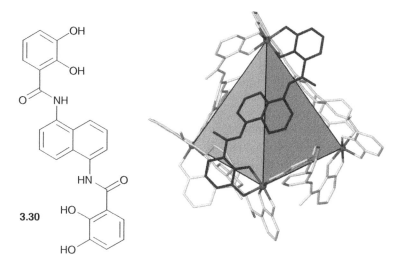

3.30

Figure 3.50 The formation of a tetrahedral capsule based around four gallium ions and four deprotonated ligands. The tetrahedral cavity is shown and one of the ligands is emphasised to show that linearity and rigidity are not essential (for clarity, counter-cations are not shown).

act as a host for cations. Various tetraalkylammonium cations (NR_4^+) are able to reside within the capsule with the flexibility of the ligand allowing the cage to expand and contract to suit guests of differing sizes. The NR_4^+ guests are also tetrahedrally disposed, highlighting the importance in complementarity between host and guest and the high degree of discrimination that can be achieved through the use of capsules. The small Me_4N^+ cation enters and leave the cavity rapidly on the ^1H NMR timescale, whereas larger cations, such as ^{i}-Pr_4N^+, exchange slowly due to their size relative to the size of the gaps in the capsule. It is also interesting to note that tetramethylammonium is completely displaced by larger guests, hence indicating that the capsule is more stable when the intracavity space is fully occupied. This principle is cemented in the anthropomorphic idiom that 'nature abhors a vacuum', *i.e.* filled space maximises van der Waals interactions between the guest and the capsule wall. The Ga_4 cage also represents an example of a highly reactive molecule being stabilised within a supramolecular container (see Chapter 2, Section 2.7.3). The cation $[Me_2C(OH)PEt_3]^+$ is highly unstable and decomposes rapidly in water. However, when included within the tetrahedral cavity and thus removed from the bulk environment, the species persists for several hours.

Molecular panelling
Molecular panelling is the formation of capsules by using ligands as the faces of the final cage, rather than as the edges.[33] Potentially, this approach can lead to more fully enclosed containers with fewer gaps through which guests can pass, thereby limiting the rate of guest exchange. Perhaps the most comprehen-

sively studied example of a panelled container is **3.31**, shown in Figure 3.48(b), and its derivatives. The tripyridyltriazine ligand is a planar, equilateral triangle, with nitrogen donors that are able to coordinate to metals, facing outwards at the corners. Such a ligand is suitable to form the triangular faces of regular poly-hedral shapes. As we saw in Figure 3.9, a divergent ligand can be capped with convergent, semi-protected metals to form discrete complexes. When the trian-gular ligand is combined with semi-protected Pd(II) ions, an octahedral struc-ture is formed in which alternate faces are occupied by the ligand. The vacant faces allow for guest exchange to occur and the large internal cavity permits multiple guests (or very large guests) to be incorporated. Slight variations in the ligand (*i.e.* the position of the nitrogen atoms) allows for other shapes to be assembled, such as hexahedra and tetrahedra. The internal cavity is hydrophobic, which drives the inclusion of organic guests in aqueous media. Molecules that contain both hydrophobic and hydrophilic groups, such as adamantanecarboxylic acid, are encapsulated with the hydrophilic groups able to protrude through the vacant walls. In fact, this remarkable capsule has been observed to show a wide variety of inclusion phenomena and intra-cavity reactions (shown in Figure 3.51), as follows:

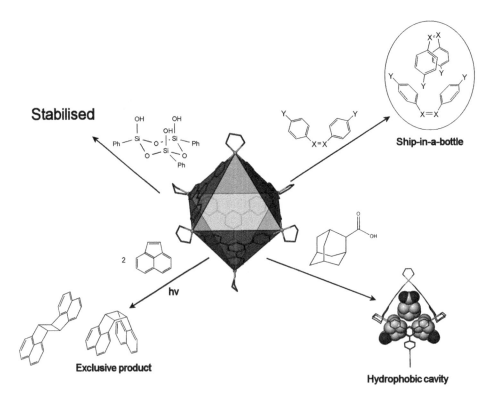

Figure 3.51   The octahedral capsule **3.31** displays many remarkable host–guest properties.

- The stabilisation of a highly reactive cyclic silanol oligomer, isolating it from the surrounding media, thus preventing hydrolysis.

- The stabilisation of the *cis*-isomers of stilbene derivatives (preventing conversion to the more stable *trans*-isomers) by the formation of dimers within the cavity, akin to building a 'ship-in-a-bottle' (the dimer must form within the capsule as it is too large to enter pre-assembled).

- Stereoselective photodimerisation of olefins as only the *syn*-isomer of the product is able to fit within the cavity – without the cage the reaction results in a mixture of the two isomers.

It is possible to create larger capsules, following the same basic design principles, simply by employing larger ligands. Typically, larger ligands also have a greater degree of flexibility associated with them and are therefore less planar. As a result, the complexes that are formed often bear only a passing resemblance to regular polyhedra and become more spherical in nature as the ligands adopt more convex geometries. Non-planar, tripodal ligands have been used to great effect in constructing nanoscale containers with roughly hexahedral

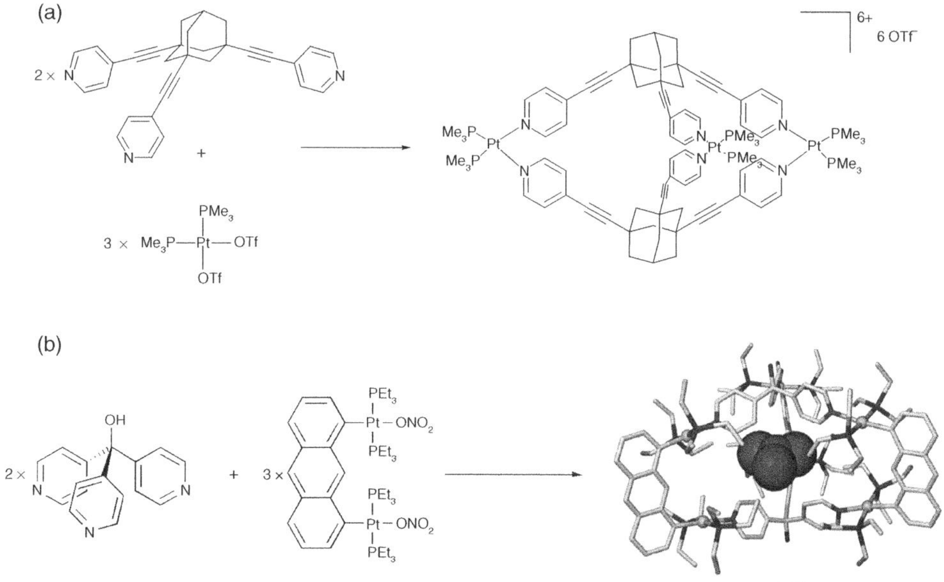

Figure 3.52 Examples of (a) a container made from tripodal, concave ligands combined with partially protected metals and (b) a cavity constructed using two different ligands containing a nitrate guest (X-ray crystal structure shown).

cavities.[34] Ligands have been synthesised around core units such as adaman-
tane and a trisubstituted carbon atom to provide the correct geometry to act
as a pseudohemispherical component. These ligands can either be used with
metal species by themselves (Figure 3.52(a)), or with other bridging species which
serve to extend the height of the cavity (Figure 3.52(b)). The cage **3.32**, shown
in Figure 3.52(b), contains a single nitrate anion as the encapsulated guest, as
determined by X-ray crystallography. The capsule shows evidence that nitrate
is required as a templating agent. This cage also demonstrates another important principle in the design of capsular species, that it is possible to utilise
more than one ligand in order to achieve a discrete container. In the example
that is shown, the cage is formed from two hemispherical multidentate ligands
and three bridging one-dimensional ligands, similar to those used in scaffold
architectures.

Perhaps one of the metal–ligand capsule systems whose interior most closely
resembles a spherical cavity is the metallo-carcerand shown in Figure 3.53.[35]
The two substituted resorcinarene ligands (**3.33**) that combine to make the
final complex are almost hemispherical and are similar to the components
used in the synthesis of purely organic carcerands and hemicarcerands
(see Chapter 2, Section 2.7.3). Two of the tridentate pendant arms that
are attached to the resorcinarene macrocycles are able to bind to Co(II)
cations, one from above and one from below, to fully satisfy the octahedral coordination sphere. The cavity is of a suitable size to hold neutral
guests, such as benzene, and, despite the cage carrying a formal negative

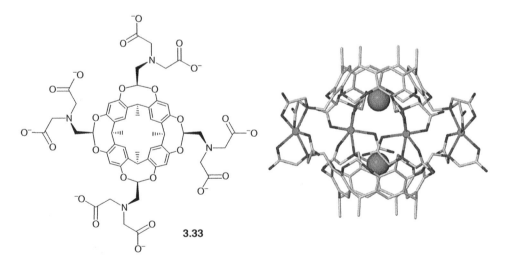

3.33

Figure 3.53 X-ray crystal structure of a metallo-carcerand, assembled from two hemispherical octa-anionic ligands, **3.33**, and four Co²⁺ ions. The cavity contains two bromide anions,
despite the −8 charge.

charge, it has been observed in the solid state to incorporate two bromide anions.

3.5.3 Hydrogen bonded capsules

Although the interactions between a metal ion and a ligand are usually stronger than hydrogen bonds, there nevertheless exists an increasing number of capsules that are exclusively held together through hydrogen bonding interactions. As with their metal-containing analogues, hydrogen bonded capsules are in equilibrium with their component parts in solution and, due to their weaker connections, are often more reliant upon the presence of a templating species within the cavity. The design principles behind capsules that are held together by hydrogen bonds are very similar to metal–ligand capsules – the components must be complementary to one another in terms of interaction centres and their spatial arrangement. Unlike the containers that we saw in the previous section, it is possible for hydrogen bonded systems to be self-complementary, *i.e.* the donor and acceptor groups are both within the same molecule which can form capsules by associating with other components of the same type. In the following discussion, we will look at some examples in two distinct categories, *i.e.* capsules created from multiple species and those created through self-recognition.

Multi-component assemblies
Hydrogen bonds are highly directional in nature and therefore offer a certain degree of predictability when designing self-assembling systems. When more than one species is used to assemble a capsule, it is usual for one component to contain groups capable of donating hydrogen bonds while the other contains acceptor atoms. There is a distinct analogy with metal–ligand systems (whereby the ligand is an electron donor and the metal an acceptor). Figure 3.54 shows the

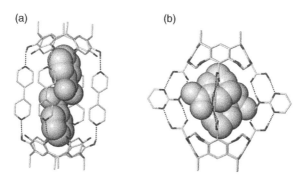

(a)   (b)

Figure 3.54 Examples of multi-component, hydrogen bonded capsules (see text for further details).

X-ray crystal structures of two examples of multi-component capsules that are held together by hydrogen bonds. Both of the assemblies are neutral, in contrast to metal-based capsules which usually carry a formal charge. Figure 3.54(a) shows the assembly formed from 4, 4'-bipyridine and a resorcinarene derivative that is held together through eight OH$\cdots$N interactions.[36] The cavity is large enough to contain two p-nitrobenzene molecules. This system has been utilised to isolate guest molecules in the solid state in order to carry out time-resolved spectroscopy of their excited states.[37] Figure 3.54(b) shows a system that incorporates donor and acceptor groups into both of the components.[38] There are mixed NH$\cdots$O and OH$\cdots$N donor/acceptor pairs arranged in a highly complementary manner akin to that observed in carboxylic acid dimers (see Chapter 4, Section 4.5.3) although in this case the interaction is between a carboxylic acid and 2-aminopyridine. The cavity within this container is of a comparable size to the previous example and is also able to incorporate two nitrobenzene molecules, although the capsule is held together through twice as many hydrogen bonds.

Although the vast majority of the examples of non-metal/ligand capsules are held together solely by hydrogen bonds, there are also a few that are held together by stronger electrostatic interactions, such as between ion-pairs. For example, it is known that under the right conditions, calix[4]arenes appended with sulfonato groups can form dimers with amidinium-substituted calixarenes.

Self-complementary capsules
Hydrogen bonded capsules need not be formed from two different species. It is common for a molecule to posses both donor and acceptor groups in close proximity (*e.g.* amides, carboxylic acids, ureas, alcohols) and appropriately functionalised molecules may thus be *self-complementary*. An impressive example of this phenomenon is shown in Figure 3.55, in which a resorcinarene with several hydroxyl substituents (**3.34**) forms a hexameric capsule held together exclusively through hydrogen bonds.[39] This extremely large capsule persists even in relatively polar media and holds eighteen methanol guests. A very closely related example in which six resorcarenes are held together by 24 Cu(II) ions is also known, highlighting the common self-assembly principles despite the different interactions involved.

Urea is well-known to form interesting structures in the solid state through self-assembly (see Chapter 4, Section 4.3.1) and urea derivatives have been widely used as self-complementary sites in the construction of container species (see key reference to section 3.5.1). Calixarenes, in particular, have been appended with urea functionalities and have been found to self-assemble into dimers in solution that are able to hold simple guest species, including ionic guests. Tertiary amines have also been used to great effect as a core unit around which pendant urea groups can be placed such as in the molecule **3.35**.[40] The compound dimerises to give a capsule with a hydrogen bonding belt around

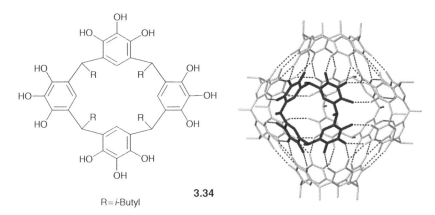

3.34

R=*i*-Butyl

Figure 3.55 A hexameric hydrogen bonded capsule (**3.34**)$_6$, with one individual component highlighted (for clarity, isobutyl groups have been omitted).

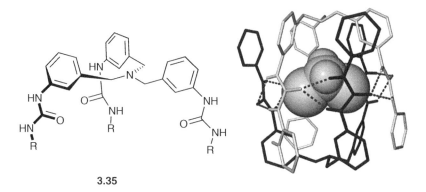

3.35

Figure 3.56 A tertiary amine (R = phenyl), **3.35**, which is capable of self-assembling to form a dimeric capsule (**3.35**)$_2$, the crystal structure of which shows an encapsulated dichloromethane guest.

the middle of the container. The urea functionalities are able to interdigitate with urea groups on an adjacent molecule through NH···O interactions (Figure 3.56).

Despite the widespread popularity of urea-based containers, the related glycoluril capsules perhaps best demonstrate the ability of molecules to self-associate (Figure 3.57).[41] Glycoluril possesses two hydrogen bond acceptor sites and four donor groups in its unsubstituted form. The glycoluril group also possesses an intrinsic curvature which facilitates the formation of closed structures. Much work has been carried out on molecules that have glycoluril groups

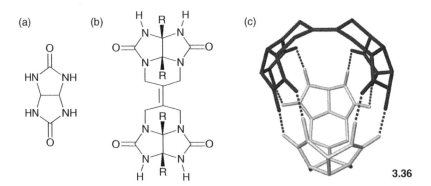

Figure 3.57 (a) The unsubstituted glycoluril moiety. (b) Its incorporation in a self-assembling molecule (R = CO$_2$Et) gives rise to (c) a dimeric capsule (for clarity, ethyl ester groups have been omitted).

appended to various different core units.[41] An example is the molecule shown in Figure 3.57(b) which contains two glycoluril substituents separated by a semi-flexible spacer. This molecule self-assembles both in solution and in the solid state, to form a pseudo-spherical dimeric capsule **3.36** (Figure 3.57(c)). The natural curvature of the monomer can be seen in the crystal structure of the dimer, giving rise to eight NH···O hydrogen bonds which hold the capsule together. The dimer has a relatively small cavity (41 Å^3) compared to other members of the glycoluril-capsule family, but nevertheless it is able to reversibly bind guests such as dichloromethane and methane in solution. Larger dimers using phthalocyanines as spacers are able to encapsulate guests such as benzene and cyclohexane and recent developments have even seen the synthesis of related chiral dimers.

Unimolecular capsules
This section has revealed numerous interesting capsules that may be formed when two or more species are joined by intermolecular hydrogen bonds. A rarer situation, however, is when a capsule can be formed by one molecule that is capable of folding in upon itself in the presence of a suitable template species. While this may not be self-assembly in its strictest sense, it is well worth pointing out that the same principle of complementary hydrogen bond donor/acceptor groups in a suitable spatial orientation can be applied. The example shown in Figure 3.58 is based around a resorcinarene with amide-functionalised arms (**3.37**).[42] The remaining free−OH group on each arm is able to hydrogen bond to the ether linkage on an adjacent arm, forming a deep bowl shape. In the presence of a suitable templating species, such as NMe$_4^+$, the molecule is able to close up and the upper end becomes sealed by amide NH···O hydrogen bonds.

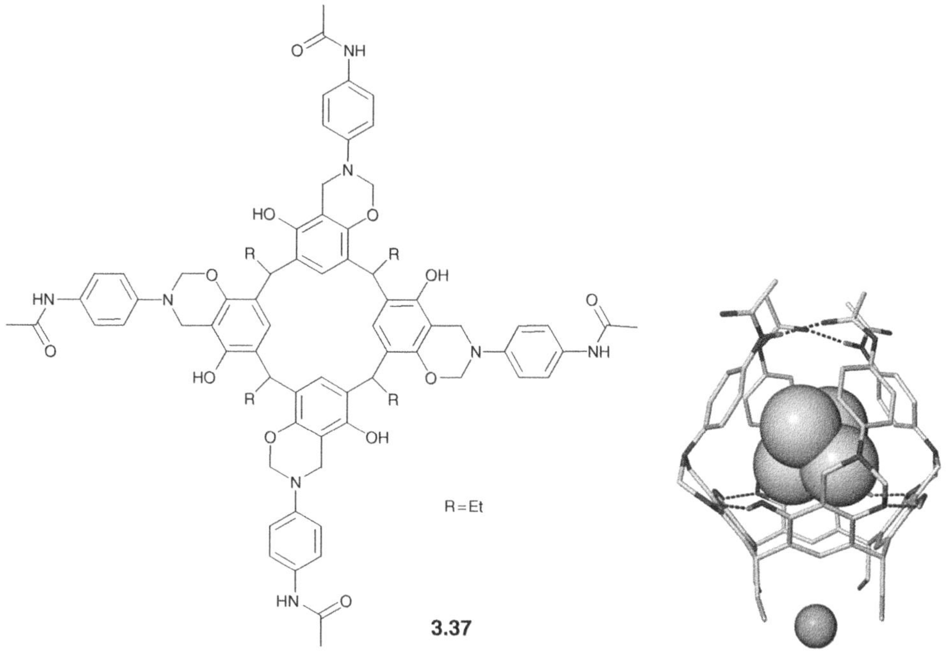

Figure 3.58 An amide-containing molecule (**3.37**) which is capable of forming a unimolecular capsule by hydrogen bonding with a templating tetramethylammonium guest and chloride counter-anion.

References

1. Venkataraman, D., Du, Y., Wilson, S. R., Hirsch, K. A., Zhang, P. and Moore, J. S., 'A coordination geometry table of the d-block elements and their ions', *J. Chem. Edu.*, 1997, **74**, 915–918.

2. Fujita, M., Sasaki, O., Mitsuhashi, T., Fujita, T., Yazaki, J., Yamaguchi, K. and Ogura, K., 'On the structure of transition-metal-linked molecular squares', *J. Chem. Soc., Chem. Commun.*, 1996, 1535–1536.

3. Butler, P. J. G., 'Self-assembly of tobacco mosaic virus: the role of an intermediate aggregate in generating both specificity and speed', *Phil. Trans. R. Soc. London, B.*, 1999, **354**, 537–550.

4. Hanan, G. S., Arana, C. R., Lehn, J.-M. and Fenske, D., 'Synthesis, structure and properties of dinuclear and trinuclear rack-type Ru-II complexes', *Angew. Chem., Int. Ed. Engl.*, 1995, **34**, 1122–1124.

5. Baxter, P. N. W., Lehn, J.-M., Kneisel, B. O., Baum, G. and Fenske, D., 'The designed self-assembly of multicomponent and multicompartmental cylindrical nanoarchitectures', *Chem. Eur. J.*, 1999, **5**, 113–120.

6. Baxter, P. N. W., Lehn, J.-M., Fischer, J. and Youinou, M.-T., 'Self-assembly and structure of a 3 × 3 inorganic grid from nine silver ions and six ligand components', *Angew. Chem., Int. Ed. Engl.*, 1994, **33**, 2284–2287.

7. Lehn, J.-M., Sauvage, J.-P., Simon, J., Ziessel, R., Piccinnileopardi, C., Germain, G., Declercq, J.-P. and Vanmeerssche, M., 'Synthesis and metal-complexes of a conformationally restricted quaterpyridine – crystal structure of its dimeric dinuclear Cu(ɪ) complex, $[Cu_2(PQP)_2]^{2+}$', *Nouv. J. Chem.*, 1983, **7**, 413–420.

8. Piguet, C., Bernardinelli, G., Bocquet, B., Schaad, O. and Williams, A. F., 'Cobalt(ɪɪɪ)/cobalt(ɪɪ) electrochemical potential controlled by steric constraints in self-assembled dinuclear triple-helical complexes', *Inorg. Chem.*, 1994, **33**, 4112–4121.

9. Constable, E. C., Ward, M. D. and Tocher, D. A., 'Spontaneous assembly of a double-helical binuclear complex of 2,2':6',2'':6'',2''':6''',2'''':6'''',2'''''-sexipyridine', *J. Am. Chem. Soc.*, 1990, **112**, 1256–1258.

10. Schweiger, M., Seidel, S. R., Arif, A. M. and Stang, P. J., 'Solution and solid-state studies of a triangle–square equilibrium: anion-induced selective crystallization in supramolecular self-assembly', *Inorg. Chem.*, 2002, **41**, 2556–2559.

11. Campos-Fernandez, C. S., Clerac, R., Koomen, J. M., Russell, D. H. and Dunbar, K. R., 'Fine-tuning the ring-size of metallacyclophanes: a rational approach to molecular pentagons', *J. Am. Chem. Soc.*, 2001, **123**, 773–774.

12. Chiu, S.-H., Rowan, S. J., Cantrill, S. J., Stoddart, J. F., White, A. J. P. and Williams, D. J., 'Post-assembly processing of [2]-rotaxanes', *Chem. Eur. J.*, 2002, **8**, 5170–5183.

13. Ashton, P. R., Fyfe, M. C. T., Hickingbottom, S. K., Stoddart, J. F., White, A. J. P. and Williams, D. J., 'Hammett correlations "beyond the molecule"', *J. Chem. Soc., Perkin Trans. 2*, 1998, 2117–2128.

14. Hogg, L., Leigh, D. A., Lusby, P. J., Morelli, A., Parsons, S. and Wong, J. K. Y., 'A simple general ligand system for assembling octahedral metal–rotaxane complexes', *Angew. Chem., Int. Ed. Engl.*, 2004, **43**, 1218–1221.

15. Sambrook, M. R., Beer, P. D., Wisner, J. A., Paul, R. L., Cowley, A. R., Szemes, F. and Drew, M. G. B., 'Anion-templated assembly of pseudorotaxanes: importance of anion template, strength of ion-pair thread association and macrocycle ring size', *J. Am. Chem. Soc.*, 2005, **127**, 2292–2302.

16. Altieri, A., Bottari, G., Dehez, F., Leigh, D. A., Wong, J. K. Y. and Zerbetto, F., 'Remarkable positional discrimination in bistable light- and heat-switchable hydrogen bonded molecular shuttles', *Angew. Chem., Int. Ed. Engl.*, 2003, **42**, 2296–2300.

17. Fujita, M., Aoyagi, M., Ibukuro, F., Ogura, K. and Yamaguchi, K., 'Made-to-order assembling of [2]-catenanes from palladium(ɪɪ)-linked rectangular molecular boxes', *J. Am. Chem. Soc.*, 1998, **120**, 611–612.

18. Cesario, M., Dietrich-Buchecker, C. O., Guilhem, J., Pascard, C. and Sauvage, J.-P., 'Molecular structure of a catenand and its copper(ɪ) catenate: complete rearrangement of the interlocked macrocyclic ligands by complexation', *J. Chem. Soc., Chem. Commun.*, 1985, 244–245.

19. Dietrich-Buchecker, C. O., Guilhem, J., Khemiss, A. K., Kintzinger, J.-P., Pascard, C. and Sauvage, J.-P., 'Molecular structure of a [3]-catenate: curling up of the interlocked system by interaction between the two copper complex subunits', *Angew. Chem., Int. Ed. Engl.*, 1987, **26**, 661–663.

20. Sambrook, M. R., Beer, P. D., Wisner, J. A., Paul, R. L. and Cowley, A. R., 'Anion-templated assembly of a [2]-catenane', *J. Am. Chem. Soc.*, 2004, **126**, 15364–15365.

21. Adams, H., Carver, F. J. and Hunter, C. A., '[2]-catenane or not [2]-catenane?', *J. Chem. Soc., Chem. Commun.*, 1995, 809–810.

22. Whang, D., Park, K.-M., Heo, J., Ashton, P. and Kim, K., 'Molecular necklace: quantitative self-assembly of a cyclic oligorotaxane from nine molecules', *J. Am. Chem. Soc.*, 1998, **120**, 4899–4900.

23. Thordarson, P., Nolte, R. J. M. and Rowan, A. E., 'Mimicking the motion of life: catalytically active rotaxanes as processive enzyme mimics', *Aust. J. Chem.*, 2004, **57**, 323–327.

24. Jiménez, M. C., Dietrich-Buchecker, C. O. and Sauvage, J.-P., 'Towards synthetic molecular muscles: contraction and stretching of a linear rotaxane dimer', *Angew. Chem., Int. Ed. Engl.*, 2000, **39**, 3284–3287.

25. Altieri, A., Gatti, F. G., Kay, E. R., Leigh, D. A., Martel, D., Paolucci, F., Slawin, A. M. Z. and Wong, J. K. Y., 'Electrochemically switchable hydrogen bonded molecular shuttles', *J. Am. Chem. Soc.*, 2003, **125**, 8644–8653.

26. Livoreil, A., Sauvage, J.-P., Armaroli, N., Balzani, V., Flamigni, L. and Ventura, B., 'Electrochemically and photochemically driven ring motions in a disymmetrical copper [2]-catenate', *J. Am. Chem. Soc.*, 1997, **119**, 12114–12124.

27. Chichak, K. S., Cantrill, S. J., Pease, A. R., Chiu, S.-H., Cave, G. W. V., Atwood, J. L. and Stoddart, J. F., 'Molecular Borromean rings', *Science*, 2004, **304**, 1308–1312.

28. Lukin, O. and Vögtle, F., 'Knotting and threading of molecules: chemistry and chirality of molecular knots and their assemblies', *Angew. Chem., Int. Ed. Engl.*, 2005, **44**, 1456–1477.

29. Dietrich-Buchecker, C. O., Rapenne, G., Sauvage, J.-P., De Cian, A. and Fischer, J., 'A dicopper(I) trefoil knot with *m*-phenylene bridges between the ligand subunits: synthesis, resolution and absolute configuration', *Chem. Eur. J.*, 1999, **5**, 1432–1439.

30. Safarowsky, O., Nieger, M., Fröhlich, R. and Vögtle, F., 'A molecular knot with twelve amide groups – one-step synthesis, crystal structure, chirality', *Angew. Chem., Int. Ed. Engl.*, 2000, **39**, 1616–1618.

31. Vilar, R., Mingos, D. M. P., White, A. J. P. and Williams, D. J., 'Anion control in the self-assembly of a cage coordination complex', *Angew. Chem., Int. Ed. Engl.*, 1998, **37**, 1258–1261.

32. Caulder, D. L., Powers, R. E., Parac, T. N. and Raymond, K. N., 'The self-assembly of a predesigned tetrahedral M_4L_6 supramolecular cluster', *Angew. Chem., Int. Ed. Engl.*, 1998, **37**, 1840–1843.

33. Fujita, M., Umemoto, K., Yoshizawa, M., Fujita, N., Kusukawa, T. and Biradha, K., 'Molecular panelling *via* coordination', *Chem. Commun.*, 2001, 509–518.

34. Kuehl, C. J., Kryschenko, Y. K., Radhakrishnan, U., Seidel, S. R., Huang, S. D. and Stang, P. J., 'Self-assembly of nanoscopic coordination cages of D3h symmetry', *Proc. Nat. Acad. Sci. USA*, 2002, **99**, 4932–4936.

35. Fox, O. D., Dalley, N. K. and Harrison, R. G., 'A metal-assembled, pH-dependent, resorcinarene-based cage molecule', *J. Am. Chem. Soc.*, 1998, **120**, 7111–7112.

36. MacGillivray, L. R., Diamente, P. R., Reid, J. L. and Ripmeester, J. A., 'Encapsulation of two aromatics by a carcerand-like capsule of nanometre-scale dimensions', *Chem. Commun.*, 2000, 359–360.

37. Coppens, P., Ma, B., Gerlits, O., Zhang, Y. and Kulshrestha, P., 'Crystal engineering, solid-state spectroscopy and time-resolved diffraction', *Cryst Eng Comm*, 2002, **4**, 302–309.

38. Kobayashi, K., Shirasaka, T., Horn, E., Furukawa, N., Yamaguchi, K. and Sakamoto, S., 'Molecular capsule constructed by multiple hydrogen bonds: self-assembly of cavitand tetracarboxylic acid with 2-aminopyrimidine', *Chem. Commun.*, 2000, 41–42.

39. Atwood, J. L., Barbour, L. J. and Jerga, A., 'Hydrogen bonded molecular capsules are stable in polar media', *Chem. Commun.*, 2001, 2376–2377.

40. Alajarín, M., Pastor, A., Orenes, R.-A. and Steed, J. W., 'Dimeric self-assembling capsules derived from the highly flexible tribenzylamine skeleton', *J. Org. Chem.*, 2002, **67**, 7091–7095.
41. Palmer, L. C. and Rebek, J., Jr, 'The ins and outs of molecular encapsulation', *Org. Biomol. Chem.*, 2004, **2**, 3051–3059.
42. Atwood, J. L., and Szumna, A., 'Hydrogen bonds seal single-molecule capsules', *J. Am. Chem. Soc.*, 2002, **124**, 10646–10647.

4

Solid state supramolecular chemistry

4.1 Introduction

Supramolecular interactions do not only occur within solution but also within solid state systems. Ordered solids, with the exception of giant covalent structures, are held together by non-covalent interactions. Although these supramolecular interactions are largely responsible for the packing of most crystals, the compounds most commonly referred to as solid state supramolecular compounds are those containing hollow cavities or channels in which guests are able to reside. This chapter initially discusses such solids – their compositions, host–guest behaviour and, in many cases, how they are of practical use.

The variety of solid state host–guest systems is still increasing. The first known examples were clathrate hydrates, discovered in 1810. These are a phase of solid water (ice), capable of holding guest species within a lattice of polygonal cavities. Clathrate hydrates are of great importance within the petroleum industry, where they can be both a source of fuel and, more often, a costly problem during natural gas extraction and transport. Clathrates of host molecules other than water also exist in which solid state networks, most often held together *via* hydrogen bonds, form cavities or channels in which suitably sized guest species are able to reside.

Zeolites and zeolite-type materials are also of use in the petroleum industry as catalytic agents for the cracking or separation of crude oil. Zeolites are inorganic-framework materials, predominantly aluminosilicates, that usually contain guests within continuous channels rather than discrete cavitites. They bear a resemblance to channel clathrate structures although the inorganic host lattice is much more robust, generally allowing guest exchange without disruption of the host framework. Zeolites are thus truly porous and this property, relatively rare

Core Concepts in Supramolecular Chemistry and Nanochemistry   Jonathan W. Steed, David R. Turner and Karl J. Wallace
© 2007 John Wiley & Sons, Ltd   ISBN: 978-0-470-85866-0 (Hardback); 978-0-470-85867-7 (Paperback)

among less robust clathrates, is responsible for many of their commercial applications. Zeolites also have common everyday uses in addition to petroleum refinement, such as odour absorption in cat litter and as water softening ion-exchange agents in detergents. It is possible to make synthetic versions of these naturally occurring minerals and to modify and enhance their properties for commercial and synthetic advantage.

There also exist many synthetic examples of microporous solids. The most common example is that of coordination polymers, or metal–organic frameworks (MOFs), which use organic ligands to bridge between metal atoms, so creating a hollow lattice network. The rigid coordination geometry around metal centres, coupled with appropriate bridging ligands, can result in intricate porous or interpenetrated assemblies, with some showing zeolite-type porosity – hence the generic term 'organic zeolites'.

In recent years, there has been an increasing trend in designed solid state structures being referred to as having been engineered. This tendency has helped the field of crystal engineering to come to prominence. By examining known structures and exploring patterns that emerge, it is possible to intuitively design new ligands and motifs to give rise to interesting new solid state structures, although it is almost impossible to second guess the exact structures that will be adopted. Such systems contain no fragments that can be classified as hosts and guests, but rather contain two or more separate species that unite to give a three-dimensional designer solid.

4.2   Zeolites

4.2.1   Zeolite structure

 Szostak, R., *Molecular Sieves*, 2nd Edn, Blackie Academic and Professional, London, UK, 1998.

Zeolites are a class of minerals which consist of an aluminosilicate framework containing cavities and channels. The pores that exist within the structure give rise to the useful properties of zeolites, such as the capacity for guest exchange and catalysis. Zeolites have a wide variety of everyday uses, from hydrocarbon cracking to water-softening agents in detergents. There are many naturally occurring zeolites, although most are synthetic, chemically tailored to suit specific applications. While the zeolite framework is not supramolecular in nature, the structure of the frameworks and the non-covalent inclusion of guest compounds is very similar to synthetic porous systems such as metal–organic frameworks (Section 4.6.2). Silica and alumina have also been used to form *biomimetic materials* around templating species (see Chapter 5, Section 5.10.5).

> **Zeolites:** Inorganic microporous solids composed of a negatively charged alumi-
> nosilicate framework in which cationic and neutral guests can reside.

Traditional zeolite frameworks consist of AlO_4 and SiO_4 tetrahedra, set in a repeating pattern to form a microporous three-dimensional structure (micropores are defined by the International Union of Pure and Applied Chemistry (IUPAC) as being $\leq 2\,nm$ in diameter). Oxygen atoms form the link between adjacent silicon and aluminium atoms (often referred to as 'T-atoms' for *T*etrahedral). The final framework structure is therefore composed of SiO_2 and AlO_2^- units and hence carries a net negative charge. This charge is balanced by cations that reside as guests within the pores alongside neutral species, such as solvents, that are also able to fit inside the cavities. The general formula for the composition of a zeolite is as follows:

$$A_{y/m}^{m+}[(SiO_2)_x \cdot (AlO_2^-)_y] \cdot zG$$

where A is the cation (of charge m) and G is the neutral guest.

The negative charge carried on the aluminium units gives rise to *Löwenstein's rule* which states that Al—O—Al linkages are forbidden. This means that the framework Si/Al ratio (x/y) must be greater than, or equal to, one. It is possible for 'T-atoms' other than silicon and aluminium to be incorporated in a zeolite-type framework (such as Ge, Ga, P and As) and these can significantly alter its properties. Such compounds have been referred to as *zeotypes* but here will be referred to under the general term of *zeolitic compounds*. Zeolites can differ both in terms of geometry (internal pore structure) and connectivity (bonding network). Zeolite topologies are identified by a three letter identification code assigned by IUPAC on recommendation by an independent board from the International Zeolite Association (IZA).[1] The difference between zeolite *topologies* and zeolite *structures* is an important one to note. The topology of a zeolite describes the spatial connectivity of the nodes, although the exact chemical make-up of the compound describes the structure. Different structures can be of the same topology. There are currently 165 recognised zeolitic topologies (as of January 2006). Three examples of zeolite structures are shown in Figure 4.1, displaying a range of cavity sizes and shapes. The accepted convention for the representation of zeolite structures is by lines representing T–O–T linkages with the 'T-atoms' residing at the vertices.

The structure of a zeolitic framework and the chemical nature of the host lattice lead to a wide variety of industrial uses, especially catalytic and screening processes within the petroleum industry. The size of the pores is one of the most crucial factors in determining zeolite use. The pore size is determined by the number of T-atoms defining the entrance (ring-size). For example, Figure 4.1(b) shows an 8-membered ring entrance at the front of the cavity. If the pore is too large or too small, then the activity of the material may be adversely

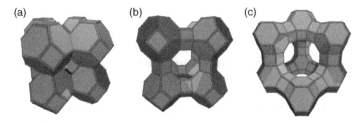

Figure 4.1 The structures of three different zeolites: (a) sodalite (SOD); (b) Linde type A (LTA); (c) faujasite (FAU). The pore size and shape can be seen to differ greatly between these structures.

affected (Table 4.1). Small-sized pores can be of use in catalytic processes involving small reactants and products, for example, the reaction between methanol and ammonia. Another example is the MTO process to produce small olefins from methanol (MTO – Methanol To Olefins). Larger pores are suitable for processes such as xylene ($C_6H_4(CH_3)_2$) separation and catalytic cracking (Section 4.2.3).

Table 4.1 Categories of pore sizes in zeolites and their uses

Ring tetrahedra	Size	Use	Occurrence	Example
4/6	Very small	Too small ($d \sim 0.2\,nm$)	Few	Afghanite
8	Small	Good for catalysing reactions of small molecules	Common	Chabazite, Linde type A
10	Medium	Useful for catalysis. Use 'pore-size enginering' to optimise after synthesis	Common	ZSM5, ZSM11
12	Large	As for medium-pore-sized zeolites	Common	Mordenite
12+	'Super-large'	—	Rare	Cloverite
Odd numbers	Various	—	Rare	SSZ-23

Zeolites, being regular crystalline solids, are composed of many identical repeating units. The arrangement of these basic building blocks gives rise to different structures. An example can be seen in the structures in Figure 4.1 which all contain the basic sodalite-β cage (Figure 4.2(a)). This basic sodalite structure only has pores composed of four-membered rings between the cages, too small for guest molecules to fit through (Figure 4.1(a)). The Linde-type and faujasite zeolites also contain double four-ring and double six-ring units in addition to the basic cage (Figures 4.2(b) and 4.2(c)), respectively, which act to space-out the sodalite cages and create larger, more useful pores.

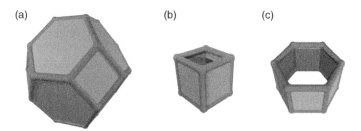

Figure 4.2 Building blocks found in (a) sodalite, (b) Linde type A and (c) faujasite zeolites, as seen in Figure 4.1.

In addition to describing the size and shapes of individual pores, zeolites can be described by the way in which the networks of pores interconnect throughout the structure to form channels. A structure in which the channels are independent of each other (*i.e.* no interconnections) is termed a one-dimensional structure (Figure 4.3(a)). A zeolite in which pores are the meeting point of two channels is termed a two-dimensional structure (Figure 4.3(b)) and those zeolites in which three channels meet together are termed three-dimensional structures (Figure 4.3(c)). These network types can have very different chemical behaviour, even if pore sizes are similar, mainly due to the vastly differing diffusion behaviour of guest species.

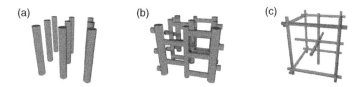

Figure 4.3 Schematic representations of (a) parallel (one-dimensional), (b) ladder-type (two-dimensional) and (c) interpenetrating (three-dimensional) channel structures. Cylinders indicate channels through the structures.

The synthesis of zeolites and their derivatives is quite unlike any traditional synthetic method. Zeolites are crystalline materials grown from a gelatinous mixture containing many different components. The slightest alteration of the gel composition can have a pronounced effect on the final product(s) and the components have complex inter-relationships. The nature of the product can also depend on other factors, such as temperature, stirring, seeding and time.

4.2.2 Zeolite composition

The size and shape of zeolite pores and channels are not the only factors that influence their use in catalysis – the chemical composition of the framework plays a major role in determining the catalytic properties. The chemical nature of the zeolite can be altered after the initial synthesis to enhance the catalytic properties for desired processes. The composition of zeolites is often referred to by the ratio of different T-atoms that are present. This is because the assembly process is essentially random and a regular distribution is not obtained. The positions of certain atoms cannot therefore be distinguished by diffraction techniques where the disorder of atom types makes accurate assignment impossible. solid state NMR spectroscopy has been of some use with this problem as ^{29}Si chemical shifts are sensitive to the neighbouring Si/Al distribution which can lead to knowledge of the ratio of various Si environments.

The T-atoms that the framework is composed of play a major part in determining the nature of the zeolite. Transition metals can be incorporated into the structure to produce catalysts for selective oxidation reactions. An example of this behaviour is the zeolite TS-1 (titanium silicate-1) which is used in industry for the production of hydroquinone and catechol from phenol and hydrogen peroxide.

The majority of the catalytic activity of zeolites comes from the acidity of the internal surfaces. There are several different aspects to zeolite acid-catalysis, some of which may be altered to change the catalytic behaviour of the microporous framework, as follows:

- Nature of acid sites (Brønsted or Lewis)

- Density/concentration of acid sites

- Strength or strength distribution

- Precise site location.

The nature of the acid sites is perhaps the most important factor in determining the activity of zeolites. Brønsted and Lewis acids behave in very different ways. Brønsted sites exist adjacent to aluminium atoms with the connecting oxygen protonated to produce a T–OH$^+$–Al linkage and are most commonly associated with catalytic activity. These sites can be detected by the use of IR spectroscopy which detects the O–H stretching band. The nature of Lewis-acid sites within a zeolitic framework is less well understood. They are thought to arise from the removal of $[(AlO)^+]_n$ units from the lattice. The density of Brønsted sites is directly proportional to the amount of aluminium in the zeolite structure. This means that zeolites with low Si/Al ratios contain more Brønsted-acid sites than those with higher ratios.

The strength of Brønsted acid sites within zeolites also varies with the Si/Al ratio and the strongest acidity lies within a narrow range. The stronger the acid site, then the greater its activity. Due to the relative electronegativity of silicon compared to aluminium, the strongest acid sites are located on isolated aluminium-containing tetrahedra, those that are not near to other aluminium centres (*i.e.* spatially, a few tetrahedra removed). Zeolites possessing a high aluminium content can be put through a post-synthetic de-alumination process to increase the Si/Al ratio, thus increasing the strength of the Brønsted-acid sites.

4.2.3 Zeolites and catalysis

 Weitkamp, J., 'Zeolites and catalysis', *Solid State Ionics*, 2000, **131**, 175–188.

The use of zeolites as catalytic materials derives from two general properties, *i.e.* the inherent acidic nature of the internal structure and the imparting of shape and size selectivity by diffusion through the porous network.[2] The acidic nature of zeolitic materials is essential for their catalytic properties. Within the confines of the zeolite framework, guest molecules are essentially in a solventless environment in very close proximity to unsolvated protons. These naked protons act as super-acidic sites and can protonate organic species such as *n*-alkanes, aromatics and waxes, which are unreactive under conventional conditions. When these guests interact with the super-acidic sites, they form reactive carbocations that can undergo rearrangement to facilitate otherwise inaccessible reaction pathways.

The usefulness of zeolites in catalysing specific processes comes mostly from the sizes of the cavities which impart a degree of selectivity over the reactions that are allowed. This selectivity derives from one of three principles, as follows:

- *Reactant selectivity*. If the diffusion of bulky species into or through the zeolite lattice is hindered, then only the smaller species will be able to react, or will at least have the chance to react first, giving a lower percentage of the product from the larger molecule.

- *Product selectivity.* When one reaction has several outcomes, such as the production of xylenes from toluene and methanol, only the smaller molecules can diffuse out of the zeolite, again reducing the percentage of the larger product that is formed as they are hindered by the pore openings.

- *Transition-state selectivity.* If there are several possible reaction pathways, it may be that one or more of them pass through a transition state that cannot be accommodated within the cavities of the zeolite.

Reactant and product selectivity are both effects that are dependant upon mass transport. These effects can be adjusted by the use of smaller or larger crystals. Larger crystals will increase the distance through which molecules must diffuse, thereby enhancing the selectivity effect. Transition state selectivity is not diffusion-controlled and therefore is not influenced by the size of the crystal.

One example of product selectivity in a zeolite system is catalysed xylene formation from toluene and methanol. An electrophilic substitution of the aromatic ring occurs within the pores of the zeolite ZSM-5, which gives a crude mixture of the three xylene isomers (*ortho, meta* and *para*). Due to the narrow, linear channels within this zeolite, it is only the *para*-isomer that is able to diffuse out of the zeolite and be isolated (Figure 4.4). The other *ortho-* and *meta*-isomers can diffuse through the pores but at much slower rates (14 and 1000 times slower, respectively). Due to these slow rates of diffusion, it is more likely that the undesirable isomers will remain within the zeolite long enough to isomerise to the *p*-xylene product.

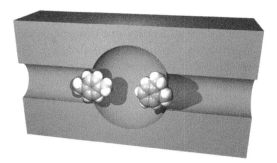

Figure 4.4 The separation of xylenes within the zeolite ZSM-5: *p*-xylene (left) can diffuse through the zeolite structure, whereas *o*-xylene (right) cannot.

Zeolites are also used in the production of gasoline. Linear *n*-alkanes are not wanted in commercial gasoline as they lower the octane number (*i.e.* produce less energy for their mass upon combustion). The ability to separate mixtures of linear and branched alkanes therefore produces a higher-value product. Highly branched alkanes diffuse though ZSM-5 slowly due to their large size, giving rise to a reactant selectivity in favour of linear species. Once inside the zeolite, larger species are also prevented from reacting due to a bulky transition state and so only the *n*-alkanes react. These straight-chain species undergo catalytic cracking to give lighter fractions which are easily separated from the mixture, leaving only the desirable branched alkanes behind. Larger-pore zeolites are also used to convert very long-chain species into gasoline and petroleum. The use of zeolites in fuel refinement represents their biggest use in terms of annual commercial value.

4.3 Clathrates

Clathrates are solid state host–guest compounds in which the guest is trapped within a void in the solid state lattice. The word 'clathrate' itself is derived from the Latin word *clathrus*, meaning 'surrounded on all sides', and was coined in the 1940s to describe clathrates of β-hydroquinone (Section 4.3.3). The term 'inclusion compound' was introduced at around the same time to describe urea channels (Section 4.3.1). These two terms are now essentially interchangeable with each other. The main structure of a clathrate compound is a framework composed of one type of molecule connected in a repeating pattern such that regular voids are obtained in which guest species of another type can reside. As the host network is composed solely of one type of molecule, it must be able to self-associate, thereby holding the crystalline structure together by virtue of non-covalent interactions with adjacent molecules. This forms structures not unlike those of zeolites but which are held together by weak interactions rather than strong covalent bonds. Clathrates form only in the presence of suitable guest species as the cavities in the structure are unstable unless occupied. Indeed, it is a favourable complementarity between the shape of the guest and that of the cavity that plays the most vital role in the formation of clathrate structures. A wide variety of supramolecular interactions can be involved in the formation of clathrate structures. For conventional clathrates, the guest does not participate in any interactions with the host structure other than weak van der Waals interactions, although there are variations on these structures in which the guest may interact more strongly with the host framework. In most cases, the clathrates are only stable when there are guests within the cavities, although there are notable exceptions to this rule, especially clathrate hydrates (Section 4.4).

Clathrates: solid state *host–guest compounds*, consisting of a network of self-associating molecules forming cavities or channels in which guest species can reside. They are usually only stable in the presence of guest species.

Clathrates are sometimes categorised according to the shapes of cavities that are formed within the host structure. For example, *cryptato-clathrates* are those which contain guests within cages and *tubulato-clathrates* are those which possess one-dimensional channels in which the guest is included. Different cavity shapes produce different properties with regards to guest-exchange rates and dynamics and will be highlighted in the examples that follow. Clathrates may also be categorised by the manner in which their host network is constructed, either by weak interactions (lattice clathrates) or by covalent bonds (macromolecular clathrates), as in the case of zeolitic materials. This section will treat clathrates exclusively as compounds that are formed by host lattices utilising weak interactions, although the similarities between these and other solid networks is evident.

The most common, and well-studied, clathrate systems are those in which water forms the host network, the *clathrate hydrates*. These are naturally occurring materials and are also of industrial importance and will be discussed separately in Section 4.4. First, we will look at a few representative examples of synthetic, organic clathrates that are created from a wide variety of molecules, each producing a unique network into which guests are included.

4.3.1   Urea/thiourea clathrates

 Harris, K. D. M., 'Urea inclusion compounds', in *Encyclopedia of Supramolecular Chemistry*, Steed, J. W. and Atwood, J. L. (Eds), Marcel Dekker, New York, NY, USA, 2004, pp. 1538–1549.

Urea and thiourea clathrates represent some of the most commonly studied synthetic clathrate systems. Both of these molecules contain acidic NH_2 groups, providing good hydrogen bond donors and a good acceptor atom (oxygen or sulfur, respectively). The presence of both donor and acceptor groups in the same molecule allows for significant self-association.

The structure of a conventional urea clathrate network comprises hollow, hexagonal, helical channels with internal diameters of 5.5–5.8 Å, in which narrow guests can reside (Figure 4.5). These channels form an interlocked honeycomb-type structure with the channels running parallel to each other. The urea molecules are arranged in a helix around the channel void and any given crystal contains only right-handed or only left-handed helices, and therefore the tubes are chiral. Each oxygen atom accepts four hydrogen bonds, maximising the possible interactions and stabilising the structure. As all of the urea hydrogen atoms are involved in interactions with adjacent host molecules, there are usually no spare sites for the guest species to interact with. The interior of the channels are effectively smooth and guest species are only retained by van der Waals forces. Common guests in these channels are long, straight alkanes with only a minimal amount of branching tolerated. Some functionalised hydrocarbons can also be accommodated into the structure such as α,ω-dihalogenoalkanes, α,ω-dicarboxylic acids and some alkanones (note that α,ω refers to the substituents being on opposite terminal positions of the hydrocarbon chain). Molecules the size of benzene or cyclohexane are generally too large to be suitable guests.

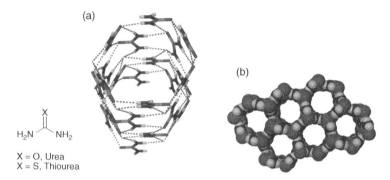

Figure 4.5 (a) The structure of a helical urea channel showing the high degree of hydrogen bonding between the urea molecules and (b) the way in which six channels pack together to produce the honeycomb structure.

As there is no specific interaction between the host and guest,[†] the complexes are often non-stoichiometric (no fixed ratio between the number of host and guest molecules) and the guests are somewhat free to move along the channel or to rotate. The relationship between the host structure and the guests in conventional urea clathrates is said to be *non-commensurate*. This means that there is no simple relationship between the repeat unit of the host (c_h) and that of the guest (c_g) (Figure 4.6). This can make characterisation difficult as the structure is often not three-dimensionally ordered (*e.g.* guests in one channel are not ordered with respect to those in the next) and so X-ray structural analysis cannot accurately locate the guest atoms. Structures in which there is a direct relationship between c_h and c_g are *commensurate*. This usually involves some kind of specific interaction between the host and the guest and is most frequently seen in unconventional urea clathrates. The guests themselves may thus display long-range ordering across the structure which is assessed by the degree of inter-tunnel ordering with a well-defined value of Δ_g (Figure 4.6).

> **Commensurate Structures:** Channel inclusion compounds in which there is a direct relationship between the repeat unit of the host structure and that of the guest. If no such simple relationship exists the structure is *non-commensurate*.

[†] There are some non-conventional urea clathrates of α,ω-diketones in which the guest does participate in interactions with the host, resulting in unusual properties, see, Brown and Hollingsworth (1995).[3]

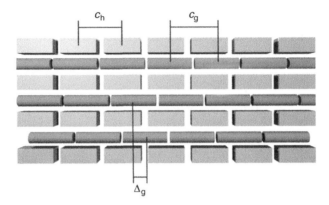

Figure 4.6 Schematic representation of a guest species (cylinders) residing in parallel guest (blocks) channels, where c_h and c_g are the periodic repeat distances of the host and guest, respectively, along the channel axis. If there is no simple relationship between these two numbers, the structure is non-commensurate. Δ_g is the inter-tunnel offset of the guest describing the ordering of guest molecules in adjacent channels.

Thiourea channels are slightly larger than their oxygen-containing analogues. This difference is due to the larger size of the sulfur atom leading to an increase in the length of all bonds involving this atom. These thiourea channels are large enough to allow some slightly branched hydrocarbon guests to reside within them, as well as larger species such as small organometallic compounds. The lack of strong interactions between the host network and the guest also allows for significant movement of the guests within the channels. The guests have a significant degree of rotational and translational freedom, which can again hinder the structure determination by diffraction methods. Guests can be readily removed from urea channels *in vacuo* which results in the collapse of the channels to give non-porous tetragonal urea which is close-packed and does not contain guest species.

Urea clathrates form readily in the presence of linear hydrocarbons. They can be highly selective, due to their restricted internal diameter, limiting the degree of branching on the hydrocarbon and excluding bulky groups. This size and shape selectivity makes *n*-alkanes the preferred guests. It is possible that, like zeolites, urea clathrates could be of use in the separation of linear and branched hydrocarbons. An interesting feature of alkane guests is the extra stabilisation that occurs with increasing chain length. It has been calculated that each extra methylene ($-CH_2-$) group which is added to the alkyl chain adds approximately $10\,kJ\,mol^{-1}$ of extra complexation enthalpy. This extra stabilisation can have important consequences on the types of guests that are included within the channels. The presence of larger groups can be tolerated as the longer chains allow for the destabilising effect of the large unit to be offset by the favourable binding of the chain. Hence, 1-phenyloctane is not included but species with longer chains, such as 1-phenylisocane, are.

The larger channels within the thiourea clathrate framework allow for the possibility of reactions within the channels, a use that the smaller, oxygen-containing analogues are unsuited for. Enclosed environments such as these channels, as well as those in other systems, such as zeolites and capsular species (see Chapter 3, Section 3.5), represent unique reaction containers whose properties cannot be mimicked in bulk-reaction conditions. The enclosed space can impart stereo- or regioselectivity upon the reactants, affecting the outcome of the reaction in a predictable manner. The cylindrical shape of the thiourea clathrate channels is ideal for promoting the formation of long, straight-chain polymers, as reactants can only interact in an 'end-to-end' manner in the narrow space. Polymer products are known to have a small molecular-weight band (*i.e.* polymers of similar lengths), representing good control.

4.3.2 Trimesic acid clathrates

Kolotuchin, S. V., Thiessen, P. A., Fenlon, E. E., Wilson, S. R., Loweth, C. J. and Zimmerman, S. C., 'Self-assembly of 1,3,5-benzenetricarboxylic (trimesic) acid and its analogues', *Chem. Eur. J.*, 1999, **5**, 2537–2547.

Trimesic acid (TMA) is the trivial name of 1,3,5-benzenetricarboxylic acid (**4.1**). Carboxylic acids are well known for their ability to self-associate, either to form a dimeric structure or, more commonly, in a conformation that avoids unfavorable secondary interactions (see Chapter 1, Section 1.3.2) and results in 'ribbons' forming throughout the structure (Figure 4.7).

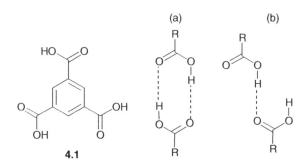

Figure 4.7 Trimesic acid (**4.1**) and the way in which carboxylic acid units can self-associate to form (a) a dimer or (b) a more common, staggered conformation.

The most interesting clathrate structures that form using TMA are those that involve one molecule of water per acid as a co-agent in the host network – therefore termed TMA monohydrates (TMA·H$_2$O). The basic structure within the clathrate system of TMA·H$_2$O is the arrangement of four TMA molecules so as to form a rectangular cavity in which guests may reside (Figure 4.8(a)).

(a)   (b)

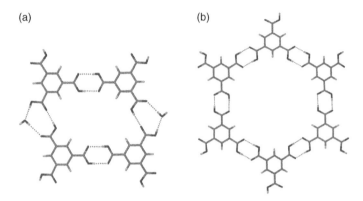

Figure 4.8   (a) Part of a planar layer of TMA hydrate showing the rectangular channel and (b) the hexagonal clathrate structure that TMA forms when in the presence of large templating guest species.

The rectangular TMA units are joined together through hydrogen bonds with water to hold together the long-range structure of the crystal. The monohydrate layers are positioned directly over one another, producing rectangular channels which suitable, neutral guests are able to fill without any specific interactions with the host network. Most TMA monohydrate structures display this stacking motif, with the exception being the structure with the empirical formula TMA·H$_2$O· [0.2 TMA] or, conventionally, TMA$_6$·5H$_2$O. In this material, TMA also acts as the guest species and the monohydrate layers are offset from each other. An infinite zig-zag strand of TMA runs through the pores, held together by carboxylic acid dimer interactions.

Clathrate structures of TMA are not, however, restricted to those containing rectangular channels. When TMA is crystallised in the presence of large species, such as branched and linear long-chain hydrocarbons or alcohols, *etc.*, the formation of a hexagonal host channel is observed (Figure 4.8(b)). This takes the form of a continuous sheet of hexagonal TMA molecules, akin to a 'chicken-wire' construction. As with the rectangular-based structures, the guests reside in the channel without partaking in specific interactions with the porous host network. One polymorph of TMA (the γ-form) contains hexagonal sheets perpendicular to one another and threaded through the hexagonal voids.

4.3.3 Hydroquinone and Dianin's compound

Flippen, J. L., Karle, J. and Karle, I. L., 'Crystal structure of a versatile organic clathrate–4-parahydroxyphenyl-2,2,4-trimethylchroman (Dianin's compound)', *J. Am. Chem. Soc.*, 1970, **92**, 3749–3755.

A series of several phenol derivatives (*e.g.* Figure 4.9) form clathrate structures incorporating hexagonal, hydrogen bonded rings between phenolic hydroxyl groups (Figure 4.10(a)) (*cf.* six-membered water rings in clathrate hydrates, see Section 4.4). This hexagonal arrangement of the hydroxyl groups forces the aryl substituents to adopt an alternating 'up–down' pattern around the hydrogen bonded ring. This results in a cavity that is lined with aryl groups and stabilised further by edge-to-face π-stacking interactions.

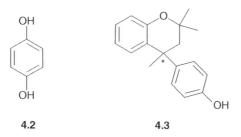

4.2 **4.3**

Figure 4.9 Molecules that form clathrate structures *via* a hexagonal motif, hydroquinone (**4.2**) and Dianin's compound (**4.3**) (note the chiral centre – marked with an asterisk).

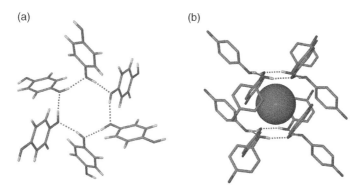

Figure 4.10 (a) The hexagonal arrangement of phenolic groups in the type-I clathrate structure of β-hydroquinone and (b) one cage with xenon encapsulated as a guest.

Hydroquinone (**4.2**) exists in multiple forms in the solid state (α, β and γ) but only β-hydroquinone is observed to form clathrates. The general formula for these clathrates is $3C_6H_4(OH)_2 \cdot xG$ (where G is a guest species and x is the 'site-occupancy factor' of this guest). The site occupancy factor, effectively what proportion of the cavities are filled, need not be unity (*i.e.* the situation in which every cage is occupied). The clathrate structure can be stable with some cavities remaining empty (*cf.* clathrate hydrates, Section 4.4). For example, with xenon as a guest, the cavity occupancy factor is 0.866 – this is just under 87 % of cavities containing a guest species. The size and shape of the cavity are highly susceptible to change, depending on the guest that is present. Small, spherical guests, such as xenon, form a highly symmetrical 'type-I' structure (Figure 4.10(b)). With slightly larger, non-spherical guests included, such as methanol or sulfur dioxide, the cavity becomes elongated and the symmetry is lowered, giving a type-II structure. Acetonitrile (CH_3CN) forms a type-III clathrate, with three individual and well-defined guest orientations. Buckminsterfullerene has also been observed as a guest in β-hydroquinone clathrates.

Dianin's compound (**4.3**) is a chiral molecule that is able to form host networks very similar to those of hydroquinone. The solid state structure forms cavities in which one enantiomer points in the opposite direction to the other around the hydrogen bonded hydroxyl ring (Figure 4.11(a)). The resolved *S*-enantiomer of this compound does not form a clathrate structure individually. The clathrates of Dianin's compound show the most versatile inclusion chemistry of any of the phenol-derived structures. A large number of different organic and inorganic guests are able to reside within the cavities, including argon, glycerol and small carbohydrates. Mostly, the host–guest ratio is 6:1, *i.e.* one guest per cavity created from six host molecules. Some large solvent molecules, such as acetone and ethanol, form 3:1 structures (two guests per cage) while methanol

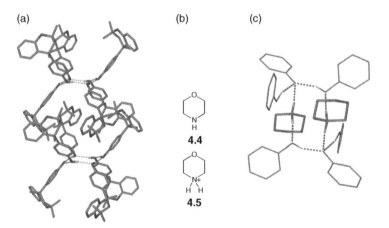

Figure 4.11   (a) An empty clathrate of Dianin's compound, (b) the structures of morpholine (**4.4**) and its protonated derivative, morpholinium (**4.5**) and (c) the hexameric hydrogen bonding in the Dianin's/morpholinium structure (for clarity, only the phenol groups of Dianin's compound are shown).

is observed to form 2:1 complexes. Piperidine (heterocyclic $C_5H_{11}N$) is the only molecule known to form a 1:1 complex inside the Dianin clathrate. Until recently, none of the known host–guest complexes of Dianin's compound have shown the guest disrupting the stable hydrogen bonding in the hexagonal motif. However, recent work examining the inclusion of morpholine (**4.4**), revealed that two guests are bound in its protonated form (morpholinium, **4.5**) after deprotonating two of the Dianins's molecules per cavity. Remarkably the overall structure is disrupted very little, as the NH_2^+ groups participate in hydrogen bonding to reform a hexameric ring, albeit of a different composition to that found in classical Dianin-type compounds (Figure 4.11(c)).[4]

4.4   Clathrate hydrates

Sloan, E. D., Jr., 'Clathrate hydrates: the other common solid water phase', *Ind. Eng. Chem. Res.*, 2000, **39**, 3123–3129.

Historically and industrially, clathrate hydrates represent the most important and well-studied class of clathrate structures. Clathrate hydrates, as their name suggests, are clathrates that are formed from water molecules resulting in cages that are capable of holding neutral guest species. The first known example was that of chlorine hydrate, discovered in 1810 by Sir Humphrey Davy and characterised by Michael Faraday in 1823. Since then, many naturally occurring and synthetic hydrates have been studied. Hydrates of natural gases are now a major area of research as they have potential uses as reserves of fuel where they occur and are also closely linked with oil production where their presence causes great problems by blocking natural gas pipelines.

> **Clathrate Hydrates:** Clathrate structures in which the framework is composed of polyhedral water cages, capable of encapsulating neutral, non-polar guest species.

4.4.1   Clathrate hydrate structure

Clathrate hydrates represent an impure phase of solid water. Water is capable of forming very strong hydrogen bonding networks as each water molecule can both donate and receive two hydrogen bonds. The framework of clathrate hydrates is composed solely of water molecules that are joined by hydrogen bonds to form polyhedral cages, forming a bulk material that resembles normal ice. These cages are only stable in the presence of guest species. Not all of the cavities have to be occupied although a significant percentage must be ($> 90\%$) to preserve the integrity of the structure. This leads to almost all clathrate hydrates having non-stoichiometric formulae in respect to water and the guest species. If all of the cavities were to be full, the mole fraction of water would be approximately 0.85 but in practice this is never the case. Without suitable guest species present during the crystallisation process, the normal hexagonal ice structure (denoted *Ih*) which contains no cavities, will form under ambient pressure. The clathrate hydrate structures are not dissimilar in terms of the interactions between water molecules with the hydrogen bonding distances between water molecules being comparable (measurable by IR spectroscopy). In addition to 'true' clathrate hydrate structures, there are other similar structures that incorporate amines or ions into the framework structure. All clathrate hydrates referred to here are true hydrates. Clathrate hydrates form under very specific conditions, usually above the melting point of *Ih* (up to 31 °C) and under elevated pressure. Clathrate hydrates occur naturally in regions of permafrost (found in some parts of Alaska and Siberia), at continental plate margins and on the beds of small seas and large lakes (such as the Black Sea and the Gulf of Mexico). Hydrate structures are composed of water polyhedra whose sides are hydrogen bonded water hexagons, pentagons and squares (Figure 4.12), unlike *Ih* which only contains hexagonal arrangements of water. The notation most commonly adopted to describe these polyhedra defines the cages by the number of each size ring in their composition. A cavity that is enclosed by twelve 5-membered rings and two 6-membered rings is a $5^{12}6^2$ cavity (note that when describing hydrates, a 5-membered ring is a ring with five water molecules, not five atoms). Different hydrate forms crystallise depending on the size of the templating guest. There are three main hydrate structures – structure I, structure II and structure H (SI, SII and SH). The main properties of these structure types are summarised in Table 4.2. Other clathrate hydrate structures are also known to exist, such as those with bromine or dimethylether, but these are much rarer. The cavities that are found in SI, SII and SH hydrates are shown in Figure 4.13

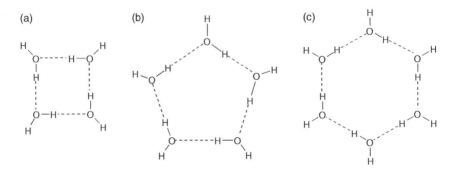

Figure 4.12 Networks of hydrogen bonds between water molecules in clathrate hydrate structures can form (a) squares, (b) pentagons and (c) hexagons.

Table 4.2 The properties of the three most common forms of clathrate hydrates

Property	SI	SII	SH
Crystal system	Primitive cubic	Face-centred cubic	Hexagonal
Space group	$Pm3n$	$Fd3m$	$P6/mmm$
Cages per unit cell	2 (5^{12})	16 (5^{12})	3 (5^{12})
	6 ($5^{12}6^2$)	8 ($5^{12}6^4$)	2 ($4^35^66^3$)
			1 ($5^{12}6^8$)
H_2O per cell	46	136	34

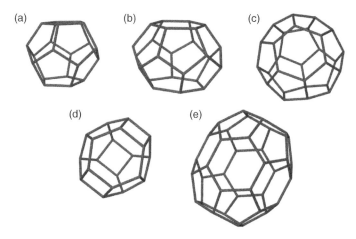

Figure 4.13 The hydrate cages found in SI, SII and SH hydrates (vertices represent oxygen atom positions): (a) 5^{12}; (b) $5^{12}6^2$; (c) $5^{12}6^4$; (d) $4^35^86^3$; (e) $5^{12}6^8$.

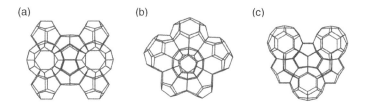

Figure 4.14 The structures of (a) SI, (b) SII and (c) SH clathrate hydrates.

and the ways in which they stack face-to-face to form the final clathrate structures are displayed in Figure 4.14.

Cavities are formed by a templation process around the guest species and therefore smaller guests tend to favour the formation of structures in which there is a higher proportion of the smaller 5^{12} cavities. The 5^{12} cages are marginally larger in type-I hydrates due to the surrounding environment and this can sometimes be a deciding factor in which structure is adopted. The shape of the cavities can also play a role in determining which guests will template which clathrate hydrate forms. The $5^{12}6^2$ cage, for example, is of an oblate nature and only CO_2 and C_2H_6 stabilise the cage as simple hydrates (only one type of guest species present in the structure) although other guests may use it in combination with 5^{12} cages in SI clathrates.

The exact sizes of the cavities are dependant on the guests that are situated within them as the cages have the ability to expand or contract slightly to maximise the stability (*i.e.* the hydrogen bonds are of variable length between structures). In general, 5^{12} cavities are able to hold molecules that are up to 5.5 Å in diameter, such as carbon dioxide. The $5^{12}6^2$ cage holds molecules up to 6 Å and the $5^{12}6^4$ cage is able to hold larger hydrocarbons up to 6.6 Å. Larger guest species (~ 8–9 Å) tend to reside in the large cavities of the SH hydrate or form less common clathrate structures. An example of a special case is dimethyl ether which forms a structure T hydrate which possesses $5^{12}6^3$, $5^{12}6^2$ and $4^15^{10}6^3$ cages, which are all capable of containing the same guest species.

The water framework of clathrate hydrates has often been studied by using solid state ^{1}H NMR spectroscopy. The framework is fluxional in nature as the water molecules are able to rotate in place (effectively switching around the directions of the hydrogen bonds) and move through the structure. NMR data show that below 50 K hydrate structures are rigid, as with *Ih*. Above this temperature, the water molecules are able to undergo reorientation and translational diffusion while still retaining the structural integrity of the clathrate structure. At 273 K, the diffusion of water molecules is found to be one hundred times greater than *Ih* and the fixed-site reorientation is twice as fast. Such rapid diffusion has an effect on the bulk dielectric constant, which is approximately half that of *Ih*.

4.4.2 Guest properties

Guest species that reside within clathrate hydrates do not participate in strong interactions with the water framework and are most often hydrophobic in nature. Molecules that are good hydrogen bond donors or acceptors (such as carboxylic acids, amines or alcohols) will disrupt the interactions between the water molecules and therefore prevent the hydrate network from forming. The most suitable guests are relatively small, non-polar species, such as halides, noble gases and small hydrocarbons.

The smallest known guest that is found within a clathrate hydrate is argon which resides in the 5^{12} cavities of the SII structure (the smallest cavities). Smaller species, such as hydrogen, helium and neon, are not observed to form clathrates under naturally occurring conditions as they are small enough to diffuse through the cage walls, so causing the cavities to fall apart. Recently however, these hydrates have been formed under extreme pressure. Some of the larger $5^{12}6^4$

cages are able to hold hydrocarbons as large as propane, *iso*-butane and *n*-butane. It is even possible for cyclopentane to be held in the SH structure if there is methane present in the other cages which is a small guest and allows for the expansion of the larger cavities. Although structures may contain more than one kind of guest species, there is only ever one guest molecule in a cavity. Only the high-pressure H_2 hydrate has formed with multiple guests residing in one cavity simultaneously. Figure 4.15 shows some common guest species and the types of cavity that they are most often found in. Many of these do not form as pure hydrate structures but as mixtures. This is particularly true of the larger species which will only form when there is a smaller secondary guest species also present.

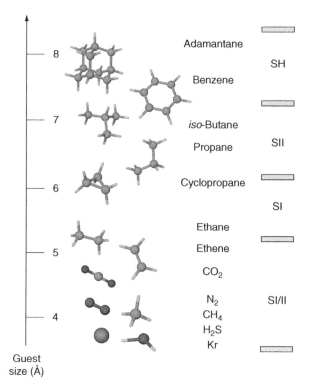

Figure 4.15 The sizes of common clathrate hydrate guests and the clathrate hydrate structures that they commonly template.

The identities of guests within clathrate hydrate structures are most often determined by spectroscopic methods, such as IR and NMR spectroscopies, as they are highly disordered within crystalline samples. This disorder is due to the lack of strong interactions to hold them in place, which allows for free rotation and translation within the cavities. It is thought that translational movement is some-what restricted, as guest species prefer to reside close to the walls of the cages rather than in the centre. There is some degree of rotational inhibition due to the dipole–dipole interactions between guest molecules in adjacent cages. Within a single cage, there is negligible rotational inhibition as the cage-water dipoles are cancelled out near the centre of the cavity.

4.4.3   Clathrate hydrates in the petroleum industry

 Koh, C. A., 'Towards a fundamental understanding of natural gas hydrates', *Chem. Soc. Rev.*, 2002, **31**, 157–167.

Clathrate hydrates of natural gases are of great importance in current research within the petroleum industry. The natural reserves are a potential source of energy, as well as a potential environmental hazard, and the accidental formation of hydrates during recovery and refinement of oil and gas is both an economical and ecological liability. Clathrate hydrates also present an opportunity for novel storage and transportation technologies.

The environments from which natural gas is usually obtained are often extremely cold and exist under high pressure. The presence of water and small hydrocarbons under such conditions is ideal for the formation of clathrate hydrates. Gas deposits become saturated with water when they are forced out from underground reservoirs. As the gas expands, the temperature in the well-head drops and hydrates start to condense and form on the sides of the pipe. These hydrates then form potentially dangerous blockages within the pipelines if pressure is allowed to build up behind the plug. Such blockages may lead to rupturing of the transport pipes with devastating environmental impact. The formation of hydrates also removes the lighter, more desirable fractions from the oil mixture, leaving behind only heavy oils and tar. The problems caused by the formation of hydrates costs the petroleum industry hundreds of millions of dollars a year and so intensive research is continuously carried out on methods for the removal and prevention of hydrates.

Possibly the most obvious method of prevention is the removal of water from the natural crude oils. However, in most situations this is not a feasible approach due to the conditions in which the extraction is carried out. Other methods of prevention rely upon altering the conditions within the pipelines away from those that are conducive to hydrate formation. For example, the pipes can be heated or the pressure within them can be reduced. However, as with drying the gas, these methods are also difficult to apply in most situations. The most commonly used approach for prevention is placing additives into the oil that stop or slow down hydrate formation. These materials are called 'hydrate inhibitors'.

Hydrate inhibitors can work in one of two ways, either as thermodynamic inhibitors or as kinetic inhibitors:

- **Thermodynamic inhibitors** are chemicals such as methanol or ethylene glycol (ethane-1,2-diol). These materials shift the hydrate formation phase boundary away from the temperature and pressure conditions of the gas-transportation process by increasing the driving force required for the formation. Thermodynamic inhibitors are used at very high concentrations, as much as 60 % methanol can be used in deep water extractions, which proves to be a very costly way of preventing the build-up of hydrates.

- **Kinetic inhibitors** are used on a smaller scale ($\sim 0.5\%$) and work by reducing the rate at which the hydrate is formed.[5] The small amounts of hydrate that do form are simply carried along in the flow and dealt with at a later stage in the refinement process. Kinetic inhibitors are usually polymeric organic materials, such as poly(vinyl pyrrolidinone) (PVP). Generally, kinetic inhibitors or other anti-agglomerants (such as tetraalkyl ammonium salts) are not used without the presence of a thermodynamic inhibitor as well.

Clathrate hydrates are generally problematic within the petroleum industry – however, the presence of natural gas hydrates also raises positive prospects of fuel resources and transport methods. At STP (0 °C, 101 kPa), the SI hydrate can contain approximately 160 times the equivalent volume of methane under the same conditions. This is enough to support its own combustion which can be observed as the paradoxical 'burning snowball'. It is estimated that around $50 \times 10^{15}\,\mathrm{m}^3$ of methane is stored in hydrate deposits in permafrost regions and in deep-sea deposits. With there being an increasing problem with fuel shortages and finding new sites for natural oil and gas, their extraction from clathrate hydrates could provide a viable alternative fuel source in years to come. It has already proved possible to reclaim useable gas from more readily available land deposits, although many of the dangers associated with recovery from marine deposits are still present in land-based systems. The high density of gas in hydrate structures also means that they are being researched as possible methods of storage, although it is not yet possible to compete economically with liquid gas.

4.4.4 Other uses and occurrences of clathrate hydrates

Although clathrate hydrates are most commonly talked about in association with the petroleum industry, they also have other effects and uses, both naturally occurring and in artificial systems.

Methane hydrate deposits in permafrost regions are not only potential fuel reserves but are also associated with aspects of global warming. It is postulated that increasing global temperatures may be speeding up the release of methane from these reservoirs, which in turn speeds up the heating process. Methane is ten times worse as a greenhouse gas than carbon dioxide. It is suggested that hydrates may have played a role in prehistoric global-warming events. Other environmental phenomena are also being blamed on hydrate deposits. Some deep-sea deposits cause massive slippage when they decompose and it is thought that these could be the cause of some large tidal-wave events. It is even suggested that gases given off when hydrates decompose could be behind the 'Bermuda Triangle phenomenon', although this has not been proven. Not all hydrate deposits have a negative consequence, for example, some deep-sea hydrate deposits act as habitats for bacteria which thrive off of methane. Clathrate hydrates are also now thought to exist (from spectroscopic data) extra-terrestrially on Mars, Saturn, Neptune and Uranus and even form parts of comets. Indeed, it has been suggested that the CO_2 clathrate may exist in winter at the Martian poles, an observation which is being put forward as evidence of extraterrestrial life.

A use of hydrates in synthetic systems is in the separation of gases by the differences in the formation conditions of their respective hydrates. For example, it is possible to separate hydrofluorocarbons and nitrogen in this manner. Pollutant streams can also be cleared up in a similar fashion. It has been shown that hydrates are able to recover carbon dioxide from a mixture of CO_2, O_2 and N_2 in high yields.

4.5   Crystal engineering

4.5.1   Concepts in crystal engineering

 Brammer, L., 'Developments in inorganic crystal engineering', *Chem. Soc. Rev.*, 2004, **33**, 476–489.

Most of the types of solid state supramolecular systems that we have seen so far have all been adapted from naturally occurring systems or molecules. The knowledge and understanding gained from the study of such systems can be taken and applied to purely synthetic systems. In this manner, it may prove possible to build up designed, functional materials from a knowledge of intermolecular interactions.

Crystal engineering is a sub-discipline within supramolecular chemistry which explores interactions within crystalline solids and attempts to apply the knowledge gained in the rational design of novel materials.[6] The distinction should be made clear at this stage that crystal engineering is *not* the same as crystal structure prediction. Structure prediction implies a totally theoretical route to determining crystal packing and symmetry (including space group) for a given molecule, a task that has yet to be accomplished reliably and is the focus of considerable current effort.

> **Crystal Engineering:** The predicted design and synthesis of novel synthetic crystalline solids through knowledge of intermolecular interactions and supramolecular synthons.

Crystal engineering relies upon systematic studies of known structures to examine recurring patterns of interactions between functional groups in order to make general predictions about likely structural features. Recurring motifs between functional groups are termed *supramolecular synthons*. The molecular building blocks of the structure are referred to as *tectons* (Figure 4.16). The recognition of supramolecular synthons comes from the deconstruction of structures to their simplest components in a manner similar to that of retrosynthetic analysis used in organic synthesis to find routes towards complex molecules. The difference between these two approaches is in the types of bonds created. Where molecular synthesis looks towards finding synthons for the construction of covalent bonds, crystal engineering and supramolecular retrosynthesis require units capable of acting together through non-covalent interactions. The deconstruction of crystals ultimately looks at the manner in which one molecule interacts with its nearest neighbours and how these interactions are utilised to hold together the long-range structure of the solid. If, for example, it is known that two types of functional groups frequently associate in the solid state, then it is reasonable to use them with the expectation that the motif will re-occur.

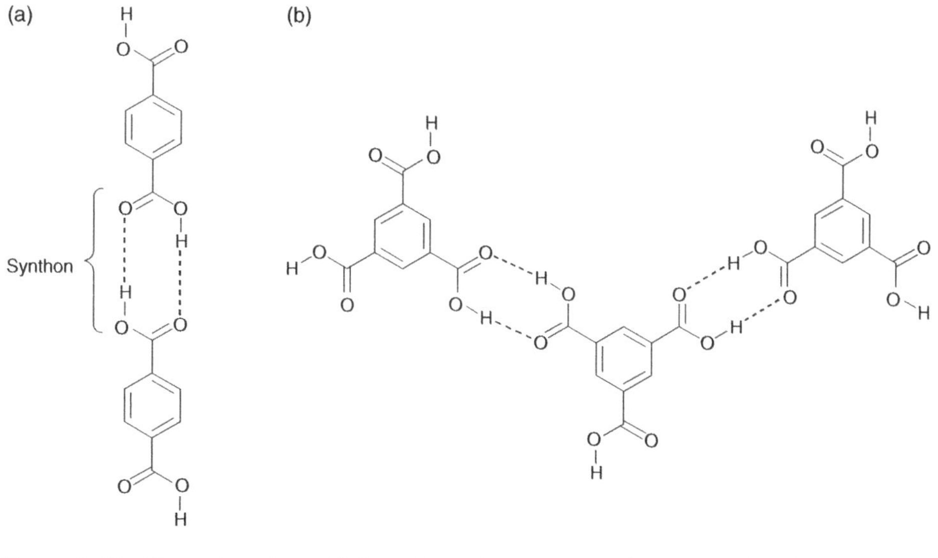

Figure 4.16   The carboxylic acid dimer is used as a supramolecular *synthon* in two systems with different *tectons*: (a) benzene-*p*-dicarboxylic acid; (b) benzene-1,3,5-tricarboxylic acid.

> **Supramolecular Synthons:** Recurring intermolecular motifs within solid state structures made up of *tectons* that may be applied towards the design of novel systems.

> **Supramolecular Tectons:** The molecular components of a crystal, held together by supramolecular synthons.

We have already seen several examples of supramolecular synthons in this chapter, such as patterns formed by carboxylic acid residues and urea functionalities (Section 4.3). The following section aims to explore in more detail how it is possible to recognise synthons and tailor their use within synthetic architectures. A particularly important source of information about the frequency of occurrence of particular synthons (and therefore their predictability) is the *Cambridge Structural Database*.

4.5.2   The Cambridge Structural Database

 Chisholm, J., Pidcock, E., van de Streek, J., Infantes, L., Motherwell, S. and Allen, F. H., 'Knowledge-based approaches to crystal design', *CrystEngComm.*, 2006, **8**, 11–28.

The Cambridge Structural Database (or CSD) is a store of known structures of small molecules (therefore excluding large biomolecules) which contain organic carbon. This includes coordination compounds, organic molecules and organometallic complexes. Related databases contain purely inorganic structures (the *Inorganic Crystal Structure Database*), protein structures (*Brookhaven Protein Data Bank*) and those of metals (*Metals Crystallographic Data File*).

The CSD is an indispensable utility for crystal-engineering studies as it is possible to search through the database of over three hundred and fifty thousand structures (as of 2006) to find intermolecular interaction patterns, as well as molecular structural data. The database contains all of the information gathered from X-ray and neutron studies of single crystals, such as atomic positions and bond lengths and angles. The availability of this data in an electronic format readily allows for statistical analysis to be undertaken so that synthons may not only be examined by their frequency of occurrence but also by directionality, geometry and implied strength.

Statistical analysis of structures allows for clear patterns to be observed for directional preferences in interactions, as well as their relative lengths and situations in which expected motifs are supplanted by other more favourable situations. Sometimes, the trends and patterns found through database studies can be rationalised computationally. The increase in computing power within the last twenty years has allowed for increasingly complex theoretical challenges to be tackled routinely. This has enabled intermolecular interactions to be modelled and for trends to be analysed in terms of energetics and allows for the theoretically most stable arrangements to be identified. However, energetic calculations are mostly conducted in gas-phase situations and in real crystals the packing effects (*i.e.* supramolecular interactions) can play a pronounced part in structure organisation, sometimes disfavouring the predicted geometries. Indeed, the difference between the calculated energies of a gas-phase-optimised structure and the structure found in the solid state calculated in the same way represents a crude measure of the minimum energy of crystal-packing forces. For this reason, the science of crystal engineering is an imprecise art and many 'engineered' structures are a result of serendipity as much as design, with the field still being described as in a 'data-collection phase'.

The Cambridge Structural Database allows for patterns to be elucidated well after the structures were originally determined. The vast majority of published structural papers contain little or no information about crystal packing through supramolecular motifs, as structural chemists are predominantly focused on intramolecular covalent connections. The X-ray crystallographic experiment yields packing information as a matter of course, however, and the ability to search through this often-overlooked information can be indispensable for determining common motifs. An example was the use of statistical analysis to clear up one of the biggest and most important supramolecular debates in the field – whether or not, CH···X hydrogen bonds exist (although pockets of resistance to this idea remain).[7] For a topic that was so controversial, no single structure could have provided conclusive evidence either for or against the argument; however, by using the CSD it was found that there are distinctly preferred length and angle distributions associated with CH···X hydrogen bonds, as there are for NH···X and OH···X interactions. Weak hydrogen bonds are now widely accepted to play significant roles within many biological systems, as well as being rife within crystal structures of synthetic compounds.

4.5.3 Crystal engineering with hydrogen bonds

Hydrogen bonds are the most prominent of supramolecular interactions, both within synthetic systems and in nature. The directionality and strength that they exhibit makes them of great use in constructing predictable and replicable structures. Within most simple, purely organic crystals, the hydrogen bond is almost exclusively responsible for ordering the structure, supported by weaker interactions such as π–π interactions and van der Waal's shape recognition (the driving force towards crystal close packing). Metal-containing systems are often very closely linked to organic structures as the external contacts of the ligands are mostly organic in nature and often the metal plays little role other than in the structure of the tecton.

The role of hydrogen bonding is of such importance that a system of nomenclature has been devised through which it is possible to describe the patterns that are observed, known as the *Graph set nomenclature*.[8] This nomenclature expresses hydrogen bond motifs in one of four categories, denoted as chains (C), rings (R), intramolecular interactions (S, for 'self') and discrete interactions (D). As well as the topology of the interaction, the number of donor and acceptor atoms are also included in the notation, with donors written as a subscript and acceptors written as a superscript. The total number of atoms within the motif, including hydrogen atoms, in the pattern is called the *degree of the pattern* and is written in brackets. Hence, $\mathbf{R}_2^2$ **(8)** denotes a ring system in which there are two hydrogen bond acceptors and two hydrogen bond donors, with 8 atoms in total, as in a carboxylic acid dimer (Figure 4.17).

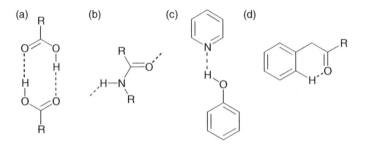

Figure 4.17 Hydrogen bond motifs described by the Graph set nomenclature: (a) $\mathbf{R}_2^2(\mathbf{8})$; (b) **C(4)**; (c) **D**; (d) **S(6)**. NB – if there is only one acceptor and one donor, then their numbers are omitted from the notation and for discrete assemblies it is not necessary to include any values if there is only one hydrogen bond.

Once motifs are established, it is also recognised that there are different levels on which the interactions can be defined. At the most basic level, all hydrogen bonds can be individually defined as discrete (**D**). This is called the *unitary* Graph set, $\mathbf{N}_1$. In systems with more than one hydrogen bond, then there is a second level of analysis called the *binary* Graph set which describes both interactions, $\mathbf{N}_2$. When more than two hydrogen bonds are present, the situation becomes increasingly complex. If three hydrogen bonds are present (*a*, *b*, and *c*) then there are three unitary graph sets (**D** for each interaction), three second-level sets ($\mathbf{N}_2$(ab), $\mathbf{N}_2$(bc) and $\mathbf{N}_2$(ac)) and a third-level set which describes all three bonds simultaneously. Usually, it is sufficient to only describe the most common or obvious features within a structure or those that play the greatest role.

In addition to this notation, there are also three general rules about hydrogen bonding put forward by Margaret Etter in 1990.[9] These are as follows:

- All good proton donors and acceptors are used in hydrogen bonding

- Six-membered-ring intramolecular hydrogen bonds form in preference to intermolecular hydrogen bonds

- The best proton donors and acceptors remaining after intramolecular hydrogen bond formation form intermolecular hydrogen bonds to one another.

Although these rules do not always hold true, they are, nevertheless, a good general guideline.

The vast range of possible interactions that are accessible in engineered systems is impossible to cover in an introductory book such as this, and the reader is strongly encouraged to use the references for further information. Below are detailed just a few of the more common hydrogen bond synthons that are observed.

Carboxylic acid dimers
Carboxylic acids contain one hydrogen bond donor group and one acceptor group. Carboxylic acid residues are therefore self-complementary as hydrogen bond acceptor–donor pairs and are able to form dimeric structures (Figure 4.18(a)). This is a remarkably robust synthon and has been observed to persist in structures despite significant changes to the molecules that the residues are attached to. A second common synthon to form from carboxylic acids is a

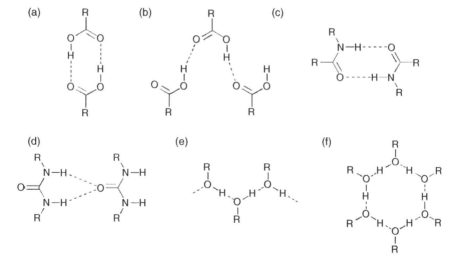

Figure 4.18 Some common supramolecular synthons based on hydrogen bonds. Carboxylic acids can form (a) discrete dimers and (b) infinite chains. (c) Amides can form dimers in a similar manner to carboxylic acids, while (d) urea may form a 'tape' motif. Alcohols may also form both (e) infinite and (f) discrete structures.

chain motif (Figure 4.18(b)), which is actually observed more often than the dimer, although not as predictably.

Amides

To a first approximation, the amide functionality may be seen to act similarly to carboxylic acids, with a strong NH donor group in place of OH and a C=O acceptor (Figure 4.18(c)). However, an additional donor exists in amide functionalities as the –CHO proton is also highly acidic in nature due to the nearby nitrogen and oxygen atoms and therefore is capable of forming strong interactions also, stronger than most C–H donors are capable of creating.

Urea is well-known to form clathrate structures (Section 4.3.1); however, it is also able to form layered structures in a 'tape' motif. The urea functionalities are able to align in such a way that each carbonyl receives two NH$\cdots$O hydrogen bonds from the molecule 'next to' it (Figure 4.18(d)). This results in the formation of $\mathbf{R}_2^1(6)$ rings running throughout the structure.

Alcohols

Alcohol functionalities contain one hydrogen bond donor and two lone pairs able to accept hydrogen bonds. It is most common to find situations in which each alcohol group donates and accepts one hydrogen bond. There are two main recurring synthons that are observed with alcohols, either infinite chains (Figure 4.18(e)) or discrete ring systems such as those in phenol derivatives (*cf.* Section 4.3.3) (Figure 4.18(f)).

The above examples are just a few in which the functionalities are self-complementary and, in the lack of any other hydrogen bonding functionalities, the final structure may well be predictable. There are many more diverse systems in which the donor and acceptor groups are different yet complementary, for example, the situation in cocrystals of alcohols and primary amines. These two functional groups represent an ideal pairing as alcohols are able to donate one hydrogen bond and receive two, whereas amines can donate two hydrogen bonds and receive only one, such a situation is referred to as 'saturated hydrogen bonding', *i.e.* all protons and lone pairs are involved in hydrogen bonding. An example of a 'chicken-wire' motif that can be formed by the cocrystallisation of an alcohol and an amine is shown in Figure 4.19.

These examples above all utilise strong hydrogen bonds; however, it is not uncommon to see systems in which the principal interactions are weaker in nature. These weaker interactions often arise from situations in which the donor group is weak, such as C–H groups, or there is an unconventional acceptor group, such as a delocalised π-system or a covalently bound halide. Weak donors are observed to interact with the more basic acceptor moieties, such as oxygen and nitrogen atoms, although there are instances of very weak C–H$\cdots$π interactions. Even very poor carbon acid donors, such as methyl groups, have been observed to form hydrogen bonds to strong acceptors.

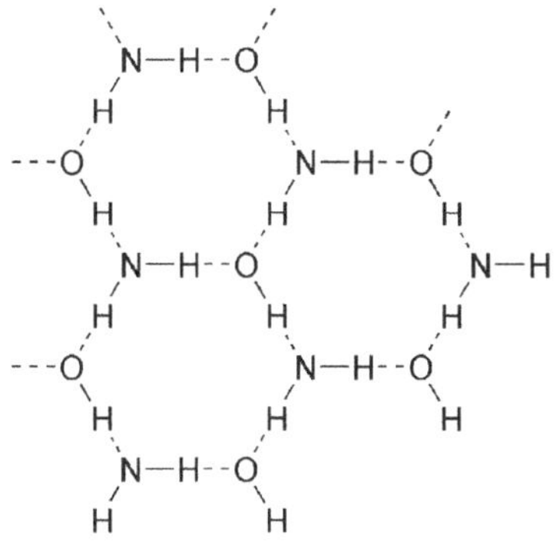

Figure 4.19 A 'chicken-wire' motif between an alcohol and an amine.

4.5.4 π-Interactions

Unlike hydrogen bonding, other intramolecular interactions are much harder to control and predict for crystal-engineering purposes. This difficulty generally arises from a lack of strength and directionality. The most commonly observed interaction other than hydrogen bonding is π–π stacking. Interactions between π-systems are most common when there are few other possible interactions that may take precedence within the structure, so allowing the structures of fused aromatic rings to display stacking between the molecules. As we saw in Chapter 1, Section 1.3.3, there are two major ways in which π-systems are able to arrange themselves, *i.e.* face-to-face and edge-to-face. Although these remain relatively unexploited in deliberately engineered systems, owing to their tendency to be over-ridden by other, stronger interactions, there are many examples of these motifs frequently recurring. Aside from the relatively simple layered structure of graphite and the herringbone structure of benzene, one well-studied synthon is the *phenyl embrace*. These motifs are commonly found between pairs of tetrahedrally disposed EAr_3 units (E = P, As, B) or the related EAr_4^+ cations and involve three phenyl rings from each group. The two EAr_3 groups are staggered with respect to each other with each of the phenyl rings, both donating and receiving in edge-to-face $\pi \cdots \pi$ interactions (Figure 4.20), with the opposing groups appearing to embrace each other. As there are a total of six phenyl rings involved, this motif is commonly referred to as the 'six-fold' phenyl embrace.[10] The large number of interactions act in a summative manner, being weak when regarded individually, and yet the phenyl embrace is a relatively strong synthon. Similar embrace motifs are seen with other aryl containing species, such as $M(2, 2'\text{-bipy})_3^{n+}$ complexes, where the strength of the embrace overcomes the repulsive cation–cation interactions.

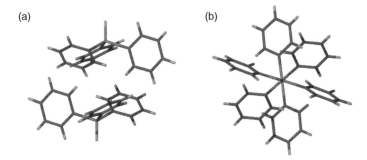

Figure 4.20   A six-fold phenyl embrace in the crystal structure of one polymorph of HGePh$_3$, seen from (a) side-on and (b) along the Ge····Ge axis.

4.5.5   Other common synthons

The section title above is somewhat misleading as, although there are many supramolecular synthons that are being identified in current research, it may be premature to label any of them as 'common' when compared to the ubiquitous nature of hydrogen bonds and coordinative interactions. As stated earlier, crystal engineering is very much a growing discipline and in many cases, although they have been statistically identified, many synthons are yet to be deliberately exploited to their full potential and are therefore awaiting widespread acceptance as genuine, reproducible supramolecular motifs.

Interactions involving halogens are frequently coming under scrutiny with C–X, and particularly M–X and X$^-$ (X = Cl, Br and I, but only rarely F) being examined as hydrogen bond acceptors, with a significant amount of data confirming this assertion.[11] Halogens bound to metals result in stronger hydrogen bonds than those attached to organic frameworks; however, the latter is still a useful synthon. A motif that is particularly noteworthy is that in which the hydrogen bond is either bifurcated or trifurcated at the donor with an MX$_2$ or MX$_3$ unit, respectively, as the acceptor species (Figure 4.21).

Interactions between pairs of halogen atoms (X···X) and between halogen atoms and electron-pair donors (such as oxygen and nitrogen) are also of current interest. This phenomenon is known as *halogen bonding*, whereby a halogen atom (usually bound to carbon/nitrogen or a metal centre) acts as an electron-pair acceptor (Figure 4.22(a)). The strongest halogen bonding is observed for iodide with a decrease in strength observed moving up the group. The strength of the interaction is closely related to the electron-withdrawing strength of the molecule to which the halogen is attached. While such interactions are rather weak (therefore readily displaced by stronger interactions), when correctly employed these synthons can give rise to predicted structures. Examples of this are 4-nitro-4'-iodobiphenyl[14] (Figure 4.22(b)), designed to display second-harmonic generation (SHG) properties due to directional stacking utilising the I···O$_2$N interaction, and the complex of 3-bromopyridine with PtCl$_2$, [PtCl$_2$(C$_5$NH$_4$Br)$_2$][15] (Figure 4.22(c)).

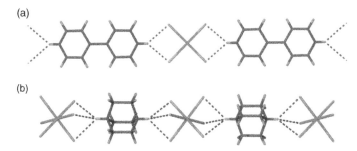

Figure 4.21 Halogens as hydrogen bond acceptors in the ribbon structures of (a) $[H_2bipy][PdCl_4]^{12}$ and (b) $[H_2DABCO][PtCl_6]^{13}$ with $NH\cdots Cl_2M$ and $NH\cdots Cl_3M$ motifs, respectively.

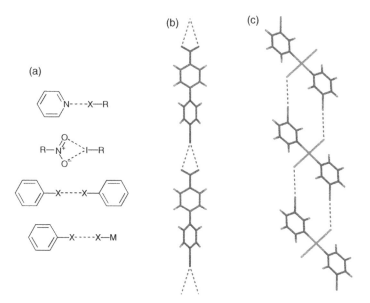

Figure 4.22 (a) Some examples of halogen bonding interactions (X = halogen) and their use within the structures of (b) 4-nitro-4'-iodobiphenyl and (c) $[PtCl_2(C_5NH_4Br)_2]$ (X = Cl, Br and I).

An even more 'exotic' interaction is found between Au(I) atoms, known as the *aurophilic interaction*.[16] The Au(I)–Au(I) interaction is of a strength comparable to a weak hydrogen bond (~ 20–$50\,kJ\,mol^{-1}$), with the two atoms within the sum of their van der Waals radii and occasionally closer together than the atoms lie in metallic gold. While this may seem strange at first, the phenomenon has been explained theoretically, occurring as a result of relativistic contraction making the $5d^{10}$ closed-shell configuration chemically active. Similar interactions also exist between silver atoms, known as *argentophilic interactions*.

One further unusual type of interaction is where a transition metal itself acts as a hydrogen bond acceptor. Metals are highly *amphoteric* in nature (*i.e.* they are able to acts as both Lewis acids and bases). Agostic interactions, those in which the σ-electron density in a C–H bond interacts with an electron-deficient metal, are well known in structural inorganic chemistry. These are classified as being three-centre (C, H and M), 2-electron bonds. There is a distinct difference between these and D–H $\cdots$ M hydrogen bonds in terms of distance and geometry. Hydrogen bonds tend towards linearity and those that are directed to a metal centre have been termed *intramolecular pseudo-agostic (IPA) interactions* (Figure 4.23(a)). Closely related to these are *intermolecular multi-centre hetero-acceptor (IMH) interactions* in which the hydrogen bond is bifurcated between the metal and another acceptor atom which is ligated to the metal (Figure 4.23(b)). This latter class of interaction is sometimes mistakenly identified as a true agostic interaction and care must be taken when classifying such situations.[17] The crucial difference is that both IPA and IMH interactions are four or more electron interactions, not two electron interactions.

(a)   (b)

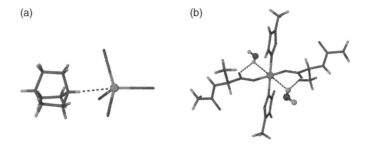

Figure 4.23   (a) An IPA interaction between a quaternary ammonium cation and $[Co(CO)_4]^-$ and (b) an IMH interaction of water to a Cu(II) carboxylate complex.

4.5.6   Solid state reactivity

Kaupp, G., 'Solid state molecular syntheses: complete reactions without auxiliaries based on the new solid state mechanism', *CrystEngComm*, 2003, **5**, 117–133.

An area of solid state supramolecular chemistry that is coming under increasing scrutiny is the ability of molecules to undergo very selective reactions in a crystalline phase. There are two general mechanisms by which these solid state reactions are able to take place. The vast majority of solid state reactions occur in a process that sees the original crystalline lattice first distorted, before various zones within the structure begin to undergo phase transitions that spread through the solid. Such reactions completely change the crystalline lattice from its initial form and the products are often of a substantially different geometry from the starting material. The other class of reactions are those that are termed *topochemical reactions*. Historically, topochemical reactions have been thought to require that the reactants are in close proximity to each other within the crystal lattice, no more than 4.2 Å apart, and that they are in the correct orientation. These assertions are known as the 'topochemical postulate'; however, there are many more exceptions to these rules than supportive examples and many topochemical reactions have been proven to show a significant change in the lattice during the reaction. Solid state reactions in which there is no molecular migration (and therefore the crystal lattice undergoes no disintegration) are said to be *topotactic*.

> **Topochemistry:** The study of reactions carried out within crystal lattices whereby adjacent molecules are able to react with each other with minimal molecular movement.

The origins of the topochemical postulate lie in the $[2+2]$ photodimerisation reaction of α-cinnamic acid ($PhCH = CHCO_2H$) in the solid state in which the double bonds of the starting material lie close to each other and are parallel in its crystalline form. In fact, the examination of solid state photodimerisation resulted in the first usage of the term 'crystal engineering' and the $[2+2]$ photodimerisation reaction remains one of the most intensely studied topochemical reaction to date. The dimerisation of α-cinnamic acid was later proved not to be a topotactic reaction. However, following from this work, attempts have been made to artificially engineer situations in which molecules can be brought into close proximity in a crystal lattice, held securely in place and reacted.[18] Two approaches have been adopted – intramolecular substitution and the use of auxiliary components. Intramolecular substitution involves the addition of functional groups onto the reactants such that they self-associate into the correct geometry for reaction. The obvious drawback with this approach is that it is no longer the desired compounds that are reacting but derivatives thereof. An example of this methodology is shown in Figure 4.24(a) with the self-association of a dicarboxylic acid (utilising a well understood supromolecular synthon) to bring two double bonds close enough to undergo a $[2+2]$ cycloaddition. The use of auxiliary species to assist in the aggregation of reactants prior to irradiation is a more attractive synthetic route. We have previously seen an example in Chapter 3 of an intra-cavity reaction that forms stereoselective products (Figure 3.51). In this case, the capsule itself is the auxiliary species. The use of cavities to conduct reactions works primarily by trapping the reacting species in a sterically well-defined enclosed space, rather than explicitly templating the reactants into the desired positions. The adaptability of both methods is poor, however, as subtle changes in the structure of the reactants can have a pronounced effect on the crystal packing, thereby preventing any reaction from occurring.

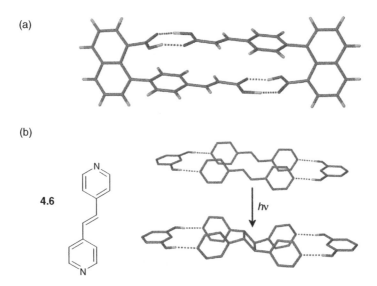

Figure 4.24 (a) The self-assembly of a dicarboxylic acid to bring two double bonds into close proximity and (b) the before and after crystal structures of a supramolecularly templated [2 + 2] photodimerisation reaction **4.6**, templated by resorcinol.

Recent efforts to template solid state reactions utilise a methodology that lies somewhere between the examples shown above. Auxiliary components are used that template the positions of the reactants through designed supramolecular synthons, rather than simply enclosing them in a cavity. Examination of the reactants and their specific demands in terms of satisfying any hydrogen bond donor/acceptor groups can lead to rational choices being made for the auxiliary templates. One of the best examples is the templation of *trans*-1,2-bis(4-pyridyl)ethylene (**4.6**) by resorcinol (1,3-dihydroxybenzene) (Figure 4.24(b)). The design of this system employs a host–guest strategy, whereby the dipyridyl molecules can be thought of as guests within the hydrogen bond donor array presented by the two resorcinol molecules. The guests are brought into a suitable proximity for reaction, according to the topochemical postulate, and upon irradiation the desired cyclobutane derivative is formed, still held within the resorcinol template. The same template has been successfully extended to reactants with terminal pyridyl groups containing multiple double bonds along their length to form products containing two or more cyclobutyl rings. This templation methodology can be extended to other systems, for example, two strands with ammonium groups at either end have been templated by polyether macrocycles (*cf.* ammonium/crown rotaxanes, see Chapter 3, Figure 3.28) and then reacted.

The standard type solid state reactivity is *mechanochemical reactivity*. This method has been referred to by as 'making crystals by smashing crystals', although in fact sometimes only a short 'grinding period' is needed to initiate a solid-phase reaction that will then proceed to completing without further mechanical force.[19] In any chemical reaction, it is necessary to bring the reactants into contact with each other. Mechanochemical reactions involve taking two solids and directly mixing them in a solvent-free environment. In practical terms, this means grinding the reactants with a pestle and mortar or in a ball-mill apparatus. Under these conditions, the surfaces of the microcrystalline reactants are brought into close contact under pressure and at increased temperature as a result of the friction generated. Sometimes, the heat that is evolved can induce localised melting of the crystalline reaction mixture so that the reaction actually occurs in a melt, albeit on a microscopic scale.

Mechanochemical Reactions: The reaction of two microcrystalline products in the solid state by direct grinding.

Mechanochemical reactions can be used to access products that cannot be synthesised by more conventional methods or, in some cases, can produce known products in a faster time. For example, the palladium-based square that we saw in Chapter 3 (Figure 3.1) takes as long as four weeks to form at 100 °C, although this can be formed in 10 min at room temperature by directly grinding the solid reactants. The products that are obtained from mechanochemical reactions are themselves microcrystalline and are therefore unsuitable for single-crystal X-ray diffraction experiments. Powder diffraction can be used to differentiate product and starting materials and identify any known product phases. Full structure determination of new solid products by powder X-ray diffraction is possible, but remains a considerable challenge. The problem of determining the structure of new solids can sometimes be circumvented by seeding the growth of single crystals suitable for single crystal X-ray crystallography with microcrystals of the mechanochemical product. Small amounts of solvent are sometimes added to mechanochemical grinding reactions, which are then termed 'kneading reactions'. Although this is not technically solid state synthesis, new crystalline products are formed, such as inclusion complexes of cyclodextrins. Another type of solvent-free method that can be used to obtain crystalline products are solid–gas reactions in which a gaseous phase of one reagent is passed through a solid sample of a second. A good example of a supramolecular mechanochemical reaction is the reaction of ferrocenedicarboxylic acid with dinitrogen bases to yield hydrogen bonded chains.[20] The reaction of the dicarboxylic acid with 1,4-diazabicyclo[2.2.2]octane, either by grinding in the solid state or by use of the amine in the vapour phase, results in the same one-dimensional chain structure (Figure 4.25).

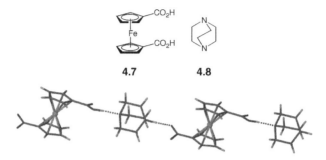

4.7   **4.8**

Figure 4.25 The hydrogen-bonded network which is formed from the mechanochemical reaction of ferrocenedicarboxylic acid (**4.7**) and 1,4-diazabicyclo[2.2.2]octane (DABCO) (**4.8**).

4.5.7 Engineering crystals

While crystal engineering refers to a systematic study of the interactions within a crystal and attempts to design solids utilising these forces, there is a trend towards 'real' crystal engineering – controlling aspects of the physical growth by utilising the known chemical similarity of the systems. An extreme example of this is the formation of biomimetic structures of templated silica and alumina (see Chapter 5, Section 5.10). Another fascinating opportunity that is made possible through templating is that of *epitaxial growth* – growing crystals around other crystals.

Crystals are described as being isomorphous with one another if the spatial arrangement of the atoms is the same, despite the chemical composition differing. Isomorphous structures are commonly found in transition metal salts, whereby the difference in structures between many of the first-row transition metals is only marginal due to size difference of the M^{2+} species. The intermolecular interactions within isomorphous crystals are therefore the same and it can reasonably be assumed that the manner of their growth and assembly is also identical. In such

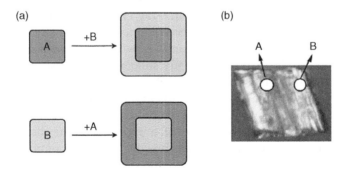

Figure 4.26 (a) Epitaxial growth occurs by one crystal growing on the surface of another, isostructural one. (b) $Co(CN)_6^{3-}$ and $Fe(CN)_6^{3-}$ salts of an organic dication can be epitaxially grown.[21] S. Ferlay and M. W. Hossieni, 'Crystalline molecular alloys', *Chem. Commun.*, 2004, 788–789. Reproduced by permission of The Royal Society of Chemistry.

cases, it is possible to grow one crystal around another to produce *crystalline molecular alloys*.[21] An example of one such system is shown in Figure 4.26, in which $Co(CN)_6^{3-}$ and $Fe(CN)_6^{3-}$ salts of an organic dication were used to form composite crystals epitaxially.

4.6 Coordination polymers

4.6.1 Introduction

James, S. L., 'Metal–organic frameworks', *Chem. Soc. Rev.*, 2003, **32**, 276–288.

Robson, R., 'A net-based approach to coordination polymers', *J. Chem. Soc., Dalton Trans.*, 2000, 3735–3744.

We saw in Chapter 3 that metals, when combined with an appropriate choice of ligand, can be used to construct complex, discrete structures. The same principles can readily be extended to give infinite structures. The use of divergent metal centres with divergent ligands (see Chapter 3, Figure 3.9(c)) should give polymeric products. A judicial selection of the components controls whether the products formed are one-, two- or three-dimensional polymers. *Coordination polymer* is the general descriptive term for an infinite array which consists of ligands bridging between metal ions, whatever their dimensionality. The term is extremely broad and encompasses a wide range of architectures from relatively simple one-dimensional chains with small bridging ligands, such as cyanide, to large mesoporous frameworks, involving metal clusters and extended organic ligands (Section 4.6.2).

> **Coordination Polymers:** Infinite arrays of metal ions connected by bridging ligands.

The transition from discrete complexes to coordination polymers can be seen as a logical progression as we change from semi-protected metal ions (see Chapter 3, Section 3.3.1) to naked ones, and in doing so increase their potential for coordination. For example, Figure 4.27 shows the progression from the Pd^{2+} square that we saw in Chapter 3, through a two-dimensional network made by utilising the axially protected octahedral Co^{2+} ion with 4, 4′-bipyridine and finally to a three-dimensional network using naked octahedral Ag^+ ions with the bridging ligand pyrazine.[††] In each step, the number of coordination sites linking to bridging ligands is increased (2, 4 and 6) but the ligand in each case is rigid and ditopic.

[††] The octahedral geometry is rarely observed for silver.

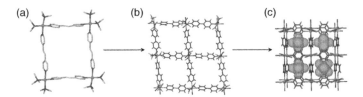

Figure 4.27 The progression from (a) discrete molecular complexes ('zero-dimensional') through (b) two-dimensional polymers using an axially blocked octahedral metal to (c) infinite three-dimensional structures containing guests.

The topologies of coordination networks, *i.e.* their connectivity (see Chapter 3, Section 3.4.1), are defined by the manner in which the individual species are joined. Coordination polymers may be deconstructed to a series of points, or nodes, that represent the molecular or ionic components. These nodes are described in terms of their connectivity to adjacent nodes. For example, the cobalt ions in Figure 4.27(b) are connected through four bridging bipyridine ligands to other cobalt atoms and are therefore said to be '4-connecting nodes'. The Ag(I) ions in Figure 4.27(c) are connecting through bridging ligands to six other Ag(I) ions and are therefore '6-connecting nodes'. Ditopic ligands, such as those in Figure 4.27, are not classified as nodes themselves as they do not affect the topology of the structure but merely serve to act as spacers between nodes.[§] Ligands that have a greater number of interaction sites can act as nodes, for example, tri(4-pyridyl)triazine (**4.9**) acts as a 3-connecting node (trigonal planar) and tetra(4-cyanophenyl)methane (**4.10**) as a 4-connecting node (tetragonal). If only one type of node is present (in terms of connectivity), then the net is uninodal; if two are present then it is binodal.

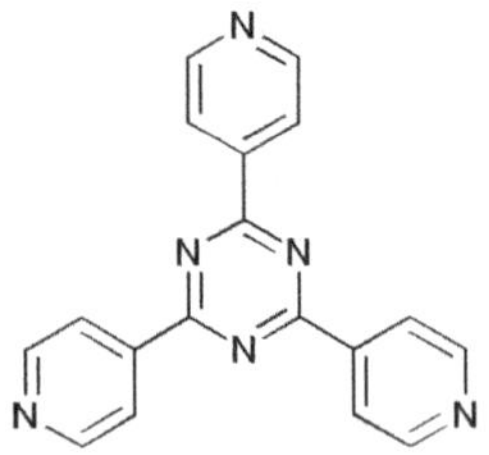

4.9 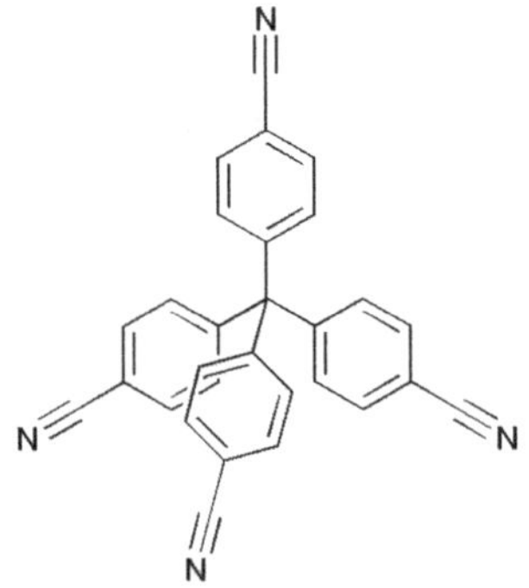   **4.10**

[§] The same principle applies for linearly 2-coordinate metals.

In addition to describing the connectivity of the nodes, it is necessary to describe the shape of the network that is formed as nodes of any given connectivity are able to form different types of network topology.[22] In order to determine the geometry of the network, we must count the smallest loop in the network to travel from one selected node back to itself. Figure 4.28(a) shows a simplistic representation of the bipyridine-based structure seen in Figure 4.27(b) with the spheres representing the 4-connecting Co^{2+} ions. If you select any sphere and find the shortest circuitous route back to it, then it is always along four edges. This common network topology, a square grid, is called a '(4,4)-sheet', using the Wells notation.[23] The nomenclature (n, p)-sheet is derived from $n =$ shortest route and $p =$ connectivity of each node. Another common two-dimensional topology is a hexagonal sheet, or (6,3)-net as it contains hexagonal holes ($n = 6$) and three connecting nodes ($p = 3$) (Figure 4.28(b)). It is important that the numbers are written in the correct order when using the Wells nomenclature, for example, a (6,3)-net is very different to a (3,6)-net. The network description of a coordination polymer is based solely upon topology and not geometry, *i.e.* two networks that may look quite different can have the same connectivity of the nodes. Figures 4.28(b) and 4.28(c) show 6,3-nets constructed using trigonal planar and T-shaped nodes, respectively, which have different geometries but the same topology. An alternate notation to the Wells nomenclature is that of Schläfli symbols. These symbols are calculated by working out the shortest circuitous routes from each possible pair of links of a node. For example, using the 4,4-net in Figure 4.28(a), four routes can be found that involve *cis*-linkages from any given node and are four linkages long. However, if the route involves *trans*-linkages, then the shortest route is six linkages long and there are two of them. The Schläfli notation is $4^4.6^2$ (where n^m has $n =$ number of links and $m =$ number of routes with n links). The net approach can also be applied to hydrogen bonding systems, for example, trimesic acid forms a clathrate with a (6,3)-net when crystallised in the presence of large guest species (see Figure 4.8(b)).

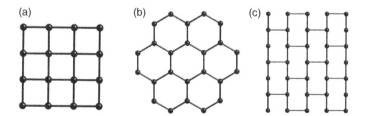

Figure 4.28 Common two-dimensional sheet networks: (a) a (4,4)-net (or $4^4.6^2$ in Schläfli notation); (b and c) (6,3)-nets (Schläfli, 6^3) using differently shaped nodes.

Three-dimensional networks are named in similar manners, although it is often harder to locate the rings in the structure. As there is a larger number of possible rings in three-dimensional structures, they are often described by simple compounds that have the same topology, for example, diamond or $ThSi_2$. Figure 4.29 shows some of the more common three-dimensional network topologies and their commonly used names.

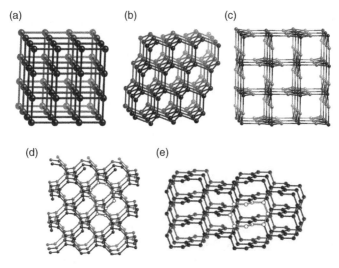

Figure 4.29 Common three-dimensional network topologies: (a) α-Po (or NaCl), uninodal, with 6-connecting octahedral nodes; (b) diamond lattice with tetrahedral nodes; (c) rutile, binodal with 3- and 6-connecting nodes. The networks (d) 10,3a and (e) 10,3b ($ThSi_2$) have the same Wells notation and Schläfli symbol (10^3) (the shortest route is shown with white nodes) but a different topology.

Transition metals are most frequently used as nodes between the organic ligands due to their predictable geometries and ligands can be utilised that link these precise coordination spheres in a well-defined way (see chapter 3, Figure 3.10). Metals such as Ag(I) that adopt linear coordination geometries are used primarily to form one-dimensional polymers with linear ligands; however, non-linear ligands can produce Ag(I)-containing structures of higher dimensionality. Similarly, square-planar metal ions generally form two-dimensional sheets when combined with two connecting ligands. The nodes most commonly used to form three-dimensional networks are tetrahedral or octahedral metals which have the potential to bind to ligands in three dimensions. Less predictable coordination geometries, such as those adopted by lanthanoid ions which have coordination numbers typically ranging from 7 to 11, can lead to more unpredictable structures and are therefore less common in the literature.

The size and shape of the spacer, or linker, that is used between the metal nodes plays an equally important role as that of the metal. Short spacers will ultimately lead to small (or non-existent) cavities when used alone; however, if used in combination with larger spacers they may be used to fine-tune structures. Examples of small ligands include halides, hydroxyl bridges and cyanide. Common medium-sized spacers include bipyridine, aromatic carboxylates and planar polynitrile ligands, such as dicyanamide, $N(CN)_2^-$, and tricyanomethanide, $C(CN)_3^-$. Longer ligands are frequently custom-synthesised and are often terminated by pyridyl, carboxylate or nitrile groups. The shape of the spacer will ultimately affect the geometry of the polymer. Linear ligands, such as those seen in Figure 4.27, result in cubic or rectangular structures when combined with metals that coordinate with 90° angles. Ligands with more complex geometries, such as trigonal ligands, will generate more complex networks; however, the geometry of the ligand is always reflected in the geometry of the product.

Another important aspect in the construction of coordination polymers is the balance of charge throughout the structure. The metal ions that are used to hold the ligands together carry a positive charge which must be countered, either by the use of anionic ligands or by non-coordinating anions that generally occupy the voids in the crystalline lattice. Neutral ligands will result in positively charge coordination networks which can be charge-balanced by non-coordinating anions, such as PF_6^-, BF_4^- and $CF_3SO_3^-$. Different shapes and sizes of anions can have an effect on the final structure that is obtained by templating channels within the network. Anionic framework ligands can produce charge-neutral networks. In some cases, it is possible to create anionic networks through the use of anionic ligands, necessitating the use of non-coordinating counter-cations in the structure. As with the use of counter-anions, the size and shape of the cation, ranging from small and spherical (Na^+) to large and structurally complex (*e.g.* tetraalkyl ammonium), can play a significant templating role in the network.

One aspect of coordination polymers that is of particular interest to supramolecular chemists is the mimicking of the structure and host–guest behaviour of natural, porous systems such as zeolites. In particular, three-dimensional coordination networks can form large, robust, mesoporous networks that closely resemble zeolites in their structure and activity, known as *metal-organic frameworks* (Section 4.6.2). Such networks are able to incorporate a large variety of guests, from simple solvent molecules to H_2 (Section 4.6.3). Coordination polymers frequently display interesting physical properties, such as nonlinear optical behaviour, long-range magnetic ordering or electrical conduction (Section 4.6.4). The deliberate attempt to synthesise structures that contain large voids can also lead to the interesting phenomenon of *interpenetration*, when one polymeric network becomes topologically entangled with one or more others, leading to fascinating solid state host–guest structures in their own right (Section 4.6.5).

4.6.2   Metal–organic frameworks

 Rowsell, J. L. C. and Yahgi, O. M., 'Metal–organic frameworks: a new class of porous materials', *Micropor. Mesopor. Mater.*, 2004, **73**, 3–14.

While zeolites represent robust, natural mesoporous solids, it is also possible to construct entirely artificial complexes that display some remarkable catalytic, guest-exchange and absorption properties. Synthetic porous networks that closely resemble zeolites in terms of their properties are termed *metal–organic frameworks* (MOFs). Other phrases, such as 'organic zeolite analogues' and 'organic–inorganic materials', are also occasionally encountered in the literature and used to describe the same types of system. The distinction between coordination polymers and MOFs can be somewhat blurred. 'Coordination polymer' is an all-encompassing term for polymeric species containing both metals and organic ligands, of which MOFs are a sub-division. MOFs are specifically three-dimensional, highly crystalline solids that are robust and usually porous. The organic ligands within MOFs can generally be synthetically modified prior to the preparation of the MOF, therefore discounting simple ligands such as cyanide and oxalate.

The design and synthesis of metal–organic frameworks is most closely related to the synthesis of discrete 'metal architectures', such as those we saw in Chapter 3, although some aspects resemble methods employed in zeolite and clathrate formation, with the use of templating species to fill the voids in the structure. While zeolites are size-restricted due to their constituent parts, artificial systems have fewer limitations. Ligands of varying shapes and sizes can be synthesised and used to connect between metal nodes to form large voids within crystalline lattices. Even though they generally lack zeolite-like thermal robustness, MOFs have other advantages over zeolites, *e.g.* they are more designable due to the use of carefully selected components and their properties are more tuneable with the ability to incorporate a wide variety of functional groups into the porous channels. In fact, the use of building blocks to build up functional materials in this manner can be readily liked to the bottom-up approach utilised in nanochemistry (Chapter 5). The structures of MOFs necessitate that a scaffolding approach be taken, similar to the manner in which capsules can be constructed (see Chapter 3, Section 3.5.2), with the ligands used forming the edges of channels, rather than the sides. By employing this method of synthesis, voids are left in the walls of the channels through which guests may pass, rather than forming structures with isolated pockets of guests (although many such structures are known).

> **Metal–Organic Frameworks (MOFs):** Porous three-dimensional coordination polymers, often with metal-cluster nodes.

Networks containing large pores are readily built by using one of two main approaches, or a combination of the two, *i.e.* the use of larger ligands or the use of larger nodes. Frequently, these structures are extensions of well-known lattice geometries. For example, the three-dimensional network formed when combining Cu(I) with 4, 4′-bipyridine, a linear ligand, closely represents a diamondoid network (*cf.* Figure 4.29(b)) in which the Cu(I) ions are the tetra-hedral nodes. Because extended ligands, rather than C–C bonds, are involved we refer to this structure as an expanded *diamondoid* network (Figure 4.30(a)). The expanded structure has the potential to leave voids into which guests could be included; however, in this case the voids are filled by neighbouring interpenetrating Cu/bipyridine networks. The geometry of the network in the Cu(I)/bipyridine case is solely a function of the metal geometry (as the ligand is merely a linear spacer); however, the use of a tetrahedral ligand, tetra(4-cyanophenyl)methane (**4.10**), leads to a similar network in which the tetrahedral vertices alternate between Cu(I) and the centre of the ligand throughout the structure (Figure 4.30(b)).

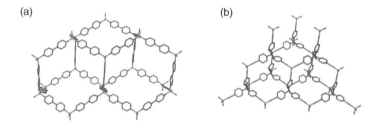

Figure 4.30 Expanded diamondoid networks in which the tetrahedral nodes are defined by (a) the Cu(I) metal alone and (b) both the Cu(I) ions and the tetrahedrally disposed ligand.

MOFs with large pores can be synthesised by what is known as *reticular synthesis* (*reticular* means something that takes the form of a periodically repeating net).[24] A combination of nodes and linkers can be said to *reticulate* when they form a three-dimensional network. This concept is an extension to the ideas first used in zeolite construction, in that pre-designed building blocks with defined geometries are used within the complexes (*cf.* Figure 4.2). For example, small metal clusters can take the place of simple metal ions. When these cluster units are used in MOF assembly they are termed *secondary building units* (SBUs). These units represent defined structural motifs, such as octahedra, tetrahedra and cuboids, and can be combined to make designed networks. Although the design of such frameworks is targeted, in that selected chemical units are used, the synthesis itself does occur *via* a self-assembly and crystallisation pathway with the final product resulting from both thermodynamic (self-assembly) considerations and the kinetic crystal nucleation and growth factors under the prevailing conditions. This process is loosely analogous to zeolite synthesis, whereby the conditions can be tuned to determine the structural elements that are present in the final structure.

In recent years, the synthesis of robust MOFs has been focused increasingly around the use of anionic carboxylate ligands rather than neutral, nitrogen-donors. An increased stability is observed due to the chelating of the ligands in combination with their negative charge. A number of reliable and repeatably observed SBUs are emerging, giving rise to formally cationic nodes with geometries ranging from square-planar to trigonal-prismatic with the geometries defined by the directions of the carboxylate 'R' groups with respect to each other (Figure 4.31). These SBUs are generated *in situ* during the formation of the MOF, often under hydrothermal conditions (heating in water under pressure), with different conditions favouring the formation of different SBUs. The challenge with many of these syntheses, however, is to ensure stability of the organic component under the prevailing conditions.

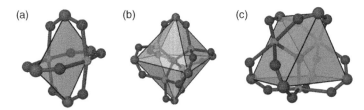

Figure 4.31 Secondary building units (SBUs) using carboxylates with rigid coordination geometries that replace metal ions as vertices in MOFs: (a) paddlewheel; (b) the octahedral basic zinc acetate SBU; (c) a trigonal prismatic oxo-centred trimer. Polyhedra use carboxylate carbon atoms as their vertices. The metal atoms are bound to only terminal ligands in addition to those shown.

An example of a framework formed in this manner is MOF-5 (Figure 4.32).[25] The framework consists of carboxylate octahedra (Figure 4.31(b)) that occupy the vertices of an infinite cubic lattice. The positively charged Zn_4 clusters are joined by benzene-1,4-dicarboxylate to form a structure that displays interesting gas-inclusion phenomena (Section 4.6.3). The structure can be further expanded by the use of larger, linear dicarboxylates ligands to yield *isoreticular structures* (structures with nets of equivalent topology). Frameworks that are constructed using metal carboxylates in this manner usually display remarkable stability, even at high temperatures ($\sim 300\,^\circ\text{C}$) and in the absence of guests. In contrast, traditional clathrates, such as urea, generally collapse or rearrange upon guest removal.

Metal–organic frameworks are largely used for their host–guest properties, such as gas adsorption, anion exchange and chemical separation, much like zeolites. The following section will deal with some of these aspects individually.

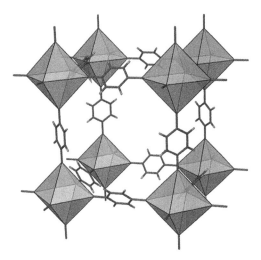

Figure 4.32 MOF-5 constructed using benzene-1,4-dicarboxylate with octahedral Zn–carboxylate clusters acting as the nodal points.

4.6.3 Guest properties of MOFs

 Kitagawa, S., Kitaura, R. and Noro, S. -I., 'Functional porous coordination polymers', *Angew. Chem., Int. Ed. Engl.*, 2004, **43**, 2334–2375.

Before discussing the uses of porous networks, it is worth defining exactly what is meant by porosity with regards to metal–organic frameworks. In order for a network to be truly porous, the lattice solvent or gas must be removable from the solid, not merely disordered within channels. The term 'open-framework' is sometimes applied, suggestive of the fact that the guests are free to enter and leave the structure. Permanent porosity refers to structures that are able to persist, with no degradation of the framework, despite the channels being completely emptied of guest species. Porosity and stability are determined by using combinations of thermogravimetric analysis, to detect solvent loss and stability as a function of temperature, and X-ray powder diffraction. The latter may be used to compare a microcrystalline solid that has been 'emptied' with the calculated powder pattern determined from the single-crystal structure of the MOF, to determine if the lattice has been compromised by removing the guests. Recently, several structures have been reported which display 'porosity without pores', in which diffusion occurs between apparently unconnected voids within crystalline lattices, although the mechanisms behind this process are less well understood.[26]

An example of a three-dimensional coordination polymer that displays guest exchange is shown in Figure 4.33.[27] The structure consists of trigonal pyramidal Ag(ɪ) nodes with 3-connecting 1,3,5-tris-(4-ethynylbenzonitrile)benzene ligands

(**4.11**), charge-balanced by trifluoromethanesulfonate ligands, also attached to the metal centres. Six interwoven networks contribute to the overall framework of the crystal, leaving large ($15 \times 22\,\text{Å}$) roughly hexagonal channels filled with disordered benzene. The enclathrated benzene exists in approximately a 2 or 3 to 1 ratio with the ligand. ^{1}H NMR spectroscopy and X-ray powder diffraction show that toluene, xylene and undecane can all replace the original benzene guests while strill retaining the integrity of the framework, with only a slight deviation from the unit cell of the original single crystal being observed. Alcohol guests, however, alter the structure, presumably through interactions with the host lattice, although it appears that the change is a subtle one and the network connectivity remains intact. The same change is observed when all guests are removed by heating, again leaving the structure intact. The empty framework absorbs guests in the vapour phase, with preference for functionalised aromatics, a selectivity presumed to be influenced by the predominantly aromatic nature of the interior of the channels.

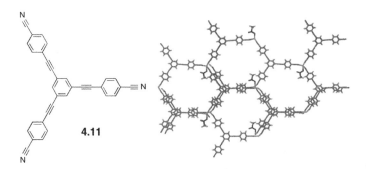

Figure 4.33 Hexagonal channels within an Ag(I)-based three-dimensional network are able to reversibly hold aromatic guest molecules (for clarity, disordered guest molecules have been omitted).

While in most cases it is desirable for the guest lattice to remain largely unchanged, by the presence of different guest molecules sometimes lattice systems are designed such that change may occur. These cases are called *dynamic frameworks*. There is significant interest in frameworks that are able to undergo reversible crystal-to-crystal transformation upon a change, or removal, of a guest. This means that as the guest is changed the structure of the lattice also undergoes a change in order to accommodate the new guest. While the coordinative bonds within MOFs are usually strong, they are often supported by weaker interactions, such as hydrogen bonds and π-interactions, that allow for a small degree of flexibility within some structures. The use of flexible ligands also allows for a small degree of freedom within some structures to accommodate guests of varying sizes and shapes. Frameworks are said to 'stretch' if a distortion occurs predominantly in one direction (often observed in cases where Jahn–Teller distortion of the metal ion, particularly Cu(II), plays a role) and 'breathe' if the expansion/contraction occurs in multiple directions. An example of a crystalline system that undergoes a crystal-to-crystal transformation is shown in Figure 4.34, containing $[Co(H_2O)_6]^{2+}$ cations and tetra(carboxyl)tetrathiafulvalene anions.[28] The initial crystalline sample, grown from aqueous media, contains disordered water within two-dimensional channels in the structure. Upon heating at 50 °C, this water is removed (observable by thermogravimetric analysis (TGA) measurements) with the sample retaining its crystalline nature. The structure of the dehydrated sample shows a significant difference in the shape of the channels. The dehydrated sample can be converted back to the original structure by rehydration at ambient temperature. Note that this material is not strictly a MOF since the components are linked by hydrogen bonding rather than metal–ligand coordination interactions.

(a)   (b)

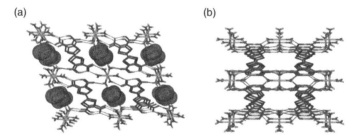

Figure 4.34 Crystal-to-crystal transformation dependent on the presence of water. The structure (a) containing enclathrated water is of lower symmetry than (b) that without it.

Possibly the most active area of research with regards to MOFs is their capacity to reversibly store small molecules, such as N_2, H_2, CH_4 and CO_2.[29] The potential exists for reversible gas storage on a commercial scale and the use of MOFs as sieves to separate small molecules, much like zeolites. Of particular importance recently have been the areas of hydrogen and methane storage, with the goal of finding a way to store and transport environmentally friendly fuels, such as H_2, under ambient conditions. Adsorption of gases can occur in one of two ways, *i.e.* chemisorption and physisorption, with the latter being the predominant method in MOF structures. Chemisorption requires that the guest molecules are attached in some chemical manner to the framework, most often by weak coordination to the metal centre or by catalytic H_2 bond-breaking to give 'spillover'.[30] Physisorption, however, is a totally physical process, akin to capillary action.

An example of a class of framework that is able to hold a large amount of methane are carboxylate-based MOFs of the formula $[Cu(O_2C-L-CO_2)(DABCO)_{0.5}]_n$ (DABCO=diazabicyclooctane, $L=C_6H_4$, $CH=CH$, $C_6H_4-C_6H_4$ and $C_6H_4-CH=CH$). In the case where $L = C_6H_4-C_6H_4$, the framework is able to hold $212\,cm^3$ (STP) g^{-1} of methane at 298 K and 35 atm. It is worth noting at this point that the majority of research in the field of gas adsorption still necessitates that pressures above atmospheric pressure be used to retain the guests inside the crystalline lattice. The realistic goal for storage is at lower pressures, rather than atmospheric pressure.

Framework MOF-5 has been found to store a number of different gases, including H_2. At 78 K and 20 atm, the framework holds around 17 molecules of hydrogen per formula unit and retains a quarter of this value at room temperature. It has been determined experimentally, using very low-temperature single-crystal X-ray diffraction (30 K), that in the case of N_2 and Ar the guests reside close to the vertices of the structure.[31]

4.6.4 Other properties of coordination polymers

Guest absorption, although of particular current interest, is not the only potential use of coordination polymers. Ordered solids are able to display interesting and useful bulk physical properties such as magnetism, conductivity and non-linear optical behaviour. The first coordination polymer to be isolated (accidentally by a dye-maker as a blue pigment in 1704) was Prussian Blue.[32] The composition remained unknown for a long time but was eventually found to be an α-Po network with the formula $Fe(III)_4[Fe(II)(CN)_6]_3 \cdot xH_2O$ ($x = 14-16$). The cyanide ligands bridge between the iron atoms to form the cubic three-dimensional network. The complex has an intense blue colour as a result of intervalence charge transfer between the iron ions. It was used as a pigment in monochrome printing processes, from where the term 'blueprint' originates. Prussian Blue has interesting bulk properties – it is electrochromic (changes colour in response to voltage) and undergoes spin-crossover (iron atoms can be changed from low-spin to high-spin by a change in temperature or by irradiation). Changes to the Prussian Blue system, such as doping with vanadium or chromium, have resulted in the formation of room-temperature magnets.

Since the initial discovery of Prussian Blue, there have been many thousands of coordination polymers synthesised with the aim of finding significant bulk properties. The properties that are of most concern are as follows:

- *Magnetism* – Coordination polymers that are able to act as magnets, by using bridging ligands to translate magnetic moments between metal centres. Magnetic materials that are constructed in this way have several potential advantages over standard magnets, *e.g.* lower density, transparency, soluble precursors and possible secondary properties.

- *Spin-crossover* – Metals in spin-crossover materials are able to be converted between their high- and low-spin states in response to an external stimulus (heat, pressure, light) leading to potential for switching applications. In some complexes, the magnetic state is different depending on whether the sample has just been cooled or heated, *i.e.* it remembers its thermal history. A porous spin-crossover material is known in which the presence of a guest species (ethanol) affects both the magnetic properties and the colour of the network, so making it a solid state sensor.[33]

- *Non-linear optical activity* – Materials that exhibit non-linear optical (NLO) behaviour are used in second-harmonic generation (SHG) which doubles the frequency of light that is passed through the sample. In order for the network to display NLO behaviour, the structure must be non-centrosymmetric (diamondoid networks are often used) and it must contain suitable chromophores.

- *Catalysis* – Porous three-dimensional networks can be viewed as zeolite mimics and can exhibits similar properties in some cases. The enclosed environment within channels, akin to zeolites (Section 4.2) and discrete capsules (see Chapter 3, Section 3.5), can provide novel reactivity or catalytic ability and can be size- and shape-selective. Coordination polymers offer an advantage over traditional zeolites in that chiral channels are able to be constructed by using chiral ligands.

- *Negative thermal expansion* – Materials that reduce in size with heating are said to display negative thermal expansion (NTE). An example of an NTE compound is $Zn(CN)_2$ in which as the temperature rises the $Zn \cdots Zn$ distance is reduced due to vibration of the CN^- ligands around the Zn–Zn vector.

4.6.5 Interpenetrating networks

 Batten, S. R., 'Topology of interpenetration', *CrystEngComm*, 2001, **3**, 67–73.

The design of channel coordination polymers and metal–organic frameworks necessitates that the polymeric network contains large voids, suitable for the inclusion of guest species. However, during the crystallisation process it sometimes occurs that one network becomes threaded through the voids by one or more other polymeric networks. These networks are said to be *interpenetrating* networks as each is threaded through the gaps left in the others. In supramolecular terms, we can say that each network is a guest of the other. It should be stressed that two networks must pass through voids in each other to be interpenetrating, while a single chain polymer passing through a channel is not interpenetrating (it is merely considered to be an included guest). In topological terms, interpenetrating networks are closely related to rotaxanes, catenanes and Borromean rings (see Chapter 3, Section 3.4), as they cannot be untangled without breaking bonds. Interpenetrating networks are also good examples of self-assembled systems.

Examples of interpenetrating networks can be discerned by examining any imaginary geometrical distortions that would allow the components to be separated, in the same way that catenanes must be constructed from inseparable components. For example, a tangled pile of spaghetti is not interpenetrating (as theoretically, if not in practice, the components are separable), whereas a chain-link fence is interpenetrating. Interpenetrating networks can be constructed from either coordination frameworks or hydrogen bonded ones.

> **Interpenetrating Networks:** Polymeric networks that pass through voids within each other to become topologically entangled such that they are inseparable without breaking bonds.

As we saw in Section 4.6.1, polymeric components are classified as one-dimensional, two-dimensional or three-dimensional, depending upon their dimensionality, where one-dimensional polymers are simple chains, two-dimensional polymers are sheets and three-dimensional polymers exist as a continuous array in all directions. Interpenetrating networks require further classification as the manner in which the component networks pass through each other can affect whether the overall structure has the same dimensionality as the individual networks or whether it is higher. For one- and two-dimensional polymers, the relative orientations of the individual networks play a large role in determining the final structure. For example, two two-dimensional networks can interpenetrate in a co-planar fashion, also producing a two-dimensional network, or they can become linked perpendicular to each other, resulting in a three-dimensional network. The nomenclature for interpenetrating networks takes the general form $m\text{D} \rightarrow n\text{D}$, whereby $m\text{D}$ is the dimensionality of the individual networks and $n\text{D}$ is that of the resultant interpenetrating system. Furthermore, the networks are termed parallel or inclined, depending on the spatial relationship of the component networks. If the polymers propagate in parallel directions or planes then they are deemed to be *parallel*; otherwise they are said to be *inclined*. The number of polymeric nets that are interpenetrating can also be defined. Two-fold interpenetration involves two nets, three-fold has three nets that are penetrating each other, and so on.

The most topologically simple interpenetrating networks are 1D → 1D parallel networks (Figure 4.35). Only two chains become threaded through holes within each other and therefore the resultant composite chain runs in the same direction as the individual components. One-dimensional chains may also form networks of higher dimensionality, for example, Figure 4.36 shows how a two-dimensional sheet arises from interpenetrated one-dimensional chains, using the 2-connecting ligand **4.14**, that propagate in different directions (1D→2D inclined interpenetration).

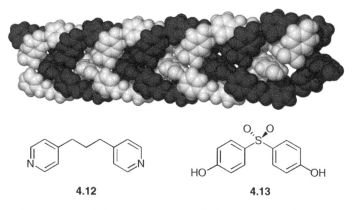

Figure 4.35 1D→1D parallel interpretation of hydrogen-bonded strands of 4, 4′-trimethylenedipyridine (**4.12**) and 4, 4′-sulfonyldiphenol (**4.13**).

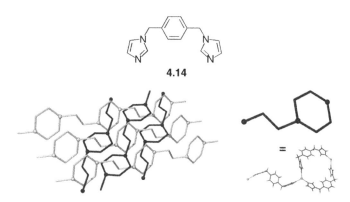

Figure 4.36 1D→2D inclined interpenetration (it is also possible for 1D→2D interpenetration to occur when one-dimensional chains are parallel).

Two-dimensional sheets may interpenetrate in one of three ways, *i.e.* 2D→2D parallel, 2D→3D parallel and 2D→3D inclined. 2D→2D parallel interpenetration can only occur when the mean planes of the intertwined layers are coincidental (*i.e.* the layers are not offset in the direction perpendicular to their propagation). A recent example of 2D→2D parallel interpenetration is seen in Figure 4.37, with the individual sheets having a (6,3)-topology. Three sheets are arranged in a Borromean fashion (see Chapter 3, Section 3.4.5) with silver atoms lying directly over each other in an argentophilic interaction, separated by only 3.06 Å.

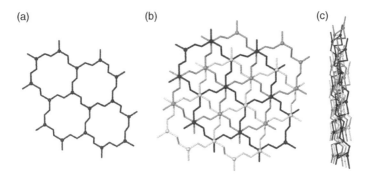

Figure 4.37 (a) A (6,3)-sheet and (b) the 2D→2D parallel Borromean interpenetrating network that it forms (c) with the mean planes of the two-dimensional sheets coincidental. The Ag atoms (spheres) lie above each other, held by argentophilic interactions.

The largest number of examples of interpenetrating networks is known for systems composed of three-dimensional nets. In these cases, there is no need for the terms 'parallel' or 'inclined', or to specify that the resultant structure is three-dimensional, as the individual polymers propagate in all directions. The wide variety of different nets that can be adopted within three-dimensional structures gives rise to many different and varied interpenetrating systems containing either identical or mixed nets. A simple example of a three-dimensional interpenetrating network, containing only two interpenetrating nets, is that of the $Co(dca)_2(pyrazine)$ system (dca $=$ dicyanamide, $N(CN)_2^-$). The individual lattices form repeating face-sharing cubes. The centre point of a cube from one lattice coincides with a vertex from the other (Figure 4.38(a)). Three-dimensional networks are also known that are self-penetrating, where the coordination network passes through gaps within itself. Figure 4.38(b) shows a partial structure of $Co(dca)(tcm)$ (tcm $=$ tricyanomethane, $C(CN)_3^-$) in which one of the tcm ligands can be seen to pass through a hexagonal void in the structure.

Although all of the examples shown above contain only one type of individual polymer in each, it is possible for interpenetrating networks to form using different nets, although this does not occur as frequently. The nets also do not have to have the same topology – it is possible for a one-dimensional net and a two-dimensional net to become interpenetrated, resulting in a three-dimensional

(a) (b)

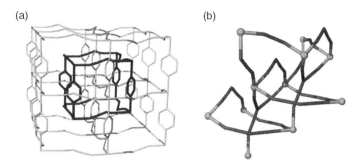

Figure 4.38 (a) A two-fold three-dimensional interpenetrating network of Co(dca)$_2$(pyrazine) and (b) part of a self-penetrating three-dimensional network Co(dca)(tcm).

network. In keeping with the nomenclature, this situation would be termed a 1D/2D→3D network.

References

1. www.iza-online.org.
2. Taguchi, A. and Schuth, F., 'Ordered mesoporous materials in catalysis', *Micropor. Mesopor. Mater.*, 2005, **77**, 1–45.
3. Brown, M. E. and Hollingsworth, M. D., 'Stress-induced domain reorientation in urea inclusion compounds', *Nature*, 1995, **376**, 323–327.
4. Lloyd, G. O., Brendenkamp, M. W. and Barbour, L. J., 'Enclathration of morpholinium cations by Dianin's compound: salt formations by partial host-to-guest proton transfer', *Chem. Commun.*, 2005, 4053–4055.
5. Kelland, M. A., 'History of the development of low dosage hydrate inhibitors', *Energ. Fuels*, 2006, **20**, 825–847.
6. Tiekink, E. R. T. and Vittal, J. J. (Eds), *Frontiers in Crystal Engineering*, John Wiley & Sons, Ltd, Chichester, UK (2006).
7. Desiraju, G., 'C – H···O and other weak hydrogen bonds. From crystal engineering to virtual screening', *Chem. Commun.*, 2005, 2995–3001.
8. Berstein, J., Davis, R. E., Shimoni, L. and Chang, N.-L., 'Patterns in hydrogen bonding: functionality and graph set analysis in crystals', *Angew. Chem., Int. Ed. Engl.*, 1994, **33**, 143–162.
9. Etter, M. C., 'Encoding and decoding hydrogen bond patterns of organic compounds', *Acc. Chem. Res.*, 1990, **23**, 120–126.
10. Dance, I. and Scudder, M., 'The sextuple phenyl embrace, a ubiquitous concerted supramolecular motif', *J. Chem. Soc., Chem. Commun.*, 1995, 1039–1040.
11. Brammer, L., Bruton, E. A. and Sherwood, P., 'Understanding the behaviour of halogens as hydrogen bond acceptors', *Cryst. Growth Des.*, 2001, **1**, 277–290.
12. Lewis, G. R. and Orpen, A. G., 'A Metal-containing synthon for crystal engineering: synthesis of the hydrogen bond ribbon polymer [4, 4′-H$_2$bipy][MCl$_4$] (M = Pd, Pt)', *J. Chem. Soc., Chem. Commun.*, 1998, 1873–1874.

13. Rivas, J. C. M. and Brammer, L., 'Self-assembly of 1-D chains of different topologies using the hydrogen bonded inorganic supramolecular synthons N–H.Cl$_2$M or N–H.Cl$_3$M', *Inorg. Chem.*, 1998, **37**, 4756–4757.

14. Sarma, J. A. R. P., Allen, F. H., Hoy, V. J., Howard, J. A. K., Thaimattam, R., Biradha, K. and Desiraju, G. R., 'Design of an SHG-active crystal, 4-iodo-4'-nitrobiphenyl: the role of supramolecular synthons', *J. Chem. Soc., Chem. Commun.*, 1997, 101–102.

15. Zordan, F., Brammer, L. and Sherwood, P., 'Supramolecular chemsitry of halogens: complementary features of inorganic (M–X) and Organic (C–X') halogens applied to M–X.X'–C halogen bond formation', *J. Am. Chem. Soc.*, 2005, **127**, 5979–5989.

16. Laguna, A., Fernández, E. J. and López-de-Luzuriaga, J. M., 'Aurophilic Interactions', in *Encyclopedia of Supramolecular Chemistry*, Steed, J. W. and Atwood, J. L. (Eds), Marcel Dekker, New York, NY, USA, 2004, pp. 82–87.

17. Braga, D., Grepioni, F., Tedesco, E., Biradha, K. and Desiraju, G. R., 'Hydrogen bonding in organometallic crystals. 6. X–H.M Hydrogen Bonds and M.(H–X) pseudo-agostic bonds', *Organometallics*, 1997, **16**, 1846–1856.

18. MacGillivray, L. R., Papaefstathiou, G. S., Friščić, T., Varshney, D. B. and Hamilton, T. D., 'Template-controlled synthesis in the solid state', *Top. Curr. Chem.*, 2004, **248**, 201–221.

19. Braga, D., Giaffreda, S. L., Grepioni, F., Pettersen, A., Maini, L., Curzi, M. and Polito, M., 'Mechanochemical preparation of molecular and supramolecular organometallic materials and coordination networks', *J. Chem. Soc., Dalton Trans.*, 2006, 1249–1263.

20. Braga, D., Maini, L., Polito, M., Mirolo, L. and Grepioni, F., 'Assembly of hybrid organic – organometallic materials through mechanochemical acid-base reactions', *Chem. Eur. J.*, 2003, **9**, 4362–4370.

21. Ferlay, S. and Hossieni, M. W., 'Crystalline molecular alloys', *Chem. Commun.*, 2004, 788–789.

22. Öhrström, L. and Larsson, K., *Molecule-Based Materials. The Structural Network Approach*, Elsevier, Amsterdam, The Netherlands, 2005.

23. Wells, A. F., *Three-Dimensional Nets and Polyhedra*, John Wiley & Sons, Inc., New York, NY, USA, 1977.

24. Yaghi, O. M., O'Keefe, M., Ockwig, N. W., Chae, H. K., Eddaoudi, M. and Kim, J., 'Reticular synthesis and the design of new materials', *Nature*, 2003, **423**, 705–714.

25. Eddaoudi, M., Li, H. L. and Yaghi, O. M., 'Highly porous and stable metal – organic frameworks: structure design and sorption properties', *J. Am. Chem. Soc.*, 2000, **122**, 1391–1397.

26. Barbour, L. J., 'Crystal porosity and the burden of proof', *Chem. Commun.*, 2006, 1163–1168.

27. Gardner, G. B., Kiang, Y.-H., Lee, S., Asgaonkar, A. and Venkataraman, D., 'Exchange properties of the three-dimensional coordination compound 1,3,5-tris (4-ethynylbenzonitrile)benzene·AgO$_3$SCF$_3$', *J. Am. Chem. Soc.*, 1996, **118**, 6946–6953.

28. Kepert, C. J., Hesek, D., Beer, P. D. and Rosseinsky, M. J., 'Desolvation of a novel microporous hydrogen bonded framework: characterization by *in situ* single-crystal and powder X-ray diffraction', *Angew. Chem., Int. Ed. Engl.*, 1998, **37**, 3158–3160.

29. Mueller, U., Schubert, M., Teich, F., Puetter, H., Schierle-Arndt, K. and Pastré, J., 'Metal – organic frameworks – prospective industrial applications', *J. Mater. Chem.*, 2006, **16**, 626–636.

30. Li, Y. W. and Yang, R. T., 'Significantly enhanced hydrogen storage in metal – organic frameworks *via* spillover', *J. Am. Chem. Soc.*, 2006, **128**, 726–727.

31. Rowsell, J. L. C., Spencer, E. C., Eckert, J., Howard, J. A. K. and Yahgi, O. M., 'Gas adsorbtion sites in a large-pore metal – organic framework', *Science*, 2005, **309**, 1350–1354.
32. Dunbar, K. R. and Heintz, R. A., 'Chemistry of transition metal cyanide compounds: modern perspectives', in *Progress in Inorganic Chemistry*, Vol. 42, Karlin, K. D. (Ed.), John Wiley & Sons, Inc., New York, NY, USA, 1997, pp. 283–392.
33. Halder, G. J., Kepert, C. J., Moubaraki, B., Murray, K. S. and Cashion, J. D., 'Guest-dependant spin crossover in a nanoporous molecular framework material', *Science*, 2002, **298**, 1762–1765.

5

Nanochemistry

5.1 Introduction

 Ozin, G. A. and Arsenault, A., *Nanochemistry: A Chemistry Approach to Nanomaterials*, The Royal Society of Chemistry, Cambridge, UK, 2005.

This chapter introduces nanometre-scale (nanoscale) synthesis and nanometre-scale materials, functional devices, switches and sensor arrays. These chemical topics form the discipline of nanochemistry, a sub-field closely related to nanotechnology. The latter is an extensive area of intense current endeavour that impinges on physics, engineering and biotechnology, as well as chemistry. Here, we give a brief introduction to nanochemistry from the view of the supramolecular chemist interested in synthesising and studying chemical aggregates on the nanoscale with the potential for the development of nanotechnological devices.[1–3]

5.1.1 What is nanotechnology?

Nanotechnology is the science of nanotechnological devices, developed from nanomaterials–materials that have spatial dimensions ranging between about 0.1 to 500 nm. Traditionally, nanotechnological devices have been prepared by the break-down of materials using techniques developed by solid state physicists. For example, a large block of silicon wafer can be reduced to smaller components by cutting, etching and slicing down to a desired size or shape. This is known as the *top-down approach*. Alternatively, the synthesis of nanostructures and nanomaterials through the utilisation of supramolecular and biomimetic materials is known as the *bottom-up approach* and forms the foundations of *nanochemistry*. Nanomaterials synthesised from the bottom-up method have novel physico-chemical proprieties that differ from the bulk material and can engender the *emergence* of novel characteristics. Both the top-down and bottom-up approaches to

Core Concepts in Supramolecular Chemistry and Nanochemistry   Jonathan W. Steed, David R. Turner and Karl J. Wallace
© 2007 John Wiley & Sons, Ltd   ISBN: 978-0-470-85866-0 (Hardback); 978-0-470-85867-7 (Paperback)

nanotechnology have interfaces with biology and biomimetic chemistry, giving rise to the field of nanobiology (Section 5.10).[2]

5.1.2 Nanotechnology: the 'top-down' approach

In 1965, Gordon E. Moore, one of the co-founders of Intel, observed that the number of transistors per square inch on an integrated circuit had doubled since the integrated circuit was first produced. From his initial observations, Moore then went on to predict that for every 18 months the complexity of an integrated circuit (generally taken to be proportional to the number of transistors) would double. This has become known as *Moore's law*. However, it is becoming more apparent that today's technologies are starting to reach their maximum capabilities and Moore's law will eventually reach its plateau. As of 2006, chip components are manufactured on the 90 nm scale, as compared to 500 nm a decade ago. Approaches based on nanotechnology are needed to produce components on the 30 nm scale and below in order to postpone the inevitable limits of Moore's law, which has become a major industrial driving force. For example, lithography and similar techniques are limited at dimensions below 100 nm. Even though this size-scale is small, it is large in comparison to the scale of atoms and 'small' molecules, the natural domain of the chemist, that have sizes in the range of tenths of nanometres.

Microfabrication is a collective term used for various kinds of *lithography* (literally, 'stone writing'). To make things smaller, scientists generally start by coating a substrate that is hard or stiff (*i.e.* if the coating is stiff, the material has a high Young's modulus) and cutting features into the coating. The most common microfabrication technique used today is photolithography, the use of UV light as a pattern source. Light carves the pattern on a photoactive organic polymer (the photoresist) which is coated onto a silicon wafer. The pattern then defines which areas of the underlying wafer are exposed for material deposition or removal. The light changes the solubility of the resist (making it either more soluble or less soluble) which can then be selectively removed by solvent. This process is used to etch integrated circuits that are used in modern computer processors and memory chips. There are two types of images that are obtained, either a *positive image* or a *negative image* (Figure 5.1). For a positive image, the resist is exposed with UV light wherever the underlying material is to be removed. In these resists, exposure to the UV light changes the chemical structure of the resist so that it becomes more soluble in the developer solvent. A classic example of a photoresist used to produce a positive image is a Bakelite-type phenol–formaldehyde resin doped with diazonaphthoquinone (DNQ), which acts as a dissolution inhibitor. The polymer is irradiated by a UV source with a mercury arc lamp in specific regions. The irradiation of the polymer produces a photochemical reaction with the DNQ which results in a transformation of the hydrophobic DNQ into a

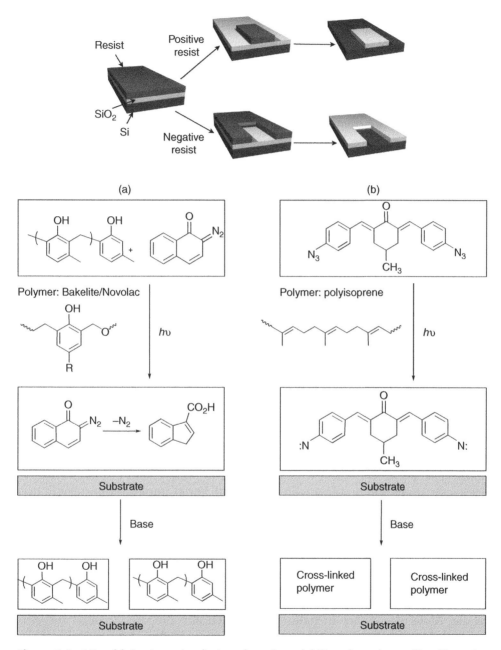

(a)

(b)

Polymer: Bakelite/Novolac

Polymer: polyisoprene

Figure 5.1 Microfabrication – irradiation alters the solubility of a polymer film. The solubility of the polymer is altered and treatment with a solvent produces a positive (a) or (b) negative image.

hydrophilic indenecarboxylic acid, *via* a Wolff rearrangement (Figure 5.1(a)). The irradiated product is then soluble in aqueous tetramethylammonium hydroxide, leaving windows of the bare underlying material. The mask, therefore, contains an exact copy of the pattern, which remains on the wafer. The disadvantage of this approach is the size of the smallest features on the image that is obtained. Diazo compounds, such as DNQ, only absorb at $\sim 350\,nm$. As a result, the optimum resolution of the smallest images is *ca.* $0.5\,\mu m$. To obtain better resolution, a shorter wavelength is required, around the $200\,nm$ range–this is the next step towards optical lithography tools for higher resolution.

Negative resists behave in just the opposite manner. Exposure to the UV light causes the negative resist to become polymerised, and more difficult to dissolve. For example, polyisoprene has a photoactive cross-linking reagent doped into the matrix, *i.e.* upon irradiation the areas exposed to the light form an insoluble polymer due to extensive cross-linking. The areas that have been masked during the process are then dissolved and washed away (Figure 5.1(b). Therefore, the negative resist remains on the surface wherever it is exposed, and the developer solution removes only the unexposed portions. Hence, masks used for negative photoresists contain the inverse (or photographic negative) of the pattern to be transferred.

5.1.3   Nanochemistry: the 'bottom-up' approach

The bottom-up approach consists of the building up of nanomaterials from smaller building blocks. This approach has given rise to nanochemistry, typically with the use of self-assembly methods (see Chapter 3) for the formation of highly ordered two- and three-dimensional nanoscale structures.

The idea of building nanomaterials atom-by-atom was first popularised by Drexler in the mid 1980s, who envisioned the construction of 'nanorobots'. These robots, nicknamed *assemblers*, were postulated to build copies of themselves by picking up atoms and building each component. For example, a nanorobot in the shape of a miniature submarine could navigate its way through the blood stream and be capable of repairing tissue or even killing cancer cells. This concept is futuristic and is in reality very unlikely, but there is still a great deal of more realistic nanotechnology of interest.

Molecules linked by supramolecular interactions are good potential building blocks for nanoscale architectures. Molecules are more convenient building blocks than atoms, as free atoms are highly reactive and difficult to generate and handle. Molecules, either in nature or synthetically prepared, exhibit distinct shapes and might already have particular properties. Molecules can self-assemble or be forced to connect to make larger structures. Nanostructures constructed by self-assembly, for example carbon nanotubes (Section 5.7), supramolecular dendrimers (Section 5.8) and large clusters are not fabricated as much as they are grown and manipulated by chemical, often non-covalent, interactions, to produce functional materials. This is the basis of the bottom-up approach and is the main focus of the remaining part of this chapter.

The preparation of nanoscale components, devices or structures *via* the bottom-up approach my be regarded as a exercise in self-assembly in which sub-nanometre scale molecules spontaneously generate nanoscale aggregates according to their intrinsic molecular programming or as a result of the influence of a template, such as a molecule, or an aggregate, such as a micelle or bilayer of surfactants or other self-assembled structures. Such approaches are often convergent and predictable. More exciting and less subject to strict control are *emergent* properties that arise from the unpredictable interaction of the components of the aggregate and evolve over time. A good example of an emergent structure in the everyday world is the complex network of tunnels and vents that regulate the environment within a termite mound. The mound is not planned by the termites and does not arise from a predictable template. It emerges from the individual, synergic efforts of the individual termites over time. Some examples of emergent nanostructures are given in Section 5.10.5.

Nanotechnology: The production of functional materials with features in the region of 0.1–500 nm in size, *i.e.* nanoscale.

Bottom-Up Approach: The synthesis of nanoscale structures and devices by chemically building up from the molecular level.

Top-Down Approach: The use of microfabrication techniques that reduce bulk material into smaller components to form nanoscale features or objects.

Emergence: The appearance of complex, unpredictable structures or properties over time.

5.2 Nanomanipulation

One of the most conceptually obvious ways to carry out chemistry on the nanoscale or to make nanoscale objects is to simply move molecules or atoms around directly.[4] Such a process is termed *nanomanipulation* and in practice it is extremely difficult to achieve. This is because it is difficult to apply the necessary force on such a small scale. There are a number of modern techniques that can achieve manipulation and on the nano- and even single-molecule scale, in particular, atomic force microscopy (AFM) and scanning tunnelling microscopy (STM), or the use of optical or magnetic 'tweezers'[†] or glass microfibres. The AFM and STM approaches are perhaps the most commonly utilised techniques. An atomic force microscope consists of a cantilever containing a sharp tip, typically made from silicon nitride, that itself has dimensions on the nanometre scale. The tip is brought into close proximity to the surface to be studied. The cantilever is deflected due to the van de Waals interactions between the tip and surface features and a laser measures the deflection. Continuous scanning results in a three-dimensional map representing the topography of the sample. The AFM technique is a versatile tool and can be used at ambient temperatures and in a liquid environment. This is a very attractive property as it means that biological samples can be investigated. For example, an AFM tip has been used to move even large flexible objects such as DNA around a mica surface to give grid and wave patterns and even write the letters 'DNA' with single DNA strands (Figure 5.2).

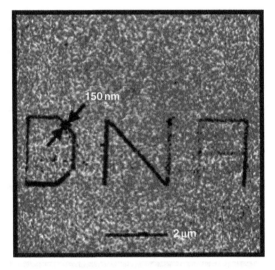

Figure 5.2   'DNA' has been spelt out on a surface with DNA strands using atomic force microscopy. Reprinted with permission from Deamers *et al.*, *Science*, **296**, 1836 © 2002 AAAS.

[†]Optical and magnetic tweezers manipulate a 'handle' in the form of a bead attached to the end of a molecule such as DNA. Optical tweezers and traps exploit the restoring force that can be exerted on a dielectric microbead by the electric-field gradients at the focus of a laser beam. In the case of magnetic tweezers, a magnetic bead is manipulated between magnetic poles.

Scanning tunnelling microscopy (STM) uses an applied voltage between the probe tip and a conducting surface. When the tip touches the surface of the specimen, a current results and is recorded as an image as the tip scans across the sample. Scanning tunnelling microscopy has been used to carry out a remarkable example of single-molecule chemistry in an analogue of the Ullman reaction.[5] The latter reaction is the copper-catalysed coupling of iodobenzene to produce biphenyl ($H_5C_6–C_6H_5$), first reported in 1904. By using a tungsten STM tip, it is possible to bring about electron-induced selective abstraction of iodine from two individual iodobenzene molecules and park the abstracted iodine atoms on an adjacent terrace on a Cu(111) surface. The STM tip is then used to tether the two phenyl groups by lateral manipulation and bring about electron-induced chemical association, resulting in the coupling to produce biphenyl. As a final proof of the new bonding the newly synthesised molecule can be pulled along with the STM tip as a single entity (Figure. 5.3).

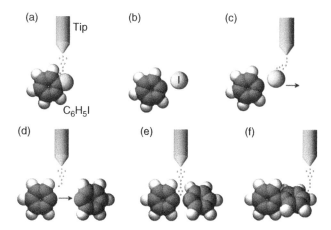

Figure 5.3 Schematic illustration of the STM tip-induced synthesis steps of a biphenyl molecule: (a, b) electron-induced selective abstraction of iodine from iodobenzene; (c) removal of the iodine atom to a terrace site Cu (111) by lateral manipulation; (d) bringing together two phenyls by lateral manipulation; (e) electron-induced chemical association of the phenyl couple to biphenyl; (f) pulling the synthesised molecule by its front end with the STM tip to confirm the association.[5]

Another, less direct, approach to manipulate molecules on a surface is through the use of chemical self-assembly principles derived from X-ray crystallographic work. A two-dimensional hydrogen bonding motif derived from crystal-engineering studies has been used to assemble a two-dimensional honeycomb lattice based on the hydrogen bonded triad motif formed between perylene tetracarboxylic acid diimide (**5.1**) and melamine (**5.2**). The honeycomb network (**5.3**) spontaneously self-assembles on a silver surface under vacuum (Scheme 5.1).[6]

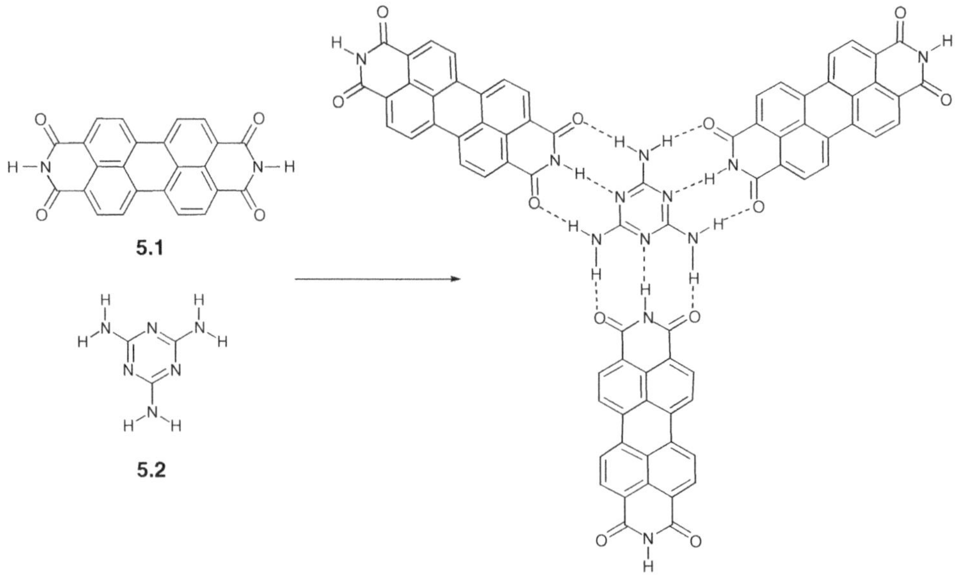

Scheme 5.1   The honeycomb self-assembled network formed between tetracarboxylic acid diimide (**5.1**) and melamine (**5.2**).

The complementary three-point hydrogen bond motif in **5.3** results in a highly symmetrical lattice. Once the lattice is formed, fullerene C_{60} can be sublimed onto the hexagonal network. The STM image shows that discrete heptamers of C_{60} assemble in the pores of the honeycomb network (Figure 5.4). The controlled location of appropriately sized molecules in the pores could prove to be useful with other guests. Potentially, these self-assembled networks could be used as nanoscale vessels for the arrangement or synthesis of a variety of nanoscale assemblies. In other nanomanipulation work, a 'V-shaped' Zn(II) bis(porphyrin) has been directly imaged by high-resolution STM which shows that it self-assembles into decaporphyrin pentagonic and dodecaporphyrin hexagonic assemblies on a gold surface. The assemblies can be fixed by alkene metathesis using Grubb's catalyst in an interesting example of self-assembly followed by covalent modification (Figure 5.5).[7] The power of STM to directly image individual molecules and nanoscale molecular assemblies in this way is likely to make it an increasingly important technique in nanocale supramolecular chemistry.

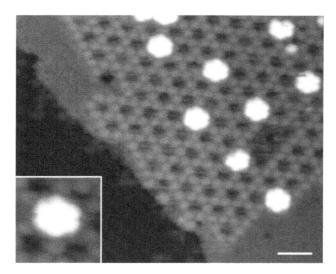

Figure 5.4 STM image of a honeycomb organic network showing fullerene heptamers residing in the cavities.[6] Reprinted with permission from Nature Publishing Group, J. A. Theobald, N. S. Oxtoby, M. A. Phillips, N. R. Champness and P. H. Beton, 'Controlling molecular deposition and layer structure with supramolecular surface assemblies', *Nature*, **424**, 1029–1031 © 2003.

5.3 Molecular devices

 Lehn, J.-M., 'Supramolecular chemistry: from molecular information towards self-organization and complex matter', *Rep. Prog. Phys.*, 2004, **67**, 249–265.

While the progress on nanomanipulation is remarkable, it is clear that making large quantities of functional multi-nanometre-scale materials is not practical on an atom-by-atom or molecule-by-molecule fashion in the immediate future. As a result, we turn our attention to directed or templated chemical methods to prepare nanoscale features and, in particular, highly complex three-dimensional nanoscale devices.[8]

One of the key goals in nanochemistry is the creation of devices that can function on the nanometre scale. The benefits that accrue with such miniaturisation include increased component density, lower costs and faster speeds, with long-term goals in molecular computing. A device can be described as an object that is invented and has a purpose. However, what is a device on the supramolecular level? Thus far, we have considered the definition of supramolecular chemistry in terms of non-covalent interactions. However, we can consider a supramolecular device to be a system made up of linked molecular components with identifiable properties that are intrinsic to each component. The interaction energy between

(a)

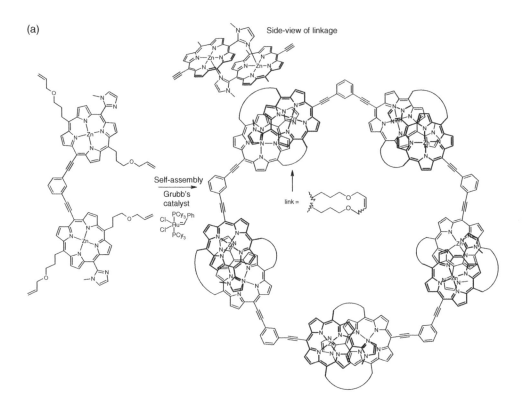

(b)

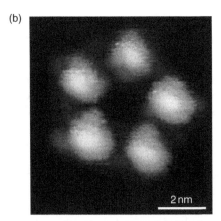

Figure 5.5 (a) Pentagonic and hexagonic assemblies based on self-assembly followed by covalent modification. (b) High-resolution STM image of one of the pentagonal structures.[7] A. Satake, H. Tanaka, F. Hajjaj, T. Kawai and Y. Kobuke, 'Single molecular observation of penta- and hexagonic assembly of bisporphyrin on a gold surface', *Chem. Commun*, 2006, 2542–2543. Reproduced by permission of The Royal Society of Chemistry.

each of the individual components is small relative to the energy of the system as a whole. This means that it is the modularity of the system which makes it supramolecular rather than the nature of the connectivity between the components, and so it is convenient to consider both covalently and non-covalently bonded molecular devices as being supramolecular.

Recent years have seen an explosion in reports of molecule-sized devices capable of switching, rectifying, performing chemical transformations or acting as molecular motors, *etc.* (some of which have been discussed in Chapter 3, Section 3.4.4) and we will highlight some of these ingenious systems in the following sections. Less well understood is the way in which such devices can be made to interact with the outside world. After all, there is no point in making a molecular bistable device (*i.e.* a molecule-sized switch) if it is impossible to read whether the switch is 'off' or 'on' (or as a molecular binary device, if it is reading '1' or '0'). Thus molecular devices are often limited by what we might loosely term their 'input–output functionality'.

In general, supramolecular devices are addressed by supplying the molecule or array of molecules with chemical energy, electrical energy or electromagnetic radiation. The most common components within a supramolecular device are those that have either photoactive moieties (groups that are able to absorb or emit light) or redox-active molecules (groups that can lose or gain electrons) within their structure. There are also biological devices and their mimics which transport ions across cell membranes, for example, whose functioning is addressed by the achievement of a *trans*-membrane ion separation.

5.3.1   Photochemical devices

The basic principle in the design of photochemical devices involves a number of sub-units, bound in a well-defined and controllable arrangement. Incident electromagnetic radiation excites a particular region of the molecule, changing the electronic interactions among the different components. This potentially gives rise to three different phenomena, namely charge-transfer (CT), energy-transfer (ET), and excimer/exciplex formation (the formation of a dimer or complex between an excited state and a ground-state chemical group) (Figure 5.6). Linking molecules together in a supramolecular array and promoting *intercomponent* processes is a particularly important concept in supramolecular chemistry and has led to the development of light-powered molecular machines, molecular switches, molecular sensors and molecular electronics.

The most common and extensively used moieties for photochemical devices are late-transition-metal centres (such as Ru(II), Os(II), Ir(III) and Re(I)) with aromatic nitrogen donor ligands coordinated to the metal, for example, 2, 2'-bipyridine (**5.4**, bpy), 1,10-phenanthroline (**5.5**, phen) and 2, 2' : 6', 2''-terpyridine (**5.6**, terpy) complexes. These types of ligand can absorb and re-emit light due to accessible π–π^* transitions. Metal-to-ligand charge-transfer (MLCT) can also occur.

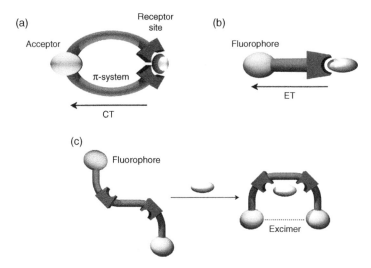

Figure 5.6 Schematic representations of the common types of molecular devices: (a) charge-transfer (CT); (b) energy-transfer; (c) excimer formation upon excitation of one fluorophore.

Complexes such as $[Ru(bpy)_3]^{2+}$ and $[Ru(phen)_3]^{2+}$ have long-lived phosphorescent states with lifetimes in the range of 10^2–10^3 ns in solution. The terpyridyl analogues are less emissive but, unlike the bpy and phen complexes, are achiral and so do not suffer from complications due to diastereoisomer formation.

5.4

5.5

5.6

5.7

5.8

5.9

5.10a, R = H
5.10b, R = F

5.11

Polynuclear metal bipyridyl-type complexes can be synthesised relatively easily by using the *complexes as metals/complexes as ligands strategy*. This approach uses metal complexes with free binding sites (*e.g.* uncomplexed N atoms on **5.9**) to bind onto metal complexes with free coordination sites or readily displaced labile ligands. Ligands **5.7** (2,5-dpp) and **5.9** (2,3-dpp) have been used to prepare multicentre and dendritic (Section 5.8) metal complexes with Ru(II), Os(II) and Re(I) centres that are terminated with capping ligands such as **5.4** and **5.8** (biq). Sophisticated arrays can be achieved by deliberately varying the metal:ligand combinations at different generations of dendrimer, thereby preparing an energy-transfer cascade from the periphery to the core of the array. For example, components that have the lowest-energy metal-to-ligand charge transfer (MLCT) absorption bands are placed at the dendrimer core and the highest-energy MLCT absorption bands are placed at the periphery of the dendrimer. This kind of planned architecture results in a controlled energy flow through the array. This process is typically known as an 'energy-transfer cascade' and many analogous systems using different metals and combinations of ligands have been employed. An example is the octanuclear complex **5.12**, based on the C,N-donor ligand **5.11**.[9] The complex contains fluorinated 2-phenylpyridine (**5.10b**) Ir(III) units on the periphery, linked *via* the bridging ligand **5.11** to similar non-fluorinated Ir(III) centres and finally to a Ru(II) tris(bpy)-derived unit. The absorption spectrum of the complex is roughly a sum of the absorption spectra of the individual components resulting in efficient light-harvesting. The emission maximum observed at 630 nm is characteristic only of the Ru(II) component, however, indicating a vectorial energy transfer from the periphery of the complex to the Ru(II) terminus. The complex is prepared by using the palladium-catalysed Suzuki reaction to give the bridging **5.11** ligand already-complexed to the metal centres in a novel variant of the complexes as metals/complexes as ligands strategy.

5.3.2 Molecular wires and rectifiers

Molecular wires are long, rod-like compounds or groups, often containing extended, conjugated π-systems, which are able to transfer electrons or energy across a significant distance, between a donor and an acceptor. In molecular wires, excitation transfer generally occurs *via* the Dexter electron-exchange mechanism that involves overlap of wavefunctions. This is a short-range excitation-transfer mechanism which operates by exchange of electrons and much of the challenge in the design of molecular wires is to extend the orbital overlap over a considerable distance to allow controlled, longer-distance electron transfer. Excitation transfer can also occur *via* the Förster mechanism, a dipole–dipole mechanism that can operate over distances as long as 10 nm. An example of a molecular wire in action is the linking of [Ru(bpy)$_3$]$^{2+}$ (donor) and [Os(bpy)$_3$]$^{2+}$ (acceptor) with

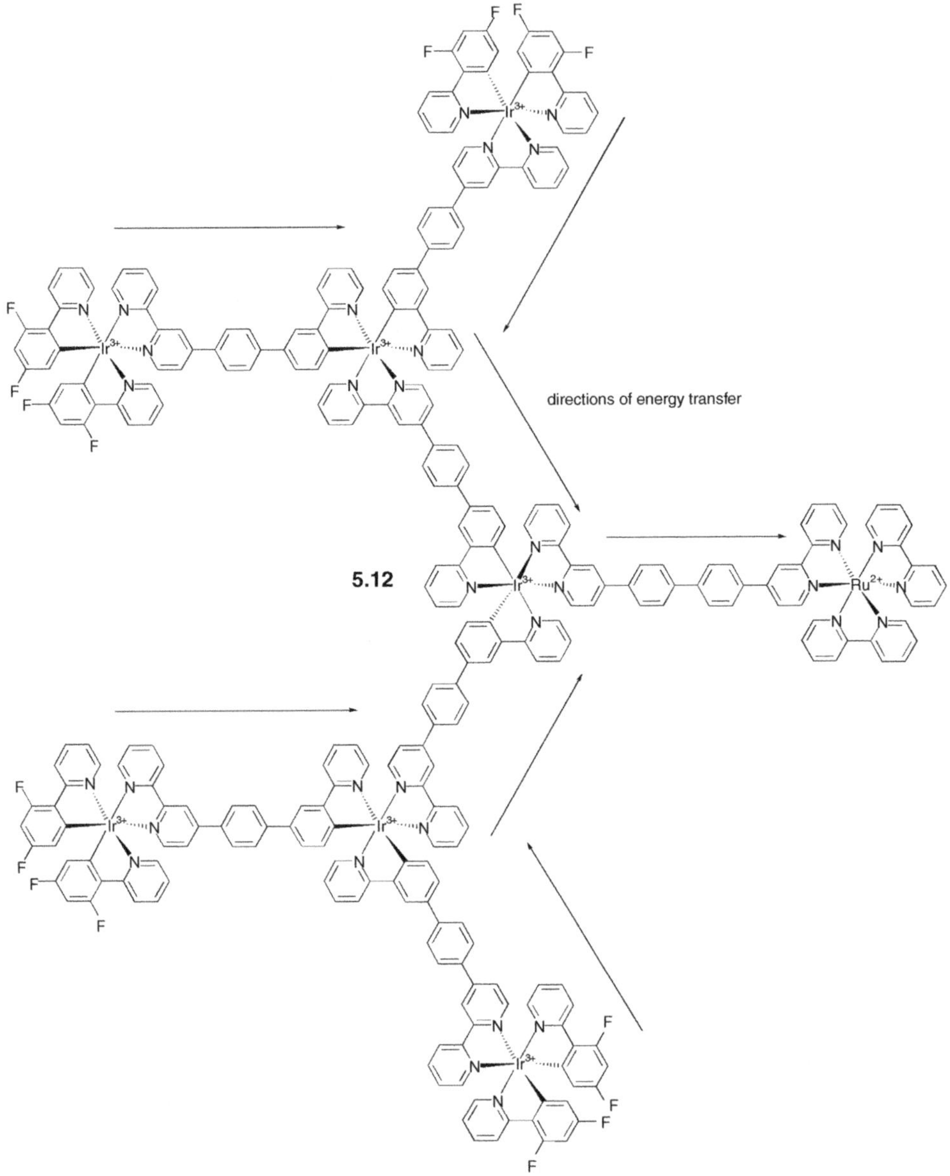

5.12

directions of energy transfer

a rigid conjugated spacer (Figure 5.7). The phosphorescence of the $[Ru(bpy)_3]^{2+}$ moiety (and of the oligophenylene spacer) is quenched by extremely rapid energy transfer to the osmium unit *via* the electron-exchange (*i.e.* Dexter) mechanism, despite the fact that the components are separated by 4.2 nm. Decreasing the

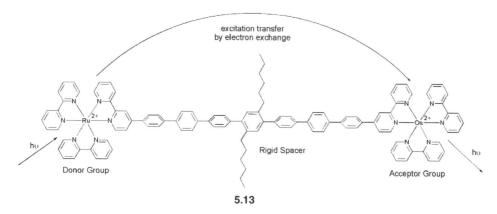

5.13

Figure 5.7 The electron-transfer between the Ru moiety donor to the Os moiety acceptor in $[Ru(bpy)_3]^{2+}$–$(ph)_7$–$[Os(bpy)_3]^{2+}$ (**5.13**).

number of oligophenylene spacer units increases the rate and efficiency of the excitation transfer process, as defined by the Dexter equation.[‡]

Molecular wires that exhibit direct-current (DC) electron conductivity have also been produced. For example, carotenoid-type compounds, termed 'caroviologens', have been used to span model membranes of dihexadecylphosphate of comparable dimensions. Conductivity enhancements of four to eight times background are observed between a internal oxidising phase containing potassium ferricyanide, $K_3[Fe(CN)_6]$, and an external reducing phase of sodium dithionite.

Unlike a wire, which allows electron flow in either direction, a device that allows the flow of electrons in only one direction is called a *rectifier*. The most common use of rectifiers is the conversion of alternating current (AC) into direct current (DC). Conventional rectifiers are made up of a contact between a *p*-type (electron-poor) and *n*-type (electron-rich) semiconductor. The rectifier operates by causing a build-up of an insulating layer in the contact region (Figure 5.8). When the *p*- and *n*-type junction is formed, the electrons flow from donor to acceptor until the charge is neutralised within the contact region and therefore no further flow is possible. However, if an external voltage is applied to the rectifier one of two things can happen, depending on the polarity of the applied potential. If the applied potential is such that the cathode is placed in contact with the *n*-type (*i.e.* electron-rich) side and the anode on the *p*-type side, the insulating layer will grow due to the additional flow of electrons until, once again, no current flow is possible. However, if the polarity is switched the insulating layer will shrink, thus allowing current to flow. If an alternating current is applied, the result is a current that flows only when the alternating cycle is of the

---

[‡]$k_{ee} = KJe^{(-2rDA/L)}$.

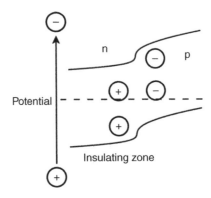

Figure 5.8 Schematic of a semiconductor p–n junction.

correct polarity–hence the conversion of AC to DC. Attempts have been made to design molecular rectifiers to mimic the semiconductior devices on a nanoscale. For example, compound **5.14** incorporates an electron donor and acceptor group separated by a carotenoid-derived molecular wire. Electron flow is in the donor to acceptor direction and *not* vice versa.

Donor group   **5.14**   Acceptor group

5.3.3 Molecular switches

In today's high-tech society, information is increasingly transported through optical networks (optical fibres and optoelectronic switches) due to the recent developments in telecommunications. Vast amounts of data can be encoded and transported over large distances. However, it is becoming apparent that the use of electrical inputs to send optical signals cannot sustain the projected amount of information that will need to be transferred in the future (of the order of terabits-per-second). Therefore, to efficiently transport information *via* optical networks, a fully optical system is needed.

Optically triggered molecular switches are a potential way of triggering optical signals. A simple photochromic process that has been extensively used in the design of molecular switches uses the dithienylcyclopentene group as the molecular switch, for example, **5.15**. This class of compound can undergo a reversible ring closure, upon irradiation with UV and visible light. In the open form, there is a high degree of conformational freedom and the thienyl moieties are able to rotate freely in solution. Due to the lack of conjugation, there is no electronic absorption in the visible region. Upon irradiation at 520 nm, ring closure takes place, the molecule becomes highly conjugated and the rotational freedom is lost, so creating a fully conjugated π-system across the bridge (Scheme 5.2). The molecule thus exhibits very different electronic properties in the 'open' and 'closed' forms.

A more complicated optically triggered molecular switch based on spiropyran (**5.16**) responds to chemical and optical inputs and produces an optical output (Scheme 5.3). Colourless spiropyran (**5.16a**) does not show any absorption at wavelengths greater than 400 nm. It switches to a purple merocyanine chromophore (**5.16b**, $\lambda_{max} = 563$ nm) upon irradiation with UV light. The merocyanine also shows an emission band at 647 nm. On treatment with acid, the merocyanine switches to a yellow–green colour and both the band at 563 nm and the emission band at 647 nm disappear to give a new absorption at 401 nm. The colourless spiropyran is regenerated when **5.16c** is irradiated with visible light, thereby completing the anticlockwise mechanism. Conversely, if acid is added to **5.16a** a yellow–green solution of **5.16c** appears again and only an absorbance band at 401 nm is seen. Upon addition of base, **5.16c** switches to the purple form **5.16b** and the clockwise mechanism is completed when **5.16b** is stored in the dark (Scheme 5.3). This system is thus a 'three-state molecular switch', whereby three input signals, (1) UV light, (2) visible light and (3) H^+, generate two output signals, (1) absorbance at 401 nm and (2) absorbance at 563 nm. This 'on–off' absorbance in response to an optical input could be used to gate optical signals within the desired time-frame for digital processing (*vide infra*).

We have already seen how rotaxanes have been used for self-assembled molecular devices (see Chapter 3, Section 3.4.4). Another interesting molecular device based on logic operations using pseudorotaxanes has been developed.[10] Logic gates are switches that use binary notation, *i.e.* the output state is a zero or

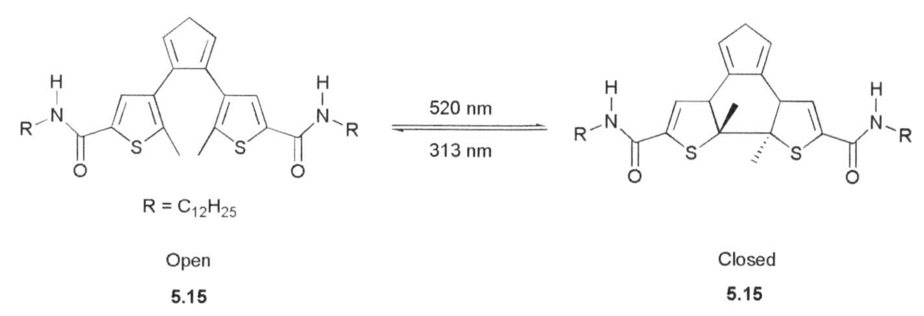

R = $C_{12}H_{25}$

Open

5.15

Closed

5.15

Scheme 5.2   Photoswitching of a dithienylcyclopentene.

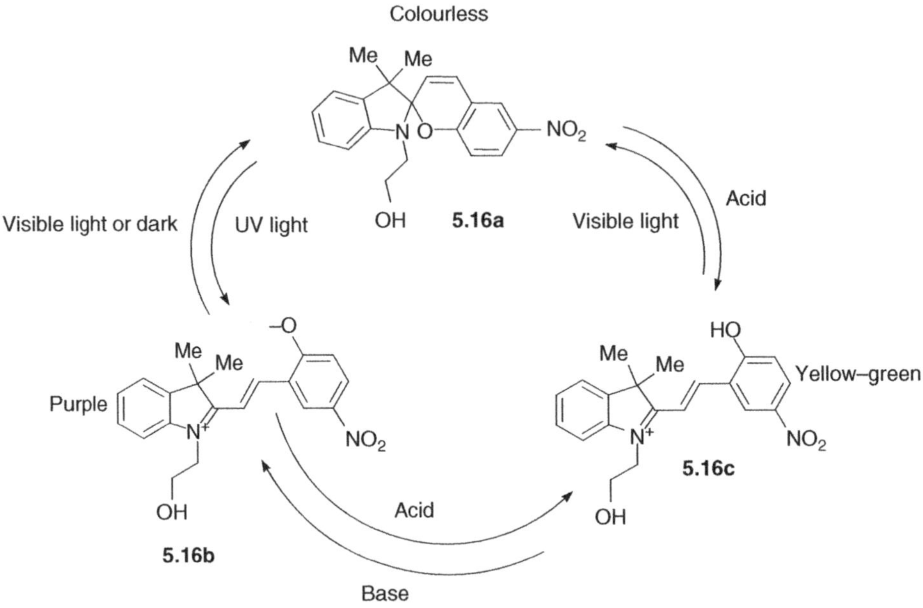

Scheme 5.3 A spiropyran optical switch turns different colours under different conditions.

one. The simplest form of logic gates are the YES and NOT gates. A YES gate passes the input bytes to the output without any change, *i.e.* input 0, output 0 and input 1, output 1. The NOT gate inverts the output, *i.e.* input 0, output 1 and input 1, output 0. However, there are more complex logic operations, such as AND, OR and XOR (eXclusive OR), which require two inputs (for the AND, gate 1 and 1 input gives a 1 output or otherwise the output is zero; an OR gate requires one or both of the two inputs to be a 1 to give a 1 output, while the XOR gate gives a 0 output for either 1 and 1 or 0 and 0 input, and a 1 output for either combination of 1 and 0 input). A nanochemical XOR gate has been prepared, based on a system using a pseudorotaxane (**5.17**). The pseudorotaxane contains a good electron acceptor (2,7-diazapyrenium, **5.18**) which interacts with the electron donating groups from an aromatic crown ether (**5.19**) and therefore the pseudorotaxane self-assembles due to the π-electron-acceptor–donor inter-action of the two groups (Scheme 5.4). This interaction is clearly observed in the UV-vis spectrum as a broad band at 400 nm assigned to the low-energy charge-transfer (CT) between the two components, and no fluorescence signal is observed. Addition of base (tributylamine) unthreads the pseudorotaxane and forms a **5.18**·(tributylamine)$_2$ complex inducing a spectral change, plus the fluo-rescence emission reappears along with a broad absorption band at around 550 nm. Upon addition of trifluoromethanesulfonic acid, the pseudorotaxane is formed again, along with a salt, tributylammonium trifluoromethanesulfonate.

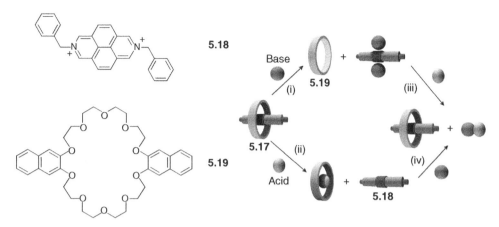

Scheme 5.4 The unthreading and threading of the pseudorotaxane, corresponding to the XOR logic function:[10] acid, CF_3SO_3H; base, $N(CH_2CH_2CH_2CH_3)_3$: (i) unthreading to give a 'locked' thread; (ii) unthreading to give a 'locked' ring; (iii and iv) unlocking and rethreading to give $HN(CH_2CH_2CH_2CH_3)_3^+CF_3SO_3^-$.

Changing the order of addition can also perform this unthreading and threading. Adding acid first protonates the crown ether and again causes the unthreading of (**5.17**).

The chemical threading and unthreading shown in Scheme 5.4 describes the input and output relationship of the XOR logic gate. For example, the strong fluorescence signal is seen only when either the amine or the acid are added, but not both (output 1 when $X = 1$ and $Y = 0$, or $X = 0$ and $Y = 1$) (see truth table, Figure 5.9).

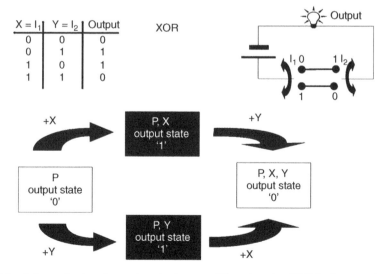

Figure 5.9 Schematic showing a chemical system (P) performing XOR logic operations under the action of two chemical inputs X and Y (acid or base in this particular example).[10]

The fluorescent signal is turned off (output 0) when both or neither the outputs are present (output 0, X = Y = 0, or X = Y = 1).

5.3.4 Molecular muscles

Muscle in the human body is responsible for movement by a complex process initiated by a potential travelling down a motor neuron. This stimulation is a mechanism that puts our limbs into automatic action to carry out a particular task. The mechanism by which an agent (*e.g.* a hand) acts upon the environment is termed an *actuator*, *i.e.* an energising process that activates a device. Mimicking biological actuating devices is an interesting goal. To achieve such a device a molecule, such as a polymer strand, is required to change shape upon stimulation. Candidate systems are *electroactive polymers* (EAPs). There are two main types of EAPs–the first group are known as *ionic* EAPs and these include ionic polymers, gels, ionomeric polymer composites, conductive polymers and carbon nano-types. All of these polymers work on the movement of free ions. The movement of these ions causes the material to bend. The disadvantage of these polymers is that they need to stay wet and are encased between two flexible coatings which require a continuous supply of electricity. If the voltage is too high, there is a danger of permanently damaging the material. The second class of EAPs are *electronic* EAPs, which include ferroelectric polymers, dielectric elastomers and electroactive graft elastomers. These polymers are stimulated by an electric field. A particular advantage of these polymers is that they can sustain high voltage and rapid movement is often observed. The usefulness of polymers containing ions originates from the strong intermolecular Coulombic interactions between the ions. There are many examples of ionic polymers, of which *ionic polymer–metal composite* (IPMC) devices are particularly interesting. The strip of IPMC bends toward the anode if it is cationic (and towards the cathode if it is anionic) when an electrical potential is applied. Therefore, by alternating the current, the strip oscillates, analogous to a muscle (Figure 5.10).[11]

A chemical example of a prototype ion-triggered molecular muscle is the extended terpyridyl ligand **5.20**.[12] In the free-state, the ligand adopts an extended conformation some 4 nm long. Upon binding Pb^{2+} ions, a 2:1 Pb:ligand complex with a coiled structure of 0.72 nm long is formed. The complex is an example of thermodynamic self-assembly (see Chapter 3, Section 3.1.2)–complexation is reversible and the Pb^{2+} ions can be removed by addition of tris(2-aminoethyl)amine (tren), regenerating the extended free ligand **5.20**. Coiling and uncoiling can be made to cycle reversibly as a function of pH. Thus, addition of triflic acid to uncoiled **5.20**/Pb^{2+}/tren induces protonation of tren, thereby freeing Pb^{2+} to bind with **5.20**, hence resulting in coiling. Alternatively, raising the pH by adding a base, such as triethylamine, deprotonates the tren, which then sequesters the Pb^{2+} ions and results in uncoiling (Figure 5.11).

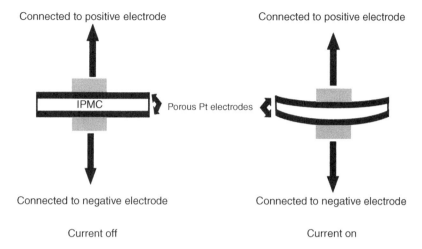

Figure 5.10 An example of a molecular muscle, *i.e.* an IPMC polymer sandwiched between porous Pt electrodes. On the application of current, the material bends and when the current is switched off the material contracts back to its original shape.

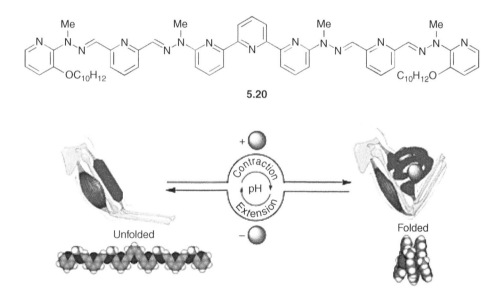

Figure 5.11 A prototype pH-triggered molecular muscle based on coiling of the ligand **5.20** by Pb^{2+}. The binding of Pb^{2+} is switched on and off according to the protonation state of the competitor ligand tren.[12] A.-M. Stadler, N. Kyritsakas and J.-M. Lehn, 'Reversible folding/unfolding of linear molecular strands into helical channel-like complexes upon proton-modulated binding and release of metal ions', *Chem. Commun*, 2004, 2024–2025. Reproduced by permission of The Royal Society of Chemistry.

5.3.5 Towards addressable nanodevices

Flood, A. H., Stoddart, J. F., Stuerman, D. W. and Heath, J. R., 'Whence molecular electronics', *Science*, 2004, **306**, 2055–2056.

A range of examples of molecular devices, such as switches, rectifiers and even transistors, some exhibiting various chemical analogues of digital logic operations or capable of carrying out mechanical operations, are now known. These systems generally operate in freely diffusing solution, however, and their outputs are communicated to the outside world by changes in bulk properties, such as fluorescence. If such systems are ever to move beyond being laboratory curiosities and fulfil their potential in molecular computing or as true nanoscale devices, they need to be addressable as single molecules as part of a robust input–output framework. To date, progress in this regard has been limited, although some operations

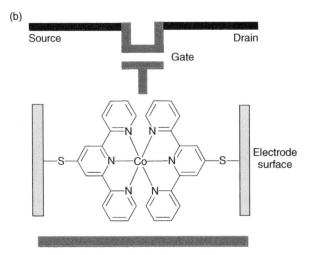

Figure 5.12 (a) A molecular rectifier connected between a thiol-coated gold tip and the surface of a gold electrode and (b) a cobalt terpyridine-based transistor.

can be performed on individual molecules (Section 5.2). This area represents a true frontier in nanoscience and there is a great deal of excitement. Crucial to progress is a fundamental understanding of the interface between electrodes and molecules and how they are linked. For example, attachment of molecules to gold electrodes *via* thiols (Section 5.4) is extremely effective, but the bond angle between the thiol and the surface has a great effect on the measured width of a molecular linker. Despite some difficulties, single-molecule electronic devices have been incorporated into more conventional electronic circuits. Figure 5.12 shows just two such simple examples, *i.e.* (a) a passive single molecule rectifier linking a thiol-coated gold tip to a gold electrode and (b) an active single-molecule transistor based on a thiol-terminated cobalt terpyridyl complex.

Another active, connected device, a molecular switch tunnel junction, has been prepared based on the switching of a two-station rotaxane. At a specific voltage, the system switches from a stable state (isomer) to a metastable one with a different conductivity. It then remains in that state until another voltage pulse returns it to its initial form (isomer).[13]

5.3.6 Addressable arrays

In recent years there has been an interest in the development of polymer-supported supramolecules for sensor arrays. Such arrays include 'electronic noses' to sense vapour analytes. There are many polymer supports that have been used to create sensor arrays, for example, conductive polymers, polymer composites, dye-doped polymer matrixes and surface acoustic wave sensors (SAWs).

The concept of identifying a specific analyte in a complex mixture of analytes, *i.e.* sensing, is not new. Historically, one of the most commonly employed techniques has exploited colloidal polymer microspheres for *latex agglutination tests* (LATs) in clinical analyses (Section 5.9.3). Commercially available LATs for more than 60 analytes are used routinely for the detection of infectious disease, illegal drugs, and even in early-stage pregnancy tests. The vast majority of these types of sensors operate on the principle of agglutination of latex particles (polymer microspheres), which occurs when the antibody-derivatised microspheres become effectively cross-linked by a foreign antigen. The process results in the microspheres becoming attached to, or unable to pass through, a filter. The dye-doped microspheres are then detected upon removal of the antigen-carrying solution. The disadvantage of this system is that LATs are difficult to array in multiple analyte detection, as the nature of the response intrinsically depends on a co-operative effect (see Chapter 1, Section 1.2.3) of the entire collection of microspheres. The idea of distinguishing one analyte from a host of others using the molecular recognition properties of a small number of molecules as part of an addressable sensing array (Section 5.3.2) is attractive but less well explored.

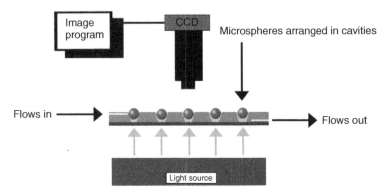

Figure 5.13 The silicon wafer array, with micromachined pyramidal wells for holding the receptor-derivatised beads. Fluid containing the experimental solution is added to the top of the array and pushed around and through the bead matrix and then out of the bottom of the pyramidal wells holding the beads.

A multianalyte chip-based sensor array system based on the very simple coloured indicator displacement assay (IDA, see Chapter 2, Section 2.4.2) has recently been developed.[14] A composite polymer microsphere (prepared from low cross-linked polystyrene and polyethylene glycol units) is functionalised with various derivatives of the receptor **5.21** (Scheme 5.5) and arranged in micromachined cavities localised on Si/SiN wafers. These cavities act as the reaction vessel and analysis chambers, analogous to the taste buds in the mammalian tongue (Figure 5.13). The analyte fluid is allowed to flow through the microreactors/analysis chambers. As the sample is introduced into the array, it flows around and over the beads and exits at the bottom of the well. Red, green and blue (RGB) transmitted light intensity values are recorded over a certain time period for each bead in the array *via* a charge-coupled device (CCD) detector. This provides a unique recognition pattern for each mixture, depending on the degree to which the analytes displace the indicator dye from the different receptors on each bead. The interesting point about the system is that each receptor bead does not have to be especially selective for an individual analyte. The pattern as a whole provides the discrimination ability.

For example, this chip-based IDA approach using receptors of type **5.21** has been used to develop an electronic array giving diagnostic patterns for adenosine 5′-triphosphate (ATP) and guanosine 5′-triphosphate (GTP). The guanidinium groups of the receptors offer a high electrostatic affinity for nucleotide triphosphate groups. Various tripeptide sequences are appended to the guanidinium sites to create a library of receptors. These scaffolds are then attached to the polymer resin bead (Scheme 5.5). The resin-bound library associates with a chromophore (fluorescein) producing an orange colour in each bead of the array. When the beads are exposed to a solution of nucleotide phosphates, the analytes displace the indicator at varying rates, resulting in a loss of colour. The RGB intensity is recorded for a library of 30 beads in the array. The uptake of analyte can be

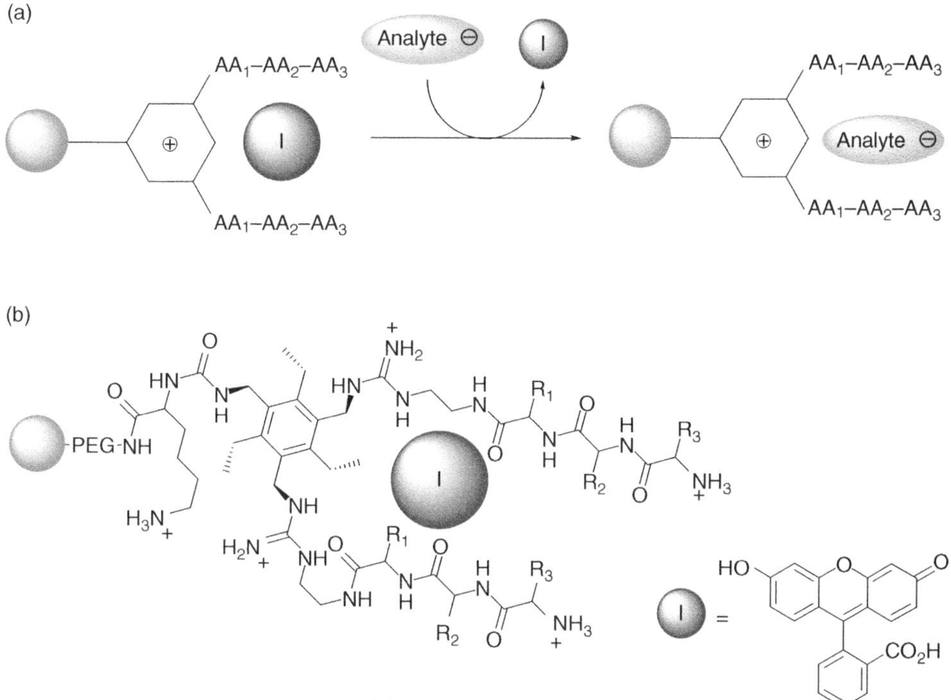

(a)

(b)

5.21

Scheme 5.5 (a) Schematic illustrating a general receptor-indicator system bound to a polymer resin and (b) an example of a resin-bound receptor and its use in IDA (AA, amino acid).

differentiated by qualitative analysis, *i.e.* visual inspection, but a more quantitative approach using pattern recognition algorithms such as Principal Component Analysis (PCA) can be employed to give higher precision.

The PCA method is a way to identify particular patterns in data and is routinely used in sensor array technology. The analysis technique highlights similarities and differences within the data. In this method, one tries to reduce the dimensionality of the data by summarising the most important (*i.e.* defining) parts while simultaneously filtering out noise. It is a powerful analytical tool for discriminating between a large numbers of very similar analytes in a complex mixture. In the example shown in Scheme 5.5, nine trials of the experiment are undertaken. To observe if there are any patterns in the data, a Principal Component (PC) axis is calculated to lie along the line of the maximum variance in the original data. Other PC axes lie along lines describing diminishing levels of variance. From this information, a 'score plot' can be calculated for the first two principal components. For this particular example, clustering of individual analytes, such as adenosine 5'-phosphate (AMP), GMP and ATP, was observed, so allowing their mutual discrimination. This 'electronic tongue' mimics the sensation of taste. The pattern created by the simultaneous response of these receptors is specific for a particular set of stimuli. For example, the mammalian tongue recognises five distinct tastes categories, sweet (carbohydrate-based), bitter (alkaloids), sour (acidity), salty (ionic) and 'umami' (savory-glutamate). Our brains experience a unique mixture of these five signals when we eat, which along with the sense of smell, gives each food its characteristic taste.

5.4 Self-assembled monolayers (SAMs)

 Ulman, A., 'Formation and structure of self-assembled monolayers', *Chem. Rev.*, 1996, **96**, 1533–1554.

Self-assembly can be used to form ordered two-dimensional monolayers by chemisorption of appropriately functionalised molecules onto substrate surfaces. Monolayers are typically formed by amphiphiles comprising long alkyl chains in conjunction with a polar head group, or long-chain molecules bearing a functional group at one end capable of surface binding. The resulting stable, ordered and dense layers have applications in anti-corrosion and wear protection, for example. Much of the early work on self-assembled monolayers (SAMs) was carried out utilising thiols on gold surfaces, and these systems continue to be widely studied (Section 5.4.2). The gold–thiolate interaction has approximately the strength of a hydrogen bond and hence the thiols have considerable surface mobility, contributing to the self-assembly process. However, there are many other systems that form SAMs, for example, siloxanes on hydroxylated surfaces and fatty acids on silver and alumina surfaces.

The most common approach to preparing SAMs is to immerse a clean substrate with a reactive surface in a solution of the coating molecules. Over time, the SAM molecule is chemisorbed to the surface, often initially in a disordered fashion but slowly self-assembling to give a more stable, ordered, close-packed monolayer. Amphiphilic SAMs are also formed at fluid interfaces (*e.g.* air–water) by a similar self-assembly process. The following sections will introduce examples of SAMs and highlight their use towards nanochemical devices.

5.4.1 Surfactants, micelles and vesicles

Surfactants, or amphiphiles, are molecules with two distinct regions that have very different solubilities. They constitute a *hydrophilic* (water-soluble) end and a *lipophilic* (organic-soluble) end that is highly *hydrophobic*. Surfactants are common in everyday life, for example, soaps (sodium dodecyl sulfate, **5.22**), shampoos (sodium lauryl sulfate) and phospholipids (phosphatidyl choline) which are the basis of biological cell membranes. Other lipids, such as fatty acids, are also part of this category. The lipophilic region is made up of long aliphatic organic

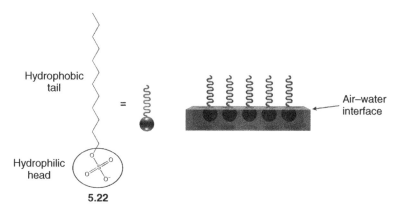

Hydrophobic tail

=

Air–water interface

Hydrophilic head

5.22

Figure 5.14 Sodium dodecyl sulfate, an amphiphile comprising a hydrophilic sulfonate head group (hydrophilic head) and the 12-carbon alkyl chain (lipophilic tail) (sodium counter-ions not shown) self-assemblies to give a monolayer at the air – water interface.

chains, such as alkanes, fluorocarbons or aromatic or other non-polar groups. The head group is made up of highly solvated hydrophilic functionalities, such as sulfonates, carboxylates, phosphonates and ammonium derivatives. It is this hydrophilic functionality that gives rise to the different classes of surfactants. Examples include anionic (sulfonate) groups which give rise to soaps and shampoos (Figure 5.14), cationic (quaternary ammonium salts), amphoteric (zwitterionic betaines) and non-ionic (fatty acids).

Hydrophobic effects influence the lipophilic portion of the amphiphile. As a consequence, there is aggregation of amphiphiles at the water–air interface in which all of the hydrophilic head groups are held together in the aqueous layer, whereas the lipophilic tails point into the air. As a consequence, a monolayer is formed, *i.e.* a layer of amphiphiles one molecule thick that line up across the water–air interface (Section 5.4.2). The degree of order of the monolayer is strongly dependent on the amount of coverage and the surface pressure. Highly ordered monolayers are generally prepared in a Langmuir trough. A sub-monolayer quantity of the water-immiscible amphiphile is spread on the water surface. The surface area is then slowly decreased by compressing the forming monolayer with a moving barrier while monitoring the surface pressure. This has the effect of causing the amphiphiles to align in the densest arrangement with the hydrophobic tails pointing into the air.

Depending on the conditions, monolayer formation is not the only behaviour observed. For example, if the concentration is raised above a certain critical concentration, the amphilphiles may close in on themselves to form spherical aggregates termed *micelles* within the bulk water. In a micelle, the hydrophobic tails are packed into the interior of the aggregate, leaving the hydrophilic head groups exposed to the solvent. The point at which these structures are formed is known as the *critical micelle concentration* (cmc). The cmc is dependent on the type of functional group and length of the tail. This is demonstrated by comparing bulky branched groups with non-branched groups, such as alkyl chains. The bulkier the group attached to the amphilphile, the more the cmc is increased, as it is more difficult to pack the tails together.

Micelles can be inverted by placing amphiphiles into organic solvents. The hydrophilic head groups coagulate within the centre of the micelle, while the lipophilic portion of the amphiphile points outward in the organic solvent. In addition to micelles, *bilayers* and *vesicles* can also form. Bilayers are commonly found in biological cells as phospholipid bilayers (Figure 5.15). A vesicle is rather like a simple biological cell membrane in that it comprises a bilayer arrangement of amphiphiles encasing an aqueous inner region.

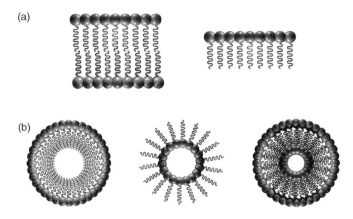

Figure 5.15 Schematics of various amphiphile aggregates: (a) bilayer and monolayer; (b) micelle, reverse micelle and vesicle.

Artificial phospholipid vesicles (liposomes) are used to transport vaccines, drugs, enzymes or other substances to target cells or organs. They also make excellent model systems for studying biological ion transport across cell membranes. The vesicles, which are several hundred nanometres in diameter, do not suffer from interference from residual natural ion-channel peptides or ionophores, unlike purified natural cells. For example, the synthetic heptapeptide **5.23** forms pores that promote chloride efflux in vesicle models. Similarly, the ion-pair receptor **2.108** can ferry NaCl from vesicles as an ion-pair ionophore (see Chapter 2, Section 2.6.2), while the 'hydraphile' **5.24** has been shown to transport Na^+ using ^{23}Na NMR spectroscopy through the bilayer walls of a vesicle model system.[15]

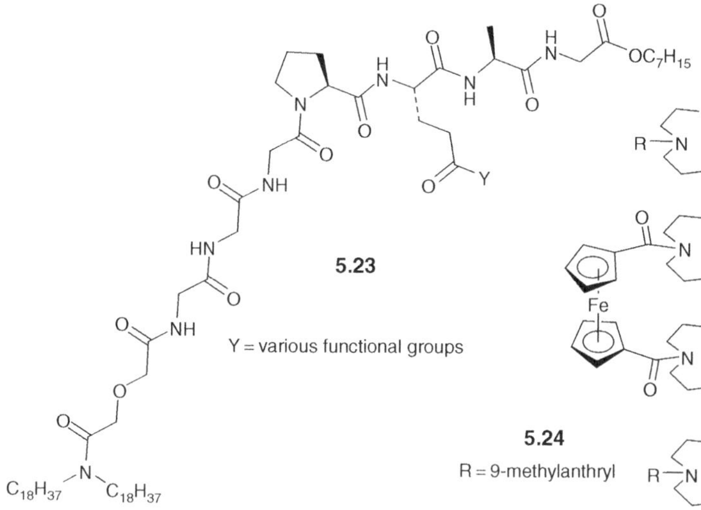

5.23

Y = various functional groups

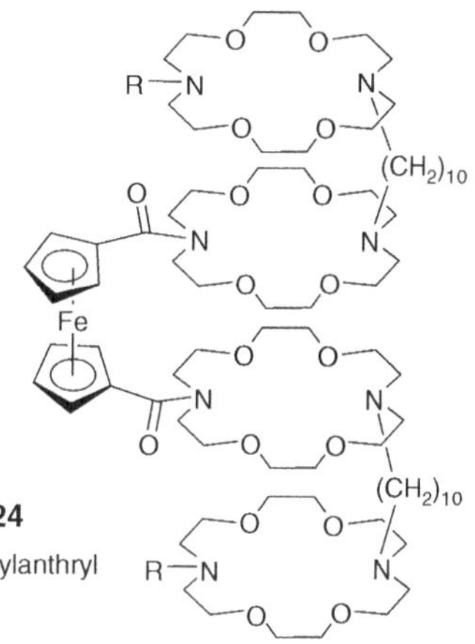

5.24

R = 9-methylanthryl

5.4.2 Langmuir–Blodgett films

In 1774, Benjamin Franklin reported the following statement to the British Royal Society:

> *At length at Clapman where there is, on the common, a large pond, which I observed to be one day very rough with the wind, I fetched out a cruet of oil, and dropped a little of it on the water. I saw it spread itself with surprising swiftness upon the surface. the oil, though not more than a teaspoonful, produced an instant calm over a space several yards square, which spread amazingly and extended itself gradually until it reached the leeside, making all that quarter of the pond, perhaps half an acre, as smooth as a looking glass.*

What was Franklin observing? It would be another 100 years until Lord Rayleigh speculated that the film was a monolayer, a single molecule thick. However, it was the work of a German, Agnes Pockles at the end of the 19th Century, who constructed a basic surface balance and used her kitchen sink to determine the water contamination of surfaces as a function of area for different oils. This work was later published in the journal *Nature* in 1891 and was used by Irwing Langmuir, whose name is now linked to the technique of forming Langmuir films and their transfer to various substrates using the Langmuir–Blodgett technique.

The original work carried out by Langmuir was the ability to form and transfer monolayers of fatty acid, ester and alcohol monolayers onto solid substrates. Several years later, Katherine Blodgett showed that multilayer films can also be formed and transferred onto solid substrates. The ability to transfer layers of organic material from a liquid phase, typically water, to a solid substrate allows the careful construction of ordered layer structures at a molecular level and the Langmuir–Blodgett technique is therefore a way to arrange molecules in organised assemblies. Highly structured, controlled multilayered thin films are prepared by the repetitive dipping of a solid substrate through a mechanically compressed monolayer spread at the gas–water interface. As the process proceeds the solid material is assembled onto the surface and this leads to a material that has a uniform thickness and laminar structure.

To understand how monolayers are formed at the interface of liquid and air, an understanding of surface tension is required. The interactions in the bulk of the liquid are balanced and experience an equal attractive force in every direction. The molecules at the interface of the surface, however, experience an imbalance of forces. Therefore, the molecules at the interface have a greater attraction to the molecules in the bulk of the liquid than the air–water interface, and thus the area of contact at the surface interface will be minimised. The result is the presence of free energy at the surface.

There are a large number of surface-active agents, commonly known as *surfactants* (Section 5.4.2), that have an amphiphilic nature and decrease the surface tension of water but are insoluble in it.[16] It is these insoluble amphiphilic monolayers that form a 'Langmuir (L) film', where this term normally describes a floating monolayer only. The nature of the amphiphilic monolayer dictates the orientation at the interface; the polar heads are immersed in the water layer and the long hydrocarbon chain points into the air or gas. For a monolayer to form, generally the hydrocarbon chain is required to be at least 12 carbon atoms long– shorter chains will tend to form different types of structures, for example, micelles (*cf.* Section 5.4.1). Conversely, if the chain is too long then the amphiphiles tend to crystallise on the surface, and a monolayer is not formed.

A 60-layer Langmuir–Blodgett film that contains an azobenzene core as a switching motif and 7,7,8,8-tetracyanoquinodimethane (TCNQ) units (**5.25**) has been prepared using this compound (Figure 5.16).[17] Upon UV irradiation at either 365 or 436 nm using a high-pressure mercury lamp, the film undergoes photoisomerisation. An intense absorption peak at 356 nm is assigned to the *trans*-isomer. On irradiation at 365 nm, the band at 356 nm is replaced by another band

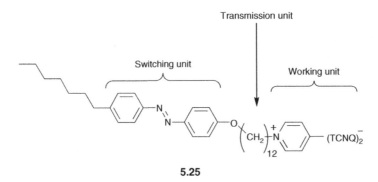

5.25

Figure 5.16 Schematic representing an organic switching device suitable for inclusion in an LB film.

appearing at 450 nm, corresponding to the *cis*-isomer. Upon irradiation at 436 nm, the *trans*-isomer is regenerated. This switching between the *trans*- and *cis*-isomers by irradiation is stable over at least ten cycles. The DC conductivity of the film is considerably higher for the *trans*-form than the *cis*-isomer. The photoisomerisation of the azobenzene in the hydrophobic part of the LB film controls the conductivity associated with the TNCQ in the hydrophilic part. This effect can be attributed to the change from *trans*-to-*cis* of the azobenzene on the ordering of the film, and hence the ease of electron-transfer between the components. A partial charge-transfer between TCNQ and *N*-dodecylpyridinium has also been observed at the air-water interface of the LB film. The film is highly conductive and therefore by changing the wavelength of irradiation the film exhibits switchable conductivity.

5.4.3 Thiol SAMs

Thiol SAMs are the most common and extensively studied monolayers. They generally involve molecules appended with a thiol group at one end interacting with a gold surface. The films are relatively straightforward to synthesise. A well-polished gold substrate is immersed into a solution of thiol and the thiols rapidly attach randomly to the substrate. Over time, the thiols align themselves onto the surface forming a monolayer according to self-assembly principles. The advantage of this technique is that no double-layer formation is observed; therefore, high-quality monolayers result if the substrate is allowed to sit in the solution for a few days. Another attractive property of thiol SAMs is the possibility of further derivatisation of the monolayer once it has formed. Many functional groups can be appended at the end of the thiol, which leads to the use of thiol SAMs in molecular recognition, as biomembrane mimics for modelling the interaction of biomolecules at different surfaces, the study of enzymes on surfaces, pH sensing devices and the preparation of molecular wires. A similar technique is used to prepare thiol-coated nanoparticles, *e.g.* as sensors (Section 5.6.3).

Tetrathiafulvalene (TTF) derivatives function as sensing motifs for cations in organic media. The bound metal–TTF complex causes polarisation of the TTF, producing a positive shift of the first oxidation potential, as measured by cyclic voltammetry. TTF-type functional groups exhibit redox chemistry in self-assembled monolayers. Therefore, there is huge potential in preparing an electrochemical SAM for the recognition of cations. For example, thiacrown ether–TTF-derived monolayers have been assembled onto gold surfaces, as demonstrated in Figure 5.17. The monolayers show a shift to varying degrees for the first oxidation potential of the TTF unit in the presence of metal cations. The smallest shifts are seen for Li^+ and K^+, 10 and 20 mV, respectively. Moderate shifts of 45 and 50 mV are observed for Na^+ and Ba^{2+} and the largest observed shift is for Ag^+ at 90 mV. The electrochemical response persists over 1000 cycles,

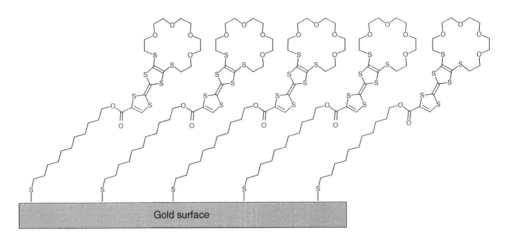

Figure 5.17 A TTF-derived molecular receptor forming a self-assembled monolayer.

indicative of the stability of the monolayers and therefore their potential use as electrochemical sensors.[18]

5.4.4 Liquid crystals

Liquid crystals are materials exhibiting one or more fluid phases with some degree of long-range order.[19] They are thus partway between a liquid and a solid. Liquid crystals exhibit fast molecular motion, but have a degree of crystal-like order. The liquid-crystal-forming-molecules (*mesogens*) are anisotropic in shape, typically either rod-shaped (*calamitic*), *ca.* 25 Å or more long, or are flat and circular (disc-shaped, *discotic*). The origin of their behaviour is a function of their anisotropic molecular shape, and the ordering of the molecules is dependent on temperature. Liquid crystals have a wide range of applications in high-tech materials, particularly liquid crystal displays (LCDs). The different mesophases formed as a function of temperature are termed *thermotropic*, in contrast to the formation of micelles that require solvent to form what is called a *lyotropic* phase (*cf.* Section 5.4.1). At higher temperatures (just below the transition to an isotropic liquid melt), mesogens form a relatively disordered *nematic* phase in which there is long-range orientational order, meaning that the molecules are all pointing in the same direction, but the mesogens are highly disordered in relation to their mutual position. As the temperature falls, molecular motion is decreased and the liquid crystal may go through different phase changes, to more ordered mesophases. In a *smectic* phase, the molecules are lined up in well-defined layers in one direction (Figure 5.18). For example, in a smectic A phase, the mesogens are aligned along the layer normal, whereas in smectic C the layers are tilted away from the molecules' long axis. There are a number of other orientations, for example, tilted phases that have hexagonal in-plane ordering are known as smectic I and smectic F. The alignment of molecules in the liquid phase gives rise to their distinctive optical properties, *vide infra*.

(a)   (b)   (c)

Figure 5.18 Mesophases at high temperature (a) nematic, and at lower temperatures, (b) smectic A and (c) smectic C.

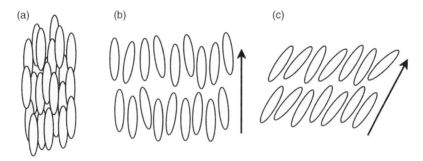

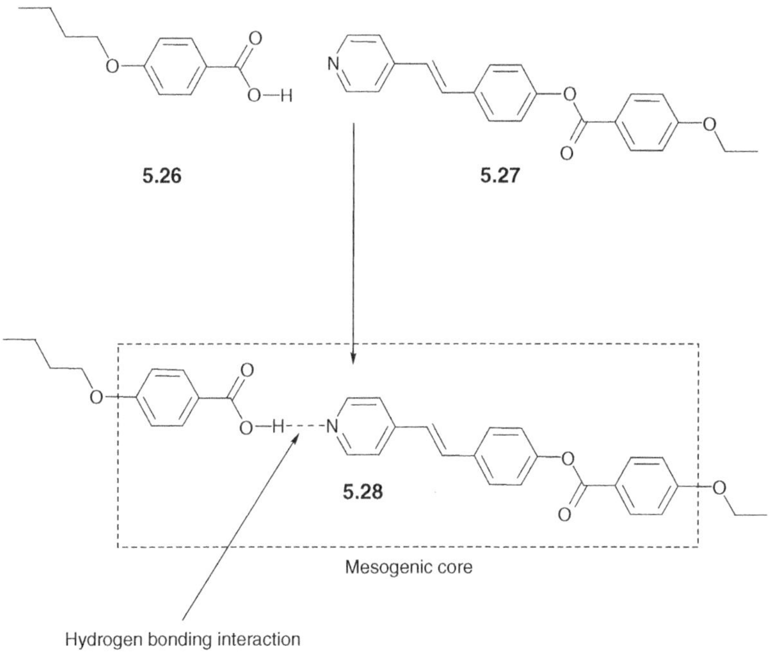

5.26   **5.27**

5.28

Mesogenic core

Hydrogen bonding interaction

Scheme 5.6   The formation of a supramolecular mesogen (**5.28**).

These self-assembled liquid crystal structures are formed by specific inter-molecular interactions, in particular, hydrogen bonding and ionic interactions (see Chapter 1, Section 1.3). One of the earliest examples of a supramolecular

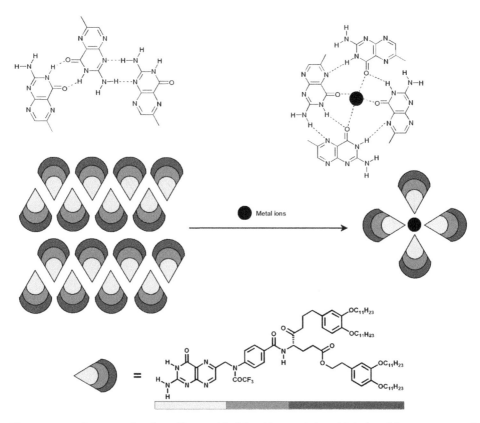

Figure 5.19 An example of a self-assembled liquid crystal, in which the ribbon structure of a folic acid derivative self-assembles into a disc-like tetramer on the addition of a metal ion, such as Na$^+$.

mesogen system was reported in 1989.[20] The mesophase is formed by hydrogen bonding interactions between 4-butoxybenzoic acid (**5.26**) and *trans*-[4-ethoxy(benzyol)oxyl]-4'-stilbazole (**5.27**) to give rod-like structures, such as **5.28** (Scheme 5.6). The two different molecules first form a *hetero* hydrogen bonding interaction between the carboxylic acid and the pyridine moieties. The high thermal stability of the molecules results in the formation of nematic phases up to 214 °C. The formation of the hydrogen bonding interaction was demonstrated by Fourier-transform infrared (FTIR) spectroscopic measurements. A one-to-one mixture of **5.26** and **5.27** shows a carbonyl stretch at 1681 cm^{-1} assigned to the C=O stretch in the hydrogen bonded dimer (C=O$\cdots$HO–). However, if a one-to-one solution is prepared in pyridine, another band at 1704 cm^{-1} is seen which is characteristic of a carbonyl stretch of the carboxylic acid group complexed with the pyridyl ring.

Since this early work, there have been numerous examples of complex supramolecular LC systems. For example, Figure 5.19 shows an LC system which

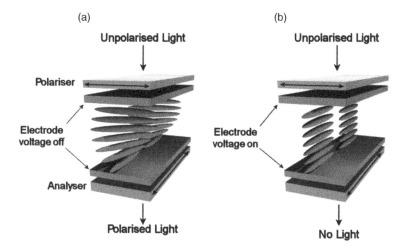

Figure 5.20 Representation of the twisted-nematic cell. In the absence of an applied field (a), light passes through and the display looks transparent to the observer. When a voltage is applied (b), light is unable to pass through and the cell is darkened.

changes from a smectic (rod-shaped) to hexagonal columnar (disc-shaped) phase on the addition of sodium ions. The oxygen atoms chelate to the metal ions and the LC reorganises into a circular geometry.[21]

The nematic phase is the most common phase used in the applications of liquid crystals, the most important of which is the liquid crystal display (LCD), used in computer monitors. The traditional LCD relies on the *twisted-nematic geometry*, whereby an electric field alters the orientation of the liquid crystal. The twisted-nematic cell is comprised of two glass slides with a transparent conducting coating, normally indium tin oxide, which act as the electrodes. The surfaces of the electrodes are in contact with the liquid crystal, which is covered by a polymer that has been 'brushed' on in one direction. This brushing technique affects the alignment of the liquid crystal. The two plates are then placed on top of each other at a 90° angle, resulting in a helical twist in the nematic phase- hence, 'twisted-nematic cell'. This twisted arrangement of the liquid crystals act as an optical waveguide and rotates the plane of polarised light by 90°, such that the light passing to the second polariser can pass through it and therefore the cell looks transparent. As a voltage is applied to the cell, the liquid crystals align themselves in the direction of the field and the twist is lost; therefore, there is no rotation of the polarised light and hence no transmission through the crossed polaroids and the display looks dark (Figure 5.20).

One of the disadvantages of the LCD display is that the response time is slow (approximately 50 μs) and the liquid-crystal phase is too symmetric to allow 'vector order'. To overcome this problem, tilted smectic phases with ferroelectric properties (polarisation can be reversed by an electric field) can be used, providing that the liquid crystal is chiral. One such ferroelectric liquid crystal is

p−decyloxybenzlidine-p'-amino-2-methylbutylcinnamte. The crystals are aligned along the molecular axis tilted from the normal layer (this is defined as the unit vector perpendicular to the plane), such that the molecules are allowed to rotate, forming a 'tilt-cone'. This alignment of the chiral molecules exhibits a spontaneous polarisation along the axis. The magnitude of polarisation is dependent on temperature. The polarisation is linear to the applied field, meaning that the ferroelectric liquid crystal can be switched quickly, between two different states, unlike a twisted-nematic cell, as the switching is a consequence of a twisting motion.[22]

5.5   Soft lithography

 Xia, Y. and Whitesides, G. M., 'Soft lithography', *Angew. Chem., Int. Ed. Engl.*, 1998, **37**, 551–575.

Soft lithography is the development of elastomeric stamps and molds by various techniques for surface patterning on the nanoscale. Many of the key strategies for the development of soft lithography arise from the bottom up approach (Section 5.1.2), in particular, the use of SAMs. There are a number of techniques employed but the most common methods are (1) microcontact printing (mCP), (2) replica moulding, (3) microtransfer moulding (mTM), (4) micromoulding in capillaries (MIMIC) and (5) solvent-assisted microcontact micromoulding (SAMIM) (Figure 5.21), with microcontact printing and replica moulding being the most common (*vide infra*). The techniques use transparent elastomeric polydimethylsiloxane (PDMS) stamps with a patterned relief on the surface to generate features in the monolayers.[23]

The soft lithography method employs the use of flexible organic molecules that have self-assembled onto a particular surface. It is the flexible nature of these organic compounds that gives the technique its name of *soft* lithography, as opposed to other microfabrication techniques that employ rigid inorganic materials. There are a number of advantages to soft lithography. The method is dependent on the self-assembly of molecules on a surface. The self-assembly process can overcome some surface defects and if any defects do occur, the procedure can be easily repeated as it is a relatively cheap technique. Unlike traditional 'hard' techniques, very small structures have been prepared (down to 30 nm), as the technique is not diffraction-limited. Different types of surfaces can also be employed which differ in size and topology. However, there are also a number of disadvantages, such as deformation of the patterns in the stamp. The technique is in a relatively infant stage and there is a lot of scrutiny to see whether materials produced by soft lithography are compatible with other integrated-circuit processes.

Typically, the stamp employed is created by casting 'prepolymers' against a master patterned by conventional lithographic techniques. This 'stamp' is then

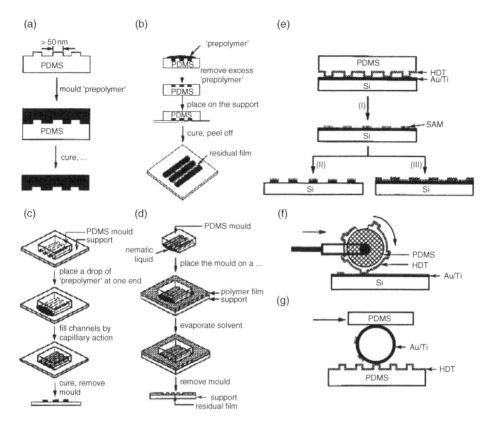

Figure 5.21 Different soft lithographic methods developed by Xia and Whitesides:[23] (a) replica moulding; (b) microtransfer moulding; (c) micromoulding in capillaries; (d) solvent-assisted microcontact moulding. Three examples of microcontact printing: (e) printing on a planar surface with a planar stamp (I), printing of the SAM (II) etching and (III) deposition; (f) printing on a planar surface with a rolling stamp; (g) printing on a non-planar surface with a rolling stamp. Reproduced with permission from Y. Xia and G. M. Whitesides, *Angew. Chem., Int. Ed. Engl.*, **37**, 550–575 © 1998 WILEY-VCH, Verlag GmbH and Co., KG Weinheim.

used to create a number of replica moulds by pressing it repeatedly into the siloxane surface.

- *Microcontact Printing* (mCP). An 'ink' of alkanethiols is spread onto a patterned PDMS stamp. The stamp is then brought into contact with the substrate, which can range from coinage metals to oxide layers. The thiol ink is transferred to the substrate where it forms a self-assembled monolayer that can act as a resist against etching.

- *Replica Moulding*. A PDMS stamp is cast against a conventionally patterned master. Polyurethane is then moulded against the secondary PDMS master.

In this way, multiple copies can be made without damaging the original master.

- *Microtransfer Moulding* (mTM). A PDMS stamp is filled with a 'prepolymer' or ceramic precursor and placed on a substrate. The material is cured and the stamp is removed. The technique generates features as small as 250 nm and is able to generate multilayer systems.

- *Micromoulding in Capillaries* (MIMIC). Continuous channels are formed when a PDMS stamp is brought into conformal contact with a solid substrate. Capillary action fills the channels with a polymer precursor. The polymer is cured and the stamp is removed.

- *Solvent-Assisted Microcontact Moulding* (SAMIM). A small amount of solvent is spread on a patterned PDMS stamp and the stamp is placed on a polymer, such as a photoresist. The solvent swells the polymer and causes it to expand to fill the surface relief of the stamp.

5.6 Nanoparticles

 Moores, A. and Goettmann, F., 'The plasmon band in noble metal nanoparticles: an introduction to theory and applications', *New J. Chem.*, 2006, **30**, 1121–1132.

Most people encounter *colloids* in everyday life, for example, in shaving foam, fog or mist, mayonnaise, and gems, such as ruby, turquoise and garnet. Colloids bridge the gap between solutions and suspensions – they are small particles that are suspended or dispersed in a bulk material and that do not separate out on standing. Colloids are molecules (or aggregates of molecules) that are larger in size than particles dissolved in solutions but smaller than those found in suspensions. They range in size between 1 and 100 nm and can exist in any of the three different physical states of matter. For example, *foam* is a gas dispersed in a liquid or solid. A *liquid aerosol* is a liquid dispersed in gas, whereas a *solid aerosol* (or smoke) is a solid dispersed in a gas. An *emulsion* is a liquid dispersed in a liquid, a *gel* is a liquid dispersed in a solid and a *sol* is a solid dispersed in a liquid or solid.

Colloidal science is a mature science in its own right but there is a striking similarity in the science of nanoparticles and colloids and more often than not they are used in the same context. Nanoparticles can be described as clusters of thousands of atoms on the size range between 1 and 100 nm. They often have specialised surface functionalities that stabilise them and prevent further aggregation into bulk materials. Below this size regime, we can think of the particles as being large molecules. Above 100 nm or so, materials start to behave in a familiar bulk fashion. Nanoparticles (or nanocrystals if they are highly ordered, faceted crystallites) exist in a size domain where their electronic properties are dominated by quantum size effects and hence are sometimes referred to as *quantum dots* (a quantum dot is generally taken to be a semiconductor nanocrystal, such as CdS, CdSe, CdTe or CdSe/ZnS).

> **Colloid**: A stable system of small particles dispersed in a different medium. A multiphase system in which one dimension of a dispersed phase is of colloidal size.

> **Nanoparticles**: Particles with controlled dimensions on the order of nanometres

5.6.1 Synthesis and derivatisation of nanoparticles

The most common nanoparticles are metallic gold nanoparticles prepared by a process called 'arrested nucleation and growth'. This occurs from the reduction of $H[AuCl_4]$ by sodium citrate or sodium borohydride. The gold nanoparticles form spheres of a size controlled by the concentration of reductant. For example, the addition of a 1 mL citrate solution to a 50 mL solution of 0.01 % $H[AuCl_4]$ gives gold nanoparticles 16 nm in diameter (measured by scanning electron microscopy (SEM), whereas, if 0.1 mL of citrate is added to the same gold solution the particle size is 147 nm. The overall charge of the gold nanoparticle is negative as a consequence of the weakly bound citrate to the surface. Citrate is a very common reducing agent and has also been used to prepare other metal nanoparticles, such as Ag, from $AgNO_3$, Pd from $H_2(PdCl_4)$ and Pt from $H_2(PtCl_6)$. It is also feasible to prepare mixed nanoparticle systems, known as composite materials, alloys or mixed-grain nanoparticles, such as magnetic FePt nanoparticles.[24] This can lead to more elaborate composites, by forming shells around the collidial seeds. For example, silver can coat itself around gold nanoparticles or can itself be capped. This method of particle growth can control size and density. These nanoparticles do not have to exist as a suspension in a liquid phase but the colloid (nanoparticle suspension) can be immobilised onto a surface of a particular substrate. This is a highly attractive and desirable property, as the chemist is able to control the growth and function of such materials. The functionalising of nanoparticles is accomplished *in situ*, *i.e.* the reagents are added together, producing an intermediate that is not isolated. For example, a reducing agent, such as sodium borohydride, reduces $H[AuCl_4]$ in the presence of (γ-mercaptopropyl)-trimethoxysilane to produce a nanoparticle containing a silane functionality (**5.29**) (Scheme 5.7).

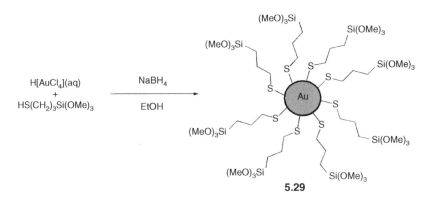

5.29

Scheme 5.7   The formation of stabilised gold nanoparticles.

This methodology has been employed to attach many functional groups onto other metal nanoparticles, including amines and carboxylic acid groups.[25]

5.6.2   Quantum size effects

The phenomenon of quantum size effects has been known since the 1960s. The quantum size effect is best observed in semiconductor nanocrystalline materials, also referred to as *quantum dots* (small regions of a material buried in another with a larger band gap). Quantum dots have electron–hole pairs known as *excitons*. The electron is excited by a photon from the valence band to the conduction band of the nanoparticle – the electron then leaves a hole of opposite charge, to which it is attracted by the Coulomb force. The exciton results from the binding of the electron with the hole. It is this feature that gives rise to the most prevalent attribute of the quantum size effect in nanoparticles – their optical features. Optical studies, using UV–Vis and fluorescence spectroscopies have shown that the energy from the electronic excitation state of an isolated molecule is greater than the inter-band transition in a macroscopic particle. This means that a size range must exist in which low-energy excitation of a crystal exhibits a characteristic wavelength for the individual particle, depending on its size. In fact, the absorption band *hypsochromically* shifts (*i.e.* moves to higher energy or is *blue-shifted*) as the particle size decreases.

5.6.3   Nanoparticles as sensors

 Davis, J. J., 'Interfacial sensing: surface-assembled molecular receptors', *Chem. Commun*, 2005, 3509–3513.

Quantum dots have had a large influence in the development of nanotechnology. They can occur spontaneously during the fabrication of monolayer formation by a technique known as *molecular beam epitaxy* or they are created by a technique called *electron beam lithography*. In the latter method, a pattern is etched onto a semiconductor chip, and a conducting metal is then deposited onto the pattern. The unique optical properties of quantum dots means that they are found in everyday applications, such as diode lasers, amplifiers and in biological systems, such as dyes. Traditionally, organic dyes have been used as sensors for the detection of diseases. The development of dye sensors often depends on complex doping of a combination of materials. However, this results in a low-efficiency output of the signal due to *self-absorption* (when an atom emits a photon, there is a chance that another atom will absorb the photon if the energy of the photon matches the electronic transition of the same atom). To address this issue, quantum dots are now being used for the detection of certain diseases, such as tumors. As a consequence of their high quantum yield, less of the material is needed *in vivo* to obtain a strong signal. Ultra-small cadmium selenide (CdSe) nanoparticle crystals have been used as white light phosphors. These nanocrystals exhibit a broad emission spectrum across the entire visible region which does not show any self-absorption, thereby increasing the efficiency of the device.[26] It is also anticipated that nanocrystals can be used to make a different kind of LED. For example, it was thought until recently thought that blue-laser technology was impossible until the first blue and violet semiconductor laser was developed by preparing gallium nitride quantum dots that spontaneously self-assemble into a film. This was a major breakthrough as it means data-storage capacity can be increased. This has consequently resulted in the development of high-density DVDs that are able to store more data than normal DVDs. This Blu-ray™ technology uses shorter wavelengths, *i.e.* 405 nm (blue lasers), whereas a normal DVD uses longer wavelengths, *i.e.* 650 nm (red lasers).

Nanoparticles have also been used as reporter groups on molecular sensors. For example, an anion sensor incorporating a zinc metalloporphyrin receptor supported on gold nanoparticles has been developed. Zinc metalloporphyrin dithiol (**5.30a**) self-assembles along with dodecane thiolate on the gold surface, forming a nanoparticle optical anion sensor (**5.30b**) (Figure 5.22). The nanoparticle shows a well-defined *plasmon band* in the visible region.[§] Upon anion binding, electromagnetic coupling perturbs the motion of the free electrons in the nanoparticle, shifting the plasmon band, hence giving optical detection. The surface pre-organisation imparted to the metalloporphyrin results in enhanced binding of anions compared to the analogous solution-phase receptor. For example, the affinities of **5.30a** and **5.30b** for Cl^- in DMSO solution are log K_a =< 2 and 4.3, respectively.[27] This is interesting as it means that highly sensitive fabricated devices can be prepared.

[§] Noble-metal nanoparticles embedded in a dielectric medium show strong absorption peaks in the visible region of the UV–vis spectrum due to the collective motion of free electrons (surface plasmon resonance).

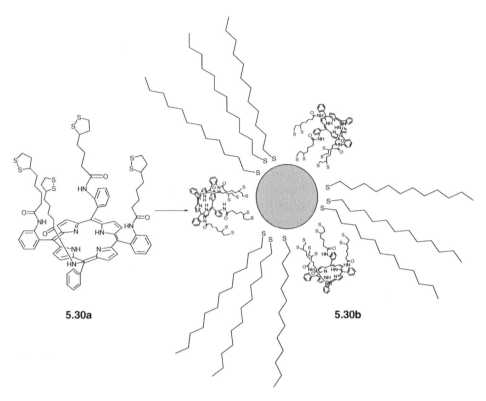

Figure 5.22   Schematic of the free porphyrin (**5.30a**) and the functionalised gold nanoparticle (**5.30b**).

Three-dimensional coating of a metal nanoparticle is analogous to the formation of a two-dimensional self-assembled monolayer on a gold surface and there are considerable parallels in the chemistry; however, the two-dimensional surface lacks the quantum size effects of the nanoparticle. There have been a number of other SAMs prepared, based on molecular receptors containing different reporter groups, coupled with a recognition motif. For example, a ferrocene-derived thiol receptor containing amide functional groups has been attached to a gold surface (Figure 5.23(a)). The recognition of an anion perturbs the electron-transfer process, either by through-space or through-bond interactions. Interestingly, it is possible to combine two-dimensional SAMs and nanoparticle chemistry to produce hierarchical nanoscale assemblies in which functionalised nanoparticles containing receptor groups are bound to a two-dimensional gold surface (Figure 5.23(b)). It is possible that enhanced sensitivity will be achieved in this way due to the large surface areas of the binding sites.

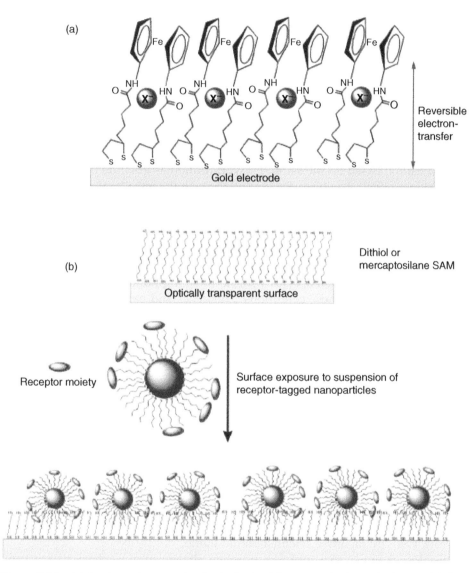

Figure 5.23 (a) A schematic example of a derivatised ferrocene thiol attached to a gold electrode. (b) A schematic highlighting that multiple-functionalised nanoparticles can be attached to a large surface to produce a highly sensitive device.[28] J. J. Davis, 'Interfacial sensing: surface-assembled molecular receptors', *Chem. Commun*, 2005, 3509–3513. Reproduced by permission of The Royal Society of Chemistry.

5.7   Fullerenes and nanotubes

Niyogi, S., Hamon, M. A., Hu, H., Zhao, B., Bhowmik, P., Sen, R., Itkis, M. E. and Haddon, R. C., 'Chemistry of single-walled carbon nanotubes', Acc. *Chem. Res.*, 2002, **35**, 1105–1113.
Diederich, F. and Gomez-Lopez, M., 'Supramolecular fullerene chemistry', *Chem. Soc. Rev.*, 1999, **28**, 263–277.

Carbon forms a number of allotropes, with graphite, diamond and fullerenes being the most stable, and ceraphite, lonsdaleite, amorphous carbon, carbon nanofoam and the most recently discovered aggregated diamond nanorods (discovered in 2005), being less well-known. Fullerenes and nanotubes are curved forms of carbon based on the same sp^2 hybridisation as graphite but differ in the fact that closed-shell structures are formed through the introduction of pentagons among the graphitic hexagonal structure. Thus, C_{60}, the most stable fullerene, requires 12 pentagons along with 20 hexagons to achieve its closed 'soccer-ball' shape. The formation of fullerenes is a remarkable example of thermodynamic self-assembly (see Chapter 3, Section 3.1.2), since under the high-energy conditions of the carbon-arc furnace even C–C bond formation becomes somewhat reversible. The closed-shell species with no dangling bonds are the most stable forms of carbon under these conditions. Fullerenes and nanotubes have become enormously popular, especially in the fields of supramolecular chemistry and nanochemistry. Fullerenes, in particular, have been used as both hosts and guests, while carbon nanotubes constitute remarkable elongated host species into which very well defined, ordered arrays of atoms can be included.[29]

5.7.1   Synthesis and structure of fullerenes and carbon nanotubes

Fullerenes were discovered in the mid 1980s by two American scientists, Robert Curl and the late Richard Smalley, along with British scientist Harold Kroto. The three went onto share the 1996 Noble Prize in Chemistry for their work in this area. Fullerene research led to the discovery of carbon nanotubes in the early 1990s by Japanese scientist Sumio Iijimo.

How does a linear structure like graphite change into a spherical structure like a fullerene or nanotube? This can be easily derived from Euler's polyhedron formula which states that exactly 12 pentagons are required to obtain a closed structure and therefore a combination of a pentagon surrounded by five hexagons is needed to form curvature of the surface in order to enclose a volume. There can be an infinite number of hexagons but only 12 pentagons are required. The stability of fullerenes is determined by the isolated pentagon (ISP) rule, which states that no two pentagons should be adjacent to one another in the structure in order to obtain a minimal local curvature and surface stress. Thus, the smallest possible fullerene is C_{20}, comprising only 12 pentagonal faces. Adding hexagons conceptually derives higher fullerenes, such as C_{60}, C_{70} and C_{84}.

5.7.2 Fullerenes as guests

Bowl-shaped molecules, such as cyclotriveratrylene (CTV) and calixarenes (see Chapter 2, Sections 2.5.2 and 2.3.6 respectively), bind fullerenes as guests and have been shown to separate C_{60} from C_{70} in a fullerene-enriched complex soot mixture (*fullerite*) formed by arc vaporisation of graphite.[30] The key to the separation technique is the ability of the larger calixarenes to selectively complex fullerenes in a toluene solution, resulting in well-defined complexes. Thus *p-t*-butylcalix[8]arene forms a one-to-one complex with C_{60} and the smaller *p-t*-butylcalix[6]arene in a double-partial-cone conformation forms one-to-two complexes with both C_{60} and C_{70}. The crystalline inclusion complexes can be separated and filtered off from a toluene solution of fullerite. Treatment of these solid inclusion complexes with chloroform dissolves the calixarene hosts, precipitating out the pure fullerenes. Figure 5.24 shows the purification of C_{60} from C_{70} and fullerite using *p-t*-butylcalix[8]arene. The C_{70} can then be obtained from the resulting C_{60}-depleted mixture using *p-t*-butylcalix[6]arene.

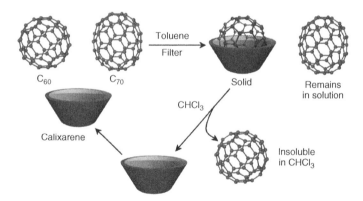

Figure 5.24 Calix[8]arene-based purification of C_{60} and C_{70}.

5.7.3 Fullerenes as hosts

Fullerenes can also act as hosts. The reactive surface of the C_{60} shell can be utilised to attach a plethora of functional groups. An elegant example uses a fullerene crown ether host as an electrochemical sensor for K^+. Compound **5.31** incorporates a dibenzo[18]crown-6 in close proximity to the C_{60} surface, acting as the binding site for a metal ion. K^+ binding within the crown ether cavity results in an anodic shift of 90 mV of the first fullerene reduction potential ($C_{60} \rightarrow C_{60}{}^{\bullet-}$).[31] The close proximity of the potassium cation to the fullerene surface makes it easier to reduce the fullerene to the mono-anion. Fullerenes can also act as hosts for atoms such as nitrogen, helium and metal atoms, which can be trapped inside them in an analogous fashion to carcerands (see Chapter 2, Section 2.7.3). Species, such as $Cu@C_{60}$ and $La@C_{82}$ (where the '@' symbol indicates inclusion of the metal inside the fullerene cage), have been prepared, generally by carbon vaporisation in the presence of metal salts. Potential applications range from tracers and spin-labels in biological systems to 'qubits' in quantum computers.

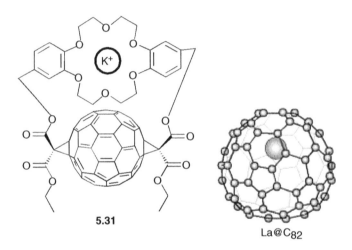

5.31

$La@C_{82}$

5.7.4   Carbon nanotubes

There are two forms of carbon nanotubes, *i.e.* single-walled carbon nanotubes (SWCNTs) and multi-walled carbon nanotubes (MWCNTs). The latter were the first to be discovered and can be described as graphite fibres (Figure 5.25). SWCNTs are closely related to fullerenes and exhibit very uniform diameters. Both kinds of nanotube can be prepared by the arc-discharge method, laser ablation or chemical vapour deposition. All of these techniques suffer from problems with the low purity of the resulting nanotubes, however. The crude nanotubes often contain species, such as polyhedral carbon particles, metal cata-lysts, amorphous carbon and smaller fullerenes. There is much research into the area of nanotube purification and many techniques have been developed to prepare uniform, clean samples, such as oxidation, acid treatment, annealing, ultrasonication, microfiltration, ferromagnetic separation, functionalisation and chromatography.

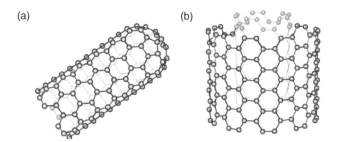

Figure 5.25 (a) A single-walled carbon nanotube (SWCNT) with a caped-end and (b) a multi-walled carbon nanotube (MWCNT).

Why have carbon nanotubes become important over the last decade, especially in nanotechnology? Nanotubes have unique electronic and mechanical properties. The C–C bond is one of the strongest bonds in nature, giving the nanotubes very high tensile strength and stiffness (a microfabricated nanotube tensile device recorded a tensile strength of some 220 times that of steel). Nanotubes also have a large Young's modulus in the axial direction (the ratio of stress to strain which indicates the elasticity of a substance). As nanotubes are very flexible, they can be used for applications in composite materials which need anisotropic properties. Nanotubes have similar electronic and chemical properties to graphite, such as high conductivity, chemical specificity and inertness. Nanotube electrical conductivity depends on their diameter and they can be either semiconducting or metallic. The conduction properties are influenced by the size of the band gap, which in turn is related to the size of the nanotube. They also have some more unusual properties due to the lattice helicity and elasticity, while they also display a large surface area that may be useful in catalytic- or absorption-type processes. Unlike graphite, the chemical reactivity of nanotubes is related to the π-orbital mismatch, caused by their curvature. Therefore, the smaller the diameter of the nanotube, then the more reactive it becomes and hence either side-wall or end-cap modification can be selectively achieved. This property may be utilised in the purification of nanotubes.

An interesting potential application of nanotubes is in energy storage, particularly hydrogen storage. Graphite, carbonaceous and carbon fibre rods are commonly used in fuel cells, batteries and other electrochemical applications. Nanotubes could also be used in such systems because of their small dimensions, smooth surface topology and specificity. Hydrogen, as an energy source, has the advantages that the only byproduct after combustion is water and hydrogen can be readily regenerated by electrolysis (as long as there is a convenient, cheap source of electrical energy!). Nanotubes have a cylindrical and hollow geometry and it has been predicted that they might be able to store liquids or gases within the cavity. However, to be commercially successful nanotube hydrogen storage materials would need to store at least 6 % hydrogen by weight and of course the problem of large-scale production of nanotubes must be addressed.

Another potential application of nanotubes is as nanoprobes in scanning probe instruments such as scanning-tunneling and atomic-force microscopes. They are unlikely to suffer damage from crashes with the target surface due to the high elasticity and could lead to improved resolution and robustness in comparison to conventional materials, such as silicon and metal tips. One of the most interesting applications of nanotubes is their use as templates to make nanowires. For this to happen, the nanotube must *not* have end caps, as is often the case in the as-synthesised material (Figure. 5.25(a)). The end caps can easily be removed chemically, however, as they are more reactive than the side-walls and can be oxidised away. For example, *quasi* one-dimensional crystals of KI have been prepared by incorporating them into SWCNTs (Figure 5.26). The lattice spacing in the included KI is substantially different from pure KI crystals. Measurements on a number of different samples of varying size show that the amount of strain in the KI, and hence its physical properties, is related to the SWCNT diameter, due to the reduced coordination of the surface atoms and proximity of the nanotube wall.[32]

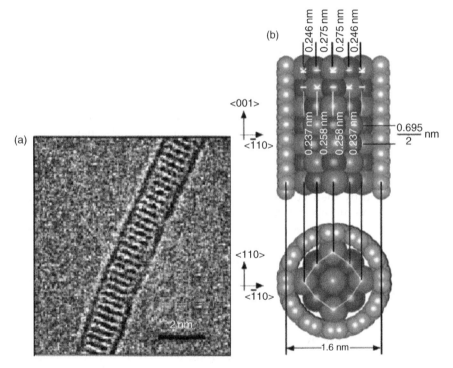

Figure 5.26 (a) High-resolution TEM image and (b) structural model of a single crystal of KI incorporated within a 1.6 nm-wide SWCNT.[32] Reprinted with permission from R. R. Meyer, J. Sloan, R. E. Dunin-Borkowski, A. I. Kirkland, M. C. Novotny, S. R. Bailey, J. L. Hutchison and M. L. H. Green, 'Discrete atom imaging of one-dimensional crystals formed within single-walled carbon nanotubes', *Science*, **289**, 1324–1326 © 2000 AAAS.

5.8 Dendrimers

Bosman, A. W., Janssen, H. M. and Meijer, E. W., 'About dendrimers: structure, physical properties and applications', *Chem. Rev.*, 1999, **99**, 1665–1688.

'Dendrimer' comes from the Greek word *dendra*, meaning tree. A dendrimer is built up from a single core in a stepwise fashion forming a highly branched oligomeric structure of nanometric dimensions and can be classified as a special kind of polymer. Dendrimers are elegant examples of using a core building block to prepare large macromolecules from the bottom-up approach. Unlike traditional polymers, the chemist can exert precise control of the molecular weight of the dendrimer because it is a chemically pure substance, *i.e.* all of the dendrimer molecules, in principle, contain the same number of repeat units. Dendrimers are generally prepared by reacting an *n*-directional core with either linear building blocks or branched building blocks to yield a first-generation dendrimer. The peripheral sites can be reacted with more building blocks to form a second-generation dendrimer and so forth (Scheme 5.8).

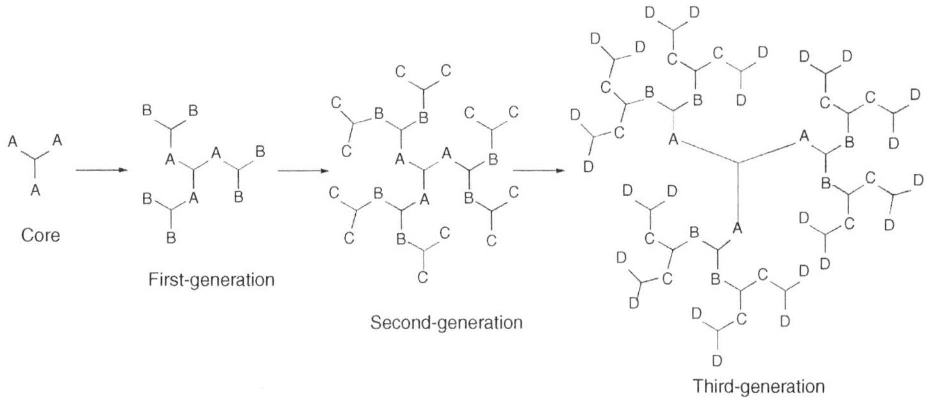

Scheme 5.8   Schematic representing the iterative preparation of a dendrimer.

More recently, alternative convergent approaches have been developed that involve attaching segments of pre-formed dendritic arms to a common core. This attachment can be carried out either covalently to give a large covalent molecule or *via* non-covalent interactions to give a supramolecular dendrimer. Convergent approaches reduce problems associated with impurities related to incomplete reaction of every site as the dendrimer grows with each generation.

Dendrimers are of interest in a variety of contexts, such as host–guest chemistry and catalysis in which the dendritic core region, which is often highly porous, can exhibit interesting host behaviour, while the densely packed outer layer acts to shield the interior region from the surrounding medium. This suggests applications in mimicking the hydrophobic pocket regions in enzymes. In medicinal applications, the feature of an outer surface comprising multiple identical binding sites has potential pharamcological importance because of amplified substrate binding. There is a plethora of both molecular and supramolecular dendrimers in the literature, which have been used in host–guest chemistry and self-assembly. Dendrimers are useful as photoactive devices, particularly in light-harvesting. The energy transfer cascade of the latter is a prevalent process which occurs in pigments of certain kinds of algae, bacteria and green plants. This is the process where absorbed photons on the photosynthetic reaction centre cascade down an energy gradient. This antenna system acts as a funnel for captured energy. A synthetic antenna dendrimer (**5.33**) has been prepared. The highly branched macromolecule is comprised of rod-like conjugated groups which are able to absorb photons. As the dendrimer is grown to form higher generations, the wavelength of the absorption maximum increases. This example of a molecular antenna is analogous to the metallodendron antenna previously discussed in Section 5.3.1.

5.33

Well-defined, tunable supramolecular dendrimer systems that form nanostructured gel phases have been reported.[33] These materials are based on a carboxylic acid dendrimer and a linear diamine (**5.34**). Gelation is optimal for the second-generation material and does not occur for the first or third-generation analogues. The sol–gel transition temperature (T_{gel}–a measure of gel stability) is also highly dependent on the diamine spacer length and decreases markedly on going from 1,12-diaminododecane to 1,6-diaminohexane. When the supramolecular acid – amine interactions are replaced by a covalent linker, gelation behaviour is also observed. Gel-phase materials are described in detail in the next section.

5.34

5.9 Fibres, gels and polymers

Terech, P. and Weiss, R. G., 'Low-molecular-mass gelators of organic liquids and the properties of their gels', *Chem. Rev.*, 1997, **97**, 3133–3160.

A fibre can be described as a compound, normally a polymer, that posses a continuous filament or a discrete elongated piece of material. This broad definition is true for both natural and synthetically prepared fibres, such as nylon, polyester and acrylic polymers. A supramolecular fibre comprises long chains of smaller molecules linked *via* non-covalent interactions, such as hydrogen bonding. The hydrogen bonding energy between a donor and acceptor group typicaly ranges from 5 to 65 kJ mol^{-1}, considerably weaker than a covalent-bonding interaction. However, if a polymer consists of an array of hydrogen bonding interactions that are aligned along the fibre axis, the polymer will show a considerable increase in tensile strength. This property is exhibited by alkylated guanosines, such as **5.35a** (Figure 5.27). This compound has been used as a building block for the preparation of a supramolecular fibre due to the extensive intermolecular hydrogen bonding interactions between the guanosine moieties. Fibres of **5.35b** have been prepared by melt-spinning using the combination of long alkylsilyl chains for flexibility and a rigid main-chain hydrogen bonding motif for strength.[34]

5.9.1 Supramolecular gels

Supramolecular gels can be described as the self-assembly of small molecules, termed low-molecular-weight gelators (LMWGs), to form extended fibres that

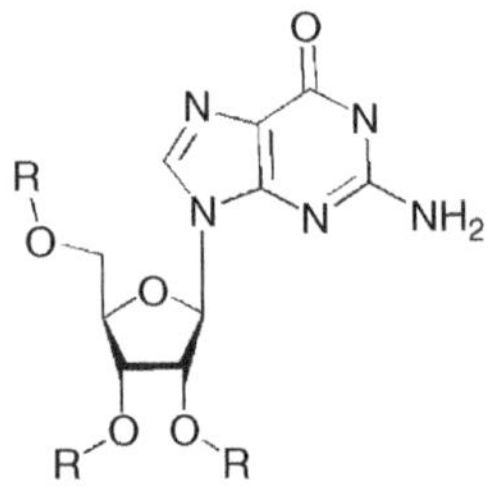

R = Si (*i*-Bu)$_2$C$_{18}$H$_{37}$ **5.35a**
R = SiMe$_2$*t*-Bu   **5.35b**

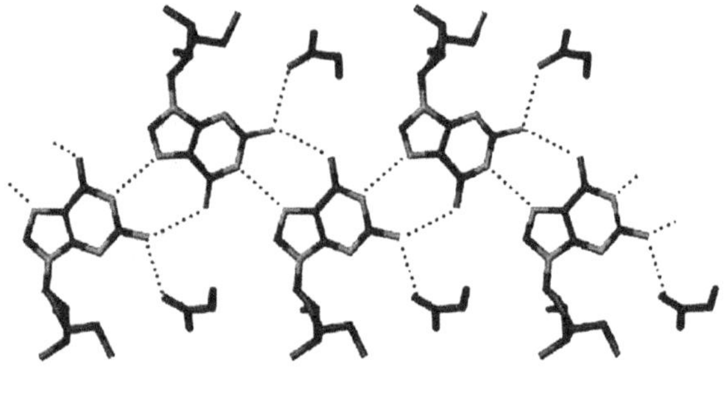

Figure 5.27 The X-ray crystallographically derived crystal packing of **5.35**, highlighting the hydrogen bonding network along the fibre axis.

interlink forming a continuous three-dimensional entangled network usually in a solvent, held together *via* non-covalent interactions (*cf.* the dendritic supramolecular gel **5.34**). The solid fibre network is highly porous and traps solvents such as water (a hydrogel), organic solvents (an organogel) or gases (an aerogel) within the fibre matrix by capillary forces. They parallel conventional gels made from polymers or natural materials, such as gelatin, which contain long covalent fibres. The most common functional groups linking together LMWGs are those containing highly directional hydrogen bonding groups, for example, peptides, carbohydrates and urea functionalities. A fibre is assembled first, which then undergoes hierarchical self-assembly, forming an entangled gel-phase network in what amounts to a kind of arrested or partial crystallisation process. Strictly speaking, this gelation process is a two-component process, as it involves two different molecules forming a complex which assembles into fibres and then into a gel network, whereas a single-component gelating process involves a single polymer which forms a gel by a combination of *intramolecular* and intermolecular interactions (see Chapter 3, Section 3.1.2).

Barbituric acid (**5.36**) and pyrimidine (**5.37**) moieties are two well-known building blocks which form mutually complementary hydrogen bonding interactions. If one face of the building blocks were deliberately blocked off, then a continuous fibre is formed in one direction, which then self-assembles to form a gel network (Scheme 5.9).[35]

Supramolecular gels can also be formed using coordination interactions and gels represent a way to control the positioning of metal ions, for example, in nanoparticle synthesis. The reaction of Fe(NO$_3$)$_3$ with 1,3,5-benzenetricarboxylic acid (BTC), for example, forms a coordination polymer gel within minutes. There is extensive cross-linking between the Fe^{3+} and the tricarboxylic acid which leads to voids within the coordination polymer structure (Scheme 5.10). This gel may be used as a polymerisation template. Methyl methacrylate may be polymerised within the gel matrix by UV irradiation of the monomer, resulting

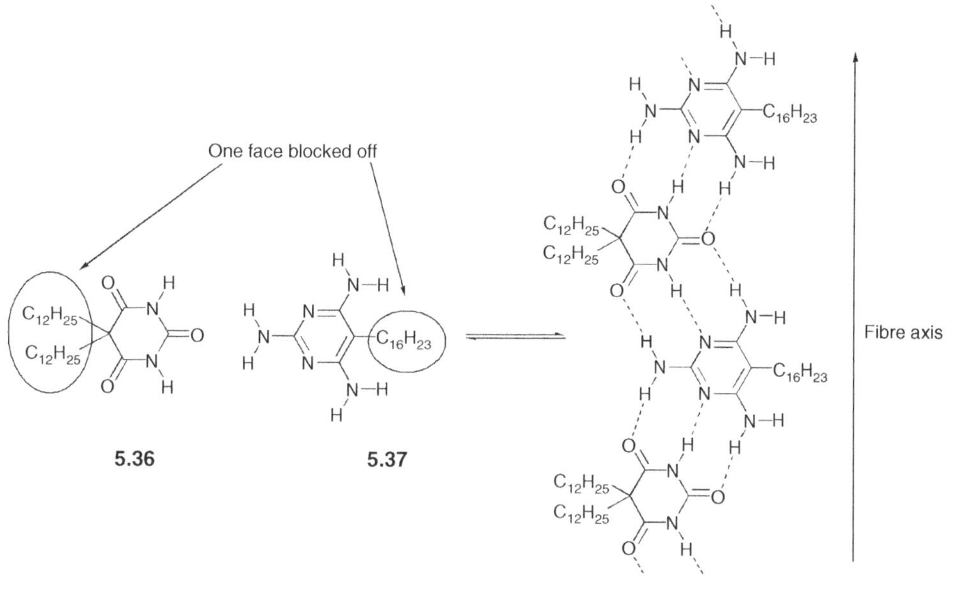

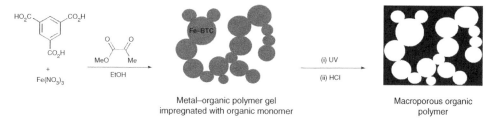

Scheme 5.9   The proposed assembly between barbituric acid (**5.36**) and pyrimidine (**5.37**) groups, giving rise to a fiberous structure and then a gel-phase material.

Scheme 5.10   Formation of a 'macroporous' organic polymer from a 'metallogel' template (BTC, benzenetricarboxylate).[36]

in extensive cross-linking to give a metallogel-templated PPMA polymer. Removal of the metallogel phase by taking advantage of its supramolecular nature produces a porous imprinted polymer in which the covalent polymer fills the former voids within the gel structure.[36] This polymer is an interesting example of a general class of *molecular-imprinted polymers* (MIPs) (see Section 5.9.2).

5.9.2   Molecular-Imprinted Polymers

 Whitcombe, M. J. and Vulfson, E. N., 'Imprinted polymers', *Adv. Mater.*, 2001, **13**, 467–478.

Molecular-imprinted polymers (MIPs) are highly crossed-linked polymer matrices that are formed by using a template. This new approach to templated polymers is highly topical. The template is removed post-synthesis, producing a cavity in the polymer. The shape and binding functionalities in the polymer are complementary to the template molecule. Recent developments in MIPs have been employed in sensor-array technology. For example, an eight-channel sensor array for biologically important amines has been developed. Once the amine templates are washed out, the polymers can be tested for their ability to differentiate between six different biological important aryl amines (Figure 5.28). The binding affinity between different amine analytes is measured by taking the constant mass of the polymer and adding it to the same concentration of each amine. The response is measured as a ratio of absorbances $(A_0 - A_i/A_0)$ before and after equilibrium between the eight different polymers and the six different analytes.

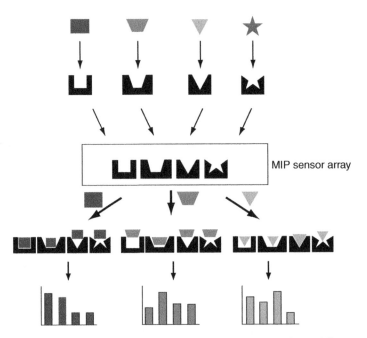

Figure 5.28   Four imprinted polymers tested in an array against three different analytes to produce different patterns of recognition.[37] N. T. Greene, S. L. Morgan and K. D. Shimizu, 'Molecularly imprinted polymer sensor arrays', *Chem. Commun*, 2004, 1172–1173. Reproduced by permission of The Royal Society of Chemistry.

The indicator of choice is structurally similar to the aryl amines, but has an intense colour with an absorption maximum at 460 nm. This band is far enough into the visible region of the spectrum such that there is no spectroscopic interference from the analytes. The affinity of the dye has been assessed, by dissolving the polymer and dye in a solution of acetonitrile. The dye absorbs into the MIP, resulting in a colourless supernatant, and the normally white polymer turns yellow. Upon the addition of the aryl amines, the bound dye is displaced back into the solution and the polymer returns to its original white colour. *Linear discriminate analysis* (LDA) can be used to give a two-dimensional plot that clearly shows binding patterns specific to particular amine analytes.[37]

5.9.3 Templated monodisperse latex

Monodisperse latex spheres of a controlled size can be arranged into three-dimensional arrays and are used as templates to prepare well-defined cavities and structures once the latex sphere template has been removed. These latex spheres are identical to one another in size and shape and are often prepared from colloidal clusters resulting from the aggregation of sol–gel colloids (Section 5.6). The spherical polymer clusters can be fabricated by the slow addition of an aqueous solution into a reservoir of hydrophobic silicone liquid, forming emulsion droplets. This produces a highly structured porous matrix with a well-defined structure upon polymerisation. The size of the droplets is controlled by the concentration of the aqueous latex, the speed at which the suspension is stirred and the ratio between the silicone liquid and latex. As the concentration of the latex spheres increases to its critical concentration, *i.e.* the concentration at which the colloidal spheres start to order themselves into a close-packed structure, the balls are filtered off and are dried, ready to be used as templates.

Latex spheres are commonly used as templates to prepare highly ordered inorganic films. For example, inorganic films have been prepared by the addition of a tetrabutyl titanate–ethanol solution to cross-linked polystyrene monodisperse spheres. The mixture is left to stand for a week to allow the titanate solution to permeate into the voids between the latex spheres. The spheres are then washed to remove the excess solution on the exterior of the spheres and left to stand in air to undergo hydrolysis to produce a polymer–titania matrix. Finally, the matrix is calcified, resulting in the removal of the polymer to give a templated rutile titania crystallite structure (Scheme 5.11). The production of materials that have specific dimensions prepared by using the latex technique has led to an array of applications, for example, the development of photonic materials, high-density magnetic data storage, microchip reactors and biosensors.

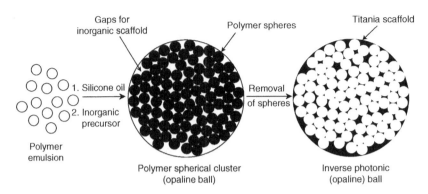

Scheme 5.11   The preparation of inverse photonic balls using polymer spheres to template to titania macroporous structure with pore diameters comparable to optical wavelengths.

5.10   Nanobiology and biomimetic chemistry

Nanobiology is the extension of nanotechnology to biology. It encompasses nanotechnological solutions to biological problems, such as diagnosis and monitoring, the study of nanoscale biological systems and phenomena, and the synthesis or construction of nanometre-scale mimics of biological entities.[38]

A major difference between nature and human design of miniature structures is that nature does not have flat surfaces to build objects from. Nature's surfaces are curved and have rounded corners, for example, proteins and biological cells. Nature has learned to work with these versatile, small and non-planar building blocks and has mastered the art of their self-assembly. Thus, nanobiology is at a fundamental level, very much more three-dimensional than the etching processes described for the nanotechnological production of printed circuit boards (Section 5.1), for example. Nanobiology, as a fusion of biology and nanotechnology, is still very much in its infancy and we will only touch upon the area here.

> **Nanobiology**: The application of nanotechnology to biological and biomimetic problems and the use of biological entities in nanotechnology.

5.10.1   Biological components in nanotechnology

   *Seeman, N. C., 'DNA in a material world', Nature, 2003, **421**, 427–431.*

Biomolecular electronics (BMEs) uses actual biological molecules in the design of nanotechnological devices. The natural self-recognition and self-assembly behaviour of biological components has the potential to make BME devices both inexpensive and easy-to-fabricate. An attractive approach is the use of natural or designed proteins that can be used for arrays of nanomachines.

Bacteria and archaea[¶] contain a crystalline single protein or glycoprotein species around their exteriors, known as 'S-layers'. S-layers exhibit either an oblique, square, or hexaganol lattice, ranging in crystallographic unit cell dimensions from 3 to 30 nm, with an S-layer thickness of 5–10 nm and pores that are evenly spaced every 2 to 8 nm (Figure 5.29). These well-defined regions of the S-layer lattice can be used to template assemblies of nanoparticles (Section 5.6). A particularly striking example is the use of bacterial S-layers as superstructures for artificial arrays. The S-layer is chemically removed and broken up into individual sub-units. One of the key features of broken, isolated and purified S-layers is their ability to self-assemble into two-dimensional arrays, either in solution or on a solid support, such as silicon wafers, metal surfaces, lipid films or liposomes. These self-assembled S-layers have been used as glucose sensors, for example,

Figure 5.29 TEM image of a freeze-etching preparation of a bacterial cell exhibiting an 'S-layer' with square (*P*4) lattice symmetry (bar = 100 nm). Reproduced from *Arch. Microbiol.*, 1986, **146**, 19–24, Figure 1a, with kind permission of Springer Science and Business Media.

[¶] Archaea comprise one of the three domains that separate all forms of life; the other two are bacteria and eukarytoa.

by binding glucose oxidisase to the S-layer and measuring the current passing through electrodes as the oxidase reacts with the glucose.

There have been numerous reports of the use of DNA as a structural scaffold, based on pioneering work by Nadrian Seeman at the University of New York. The idea is to unravel the two strands of DNA and then stick them together with matched ends of other DNA molecules that are complementary with other specific sequences of nucleic acid, thereby forming a jigsaw puzzle. The process of joining different DNA molecules together is not new and is the basis of evolution that requires genetic change. Organisms do this all of the time in meiosis and other cell division processes. During the cell division, a temporally X-shaped structure is formed, called the *Holliday junction*. With complementary binding sites, it is possible to have Holliday junctions at either end of the DNA strand to produce either a two-or three-dimensional array. Normally, DNA chromosomes are flexible and are twisted back into bundles of a densely packed hierarchy of helixes and are not a rigid array. However, Seeman has been able to prepare two- and three-dimensional arrays of DNA by allowing the DNA strands to attach to each other twice at these junctions. The resulting 'double-crossover' DNA is very stiff and rigid.

One potentially attractive property of DNA is its potential to make *DNA computers* in which DNA strands might process data in a similar way to electronic computers. The ability of DNA to act as a computer was first demonstrated by Leonard Adleman, who used it to solve a variant of the 'travelling salesman' problem.[//] Adleman used strands of DNA to represent the roads which were connected *via* base pairings, which represent the cities. By mixing the strands together and connecting the 'cities' by 'roads', he was able to show that the strand could self-assemble to solve the problem.

DNA is not the only natural material used to prepare molecular devices. Metalloproteins, such as azuarin and plastocyanin, contain copper atoms in their protein scaffold, acting as redox-active sites. These blue copper proteins have been used as candidates for biomolecular nanoelectronic applications, due to their natural electron-transfer activity in biological environments. For example, azuarin contains 2 cysteine groups and 12 amino acids that can act as 'handles' to bind onto surfaces, forming layers of proteins on semiconductors, oxides and other metal surfaces. These arrays have potential applications as devices, by tuning their redox-state with the aid of an external voltage. Azuarin proteins have been immobilised on surfaces in both ordered and disordered self-assembled arrangements. A current is then applied to the surface and both the ordered and disordered arrays show diode-behaviour. The large current value obtained from the current–voltage curves indicates an electron-transfer-process mechanism. The ordered self-assembled azuarin device shows a ten times larger current than the disordered one and acts as a better diode.

[//] The objective is to find the shortest path a 'travelling salesman' could follow in order to visit each of a set of geographically distributed locations (cities), while visiting each location only once.

5.10.2 Nanoparticles in medicine

 Parak, W. J., Pellegrino, T. and Plank, C., 'Labelling of cells with quantum dots', *Nanotechnology*, 2005, **16**, R9–R25.

Among interesting recent developments is the use of nanoparticles (Section 5.6) in a medical imaging and diagnostic role. For example, quantum-dot fluorescent semiconductor nanocrystals have been used to selectively image and differentiate tumour vessels from normal tissue, as well as to assess the ability of different particles to access tumours. A key feature is the water solubilisation of the nanoparticles by use of a hydrophilic coating, such as an amphiphilic polymer or surfactant. Nanoparticles bind well to thiol or amino groups in proteins and derivatised nucleic acids. This property has been used to assemble regular arrays of gold nanoparticles on nucleic acid scaffolds, for example. Nanoparticles have the advantage of being much more robust than organic dyes and so do not suffer from bleaching. They are also very readily imaged by electron microscopy. Gold nanoparticles of size 10–40 nm tagged to antibodies are routinely used in histochemistry for the biospecific labelling of certain regions of tissue samples.

Nanoparticles may also have a therapeutic role. Silver nanoparticles capped with a carbon-capping agent (the capping agent is used to control the size of the particle) have been used as treatment for viruses such as HIV. These silver particles were able to kill HIV-1 cells in 24 h. It is believed that the nanoparticles are bound to the gp120 glycoprotein knobs on HIV-1 through the sulfur residues. These silver nanoparticles are also effective towards other viruses and superbugs, such as *Methicillin Resistant Staphylococcus Aureus* (MRSA). However, there is current concern regarding using nanoparticles *in vivo* due to their potential toxicity.

5.10.3 Bio-inspired materials

An elegant example of how researchers are using nature to imitate biology on the sub-micrometre level is the mimicking of biological materials that have extraordinary properties. The gecko, for example, has the ability to walk upside down on surfaces that seem to be smooth. The reason that geckos' feet are able to stick to surfaces is the presence of 200 nm-thick microscopic elastic keratin hairs, called *setae*, covering their toes. The ends of these hairs split further into 'spatulas' and it is these tiny stalks that are responsible for the interaction on the surface (Figure 5.30). van der Waals forces cause hairs of this small size to stick to surfaces, as verified by model gecko spatulae made of polymers such as polyester.[39] The hairs on the soles of gecko feet are so densely packed that the overall structure exerts a force on the scale of $10 \, \text{N cm}^{-2}$ and they have the proper aspect ratio, thickness, stiffness and structure.[40]

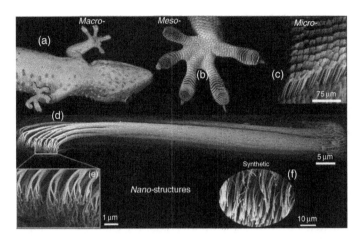

Figure 5.30 Structural hierarchy of the gecko adhesive system: (a) ventral view of a tokay gecko (*Gekko gecko*) climbing a vertical glass surface; (b) ventral view of the foot of a tokay gecko, showing a mesoscale array of seta-bearing 'scansors' (adhesive lamellae); (c) microscale array of setae are arranged in a nearly grid-like pattern on the ventral surface of each scansor – in this scanning electron micrograph, each diamond-shaped structure is the branched end of a group of four setae, clustered together in a tetrad; (d) micrograph of a single gecko seta assembled from a montage of five cryo-SEM images – note that the individual keratin fibrils comprise the setal shaft; (e) nanoscale array of hundreds of spatular tips of a single gecko seta; (f) synthetic spatulae, fabricated from polyimide at UC Berkeley, California, USA in the laboratory of Ronald Fearing, using nanomoulding. Images (a)–(d) provided by courtesy of Professor Kellar Autumn, Lewis and Clark College, Portland, Oregon, USA.

There have been numerous attempts to prepare synthetic systems analogous to geckos' feet, for example, by fabricating arrays of fibrous plastic pillars. The synthetic hairs have been found to produce a 'sticky tape'. Polymer surfaces with multi-walled carbon nanotubes (MWCNTs, see Section 5.7) have been produced which exhibit strong nanometre-level adhesion ('dry adhesives') and on the nanoscale appear more effective than the gecko foot hairs. This work has not yet been translated to the macroscale, however! The growth of the MWCNT is achieved by vapour deposition onto either quartz or silicon substrates. A gaseous mixture of ferrocene (as a catalyst) and xylene as a carbon source is heated to 150 °C and passed over the substrate, itself heated to 800 °C in a furnace. This method selectively grows the MWCNT on the oxide layer of the substrate, with remarkable control of thickness and length. The oxide layer can be patterned by photolithography (Section 5.1.2) to create various patterns of MWCNTs (Figure 5.31(a)).

The typical lengths of the MWCNTs are approximately 65 μm and have a diameter of 10–20 nm. The MWCNTs are then embedded and stabilised in a poly (methyl methacrylate) (PMMA) matrix. This matrix is then peeled off the silicon to form a very smooth surface, which is etched with a good solvent, typically acetone, to produce brushes of MWNTs, analogous to the spatula of the

Figure 5.31 SEM images of vertically aligned MWCNTs (a) grown on silicon and (b) trans-ferred into a PMMA matrix and then exposed on the surface after etching.[41] B. Yurdumakan, N. R. Raravikar, P. M. Ajayan and A. Dhinojwala, 'Synthetic gecko foot hairs from multi-walled carbon nanotubes', *Chem. Commun.*, 2005, 3799–3801. Reproduced by permission of The Royal Society of Chemistry.

gecko's toes (see Figure 5.30). The adhesion forces and elastic properties of the MWNT brushes are measured by multimode scanning probe microscopy, a tool typically used for measuring the properties of carbon nanotubes. The minimum force/area for the bundles of nanotubes was calculated to be 1.6×10^{-2} nN nm^{-2}, whereas a typical adhesive force for a gecko's foot is 10^{-4} nN nm^{-2}. This increase is attributed to the van der Waals forces originating from contact with multiple nanotubes or from large contact areas with single nanotubes.

5.10.4 An inorganic cell

In terms of nanoscale biomimetic chemistry, recent work on very large entirely inorganic spherical capsules based on the (pentagon)$_{12}$(linker)$_{30}$ type, such as the polyoxomolybdate $[P_{12}\{Mo_2O_4(MeCO_2)\}_{30}]^{42-}$ (Figure 5.32, where P is the pentag-onal cluster $\{(Mo)Mo_5O_{21}(H_2O)_6\}^{6-}$), has shown that these entirely synthetic, *abiotic* containers can function as artificial cells (or 'nano-chromatographs' for the separation of metal ions, depending on one's point of view).[42] These inorganic capsules may be very readily prepared on a kg scale from the reduction of Mo(VI) solutions with various reducing agents to give what has been traditionally termed 'molybdenum blue' because of the colour imparted by delocalised electrons. The capsule size and shape can be tailored by careful choice of conditions and the largest such hollow sphere comprises 368 Mo atoms and has an internal cavity 2.5 nm wide and 4 nm long, containing about 400 structured water molecules in a confined environment. Metal cations are able to selectively diffuse in and out

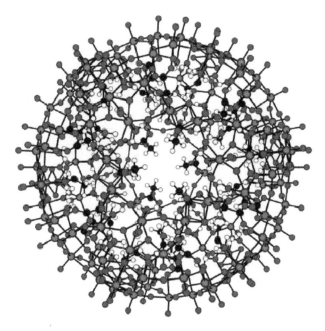

Figure 5.32 The polyoxomolybdate nanocluster $[P_{12}\{Mo_2O_4(MeCO_2)\}_{30}]^{42-}$ (where $P = \{(Mo)Mo_5O_{21}(H_2O)_6\}^6$) which behaves as an artificial cell, showing the pore opening in the centre. Reprinted from A. Müller, H. Bogge and M. Henry, 'Coordination chemistry under confined conditions: a simplified illustrative view', *C. R. Chim.*, **8**, 47–56, © (2005), with permission from Elsevier Science.

of the holes created by the carboxylate substituents, with a remarkable dependence on the ability of the water confined within the capsule to coordinate to the cation. The confined environment of the inorganic cell exhibits some remarkable properties, while the polyoxomolybdate cell wall exhibits some features in common with cation transport through biological ion channels. The adaptation of the confined structured water to the ingress of metal ions suggests insights into the way in which a cell converts an incoming chemical signal into a response.

5.10.5   Templated biomimetic materials

 Ozin, G. A. and Arsenault, A., 'Morphogensis of biomineral and morphosynthesis of biomimetic forms', *Acc. Chem. Res.*, 1997, **30**, 17–27.

In Chapter 4, we looked at zeolites and metal–organic frameworks (MOFs) (Section 4.6.2). Recent work in 'soft' (*i.e.* organic) nanochemistry utilises templating of microporous silica and alumina by organic molecules, in particular, surfactants (Section 5.4), to prepare very elaborate morphologies.

In 1887, the German biologist and philosopher Ernst Haeckel popularised in Germany Charles Darwin's work on evolution. Although Haeckel's work on evolution was important, he is most famous for his work on diatoms and radiolaria, where he catalogued thousands of species. These organisms are single-celled algae, which are known for their delicate appearance, geometrically regular shapes and elaborate patterns (Figure 5.33).

(a)   (b)   (c)

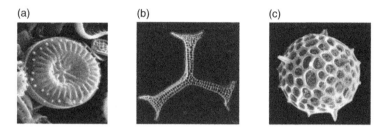

Figure 5.33 SEM images of (a) diatom and (b, c) radiolarian species, found in Southern Spain, the Arabian Sea and the NW Indian Ocean, respectively. Image (a) reproduced by permission of Dr P R Sweets, University of Indianapolis, with images (b and c) provided by courtesy of the UCL Micropalaeontology Unit.

The design of hierarchical templated inorganic structures containing porous networks has been influenced by these diatom and radiolarian microporous silica microskeletal structures. Work carried out in the 1990s by a team at Mobile showed that supramolecular assemblies of amphiphilic molecules, in the form of micelles, lamellae and bicontinuous phases[†] (Section 5.4) were able to assemble with inorganic materials, in particular, silica. This missing link using 'soft' chemistry, such as the integration of amphilphilic molecules as templates with silica, led to the creation of a plethora of abiotic microskeletal structures, like the diatoms and radiolaria found in nature. These materials are formed due to the templating behaviour of surfactant–cosurfactant-based supramolecular assemblies. Mixing of different ratios of tetraethylene glycol (TEG) and an amphilic alkylamine ($C_{10}H_{21}NH_2$) with phosphoric acid results in immediate gelation. The mixture can be identified by X-ray crystallography as a crystalline 'mesolamellar' alkylammonium dihydrogen phosphate phase, $[C_{10}H_{21}NH_3]^+[H_2PO_4]^-$, linked by an extensive hydrogen bonding array. On addition of alumina, phase-separation occurs, forming (TEG)Al(III) ionophores throughout the vesicle structure. A proposed mechanism for the templating is shown in Figure 5.34.[43] This

---

[†] A bicontinuous phase occurs when the amount of water and oil are approximately equal, forming a continuous phase. The analogy with a sponge clearly demonstrates the principle: as a liquid is poured into a sponge-like material, the sponge contains a continuous phase of the liquid, in the same way the material that the sponge is made from also forms a continuous phase–hence biocontinuous phases are also called sponge phases.

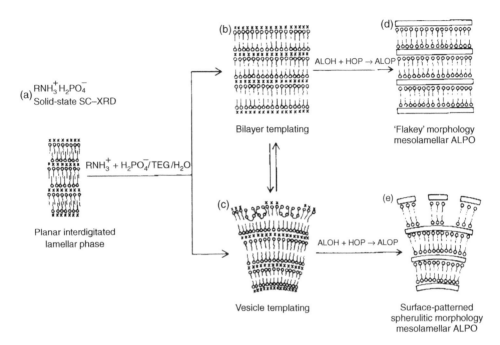

Figure 5.34 The proposed template patterning mechanism in the formation of biomimetic microskeletal structures, where the circles with tails represent the cationic surfactant, the crosses represent the anionic dihydrogen phosphate counter-ions and the connected circles represent the tetraethylene glycol (TEG) molecules.[43] Reprinted with permission from Nature Publishing Group, S. Oliver, A. Kuperman, N. Coombs, A. Lough and G. A. Ozin, 'Lamellar aluminophosphates with surface patterns that mimic diatom and radiolarian microskeletons', *Nature*, **378**, 47–50 © 1995.

system helps to explain the morphologies observed in natural microskeletal structures. This initial experiment opened the door for the use of small-molecule organic surfactant-based vesicles that can co-assemble inorganic minerals to form composite materials, producing a plethora of morphologies, such as hexagonal, gradient, polydispersion, shear, shrink and mesh patterns.

Microporous alumina composites are not the only materials that have been developed. There are also a number of shapes and surface patterns of mesoporous silica materials utilising surfactant assemblies as supramolecular templates. These silica bodies have been prepared under quiescent aqueous acidic conditions, *i.e.* the mixture is unstirred, using cetyltrimethylammonium chloride (CTACl) as the surfactant template and tetraethylorthosilicate (TEOS) as the silica source. The morphologies of these materials determined by SEM reveal a diverse range of topologies (Figure 5.35). These topologies are said to be *emergent*, because they are not predicted and evolve over time (Section 5.1).

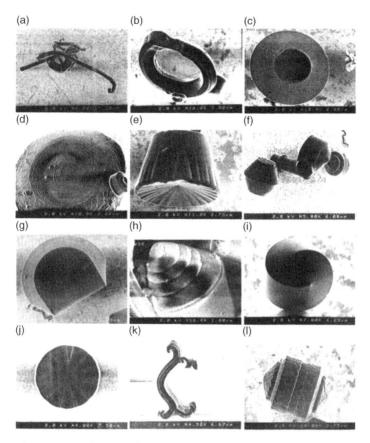

Figure 5.35 SEM images showing diverse mesoporous silica shapes and patterns produced by surfactant templating: (a) rope; (b) toroid; (c) discoid; (d) pinwheel; (e) wheel; (f) gyroid; (g) bagel; (h) shell; (i) knot; (j) clock; (k) 'eccentric 1'; (l) eccentric.[44] Reprinted with permission from Nature Publishing Group, H. Yang, N. Coombs and G. A. Ozin, 'Morphogenesis of shapes and surface patterns in mesoporous silica', *Nature*, **386**, 692–695 © 1997.

References

1. Chandross, E. A. and Miller, R. D., 'Nanostructures: introduction', *Chem. Rev.*, 1999, **99**, 1641–1642.
2. Joachim, C., 'To be nano or not to be nano?', *Nature Mater.* **4**, 2005, 107–109.
3. Madou, M., 'Nanotechnology: dry versus wet engineering?', *Anal. Bioanal. Chem.*, 2006, **384**, 4–6.
4. Wouters, D. and Schubert, U. S., 'Nanolithography and nanochemistry: probe-related patterning techniques and chemical modification for nanometer-sized devices', *Angew. Chem., Int. Ed. Engl.*, 2004, **43**, 2480–2495.

5. Hla, S. W., Bartels, L., Meyer, G. and Rieder, K. H., 'Inducing all steps of a chemical reaction with the scanning tunneling microscope tip: towards single molecule engineering', *Phys. Rev. Lett.*, 2000, **85**, 2777–2780.

6. Theobald, J. A., Oxtoby, N. S., Phillips, M. A., Champness, N. R. and Beton, P. H., 'Controlling molecular deposition and layer structure with supramolecular surface assemblies', *Nature*, 2003, **424**, 1029–1031.

7. Satake, A., Tanaka, H., Hajjaj, F., Kawai, T. and Kobuke, Y., 'Single molecular observation of penta- and hexagonic assembly of bisporphyrin on a gold surface', *Chem. Commun.*, 2006, 2542–2543.

8. Shipway, A. N., Katz, E. and Wilmer, I., 'Nanoparticle arrays on surfaces for electronic, optical and sensor applications', *Chem. Phys. Chem.*, 2000, **1**, 18–52.

9. Arm, K. J. and Williams, J. A. G., 'Boronic acid-substituted metal complexes: versatile building blocks for the synthesis of multimetallic assemblies', *Chem. Commun.*, 2005, 230–232.

10. Credi, A., Balzani, V., Langford, S. J. and Stoddart, J. F., 'Logic operations at the molecular level. An XOR gate based on a molecular machine', *J. Am. Chem. Soc.*, 1997, **119**, 2679–2681.

11. Shahinpoor, M. and Kim, K. J., 'The effect of surface-electrode resistance on the performance of ionic polymer–metal composite (IPMC) artificial muscles', *Smart Mater. Struct.*, 2000, **9**, 543–551.

12. Stadler, A.-M., Kyritsakas, N. and Lehn, J.-M., 'Reversible folding/unfolding of linear molecular strands into helical channel-like complexes upon proton-modulated binding and release of metal ions', *Chem. Commun.*, 2004, 2024–2025.

13. Heath, J. R., Stoddart, J. F. and Williams, R. S., 'More on molecular electronics', *Science*, 2004, **303**, 1136–1137.

14. Wright, A. T. and Anslyn, E. V., 'Differential receptor arrays and assays for solution-based molecular recognition', *Chem. Soc. Rev.*, 2006, **35**, 14–28.

15. Hall, A. C., Suarez, C., Hom-Choudhury, A., Manu, A. N. A., Hall, C. D., Kirkovits, G. J. and Ghiriviga, I., 'Cation transport by a redox-active synthetic ion channel', *Org. Biomol. Chem.*, 2003, **1**, 2973–2982.

16. Schmittel, M., Lal, M., Graf, K., Jeschke, G., Suskec, I. and Salbeckc, J., '*N*, *N*′-Dimethyl-2,3-dialkylpyrazinium salts as redox-switchable surfactants? Redox, spectral, EPR and surfactant properties', *Chem. Commun.*, 2005, 5650–5652.

17. Tachibana, H., Azumi, R., Nakamura, T., Matsumoto, M. and Kawabata, Y., 'New types of photochemical switching phenomena in Langmuir–Blodgett Films', *Chem. Lett.*, 1992, **21**, 173–176.

18. Moore, A. J., Goldenberg, L. M., Bryce, M. R., Petty, M. C., Moloney, J., Howard, J. A. K., Joyce, M. J. and Port, S. N., 'New crown annelated tetrathiafulvalenes: Synthesis, electrochemistry, self-assembly of thiol derivatives and metal cation recognition', *J. Org. Chem.*, 2000, **65**, 8269–8276.

19. Kato, T., Mizoshita, N. and Kishimoto, K., 'Functional liquid-crystalline assemblies: self-organised soft-materials', *Angew. Chem., Int. Ed. Engl.*, 2006, **45**, 38–68.

20. Kato, T. and Frechet, J. M. J., 'New approach to mesophase stabilization through hydrogen bonding molecular-interactions in binary-mixtures', *J. Am. Chem. Soc.*, 1989, **111**, 8533–8534.

21. Pieraccini, S., Gottarelli, G., Mariani, P., Masiero, S., Saturni, L. and Spada, G. P., 'Columnar lyomesophases formed in hydrocarbon solvents by chiral lipophilic guanosine–alkali metal complexes', *Chirality*, 2001, **13**, 7–12.

22. Smarsly, B. and Kaper, H., 'Liquid inorganic–organic nanocomposites: novel electrolytes and ferrofluids', *Angew. Chem., Int. Ed. Engl.*, 2005, **44**, 3809–3811.

23. Xia, Y. and Whitesides, G. M., 'Soft lithography', *Angew. Chem., Int. Ed. Engl.*, 1998, **37**, 550–575.

24. Howard, L. E. M., Nguyen, H. L., Giblin, S. R., Tanner, B. K., Terry, I., Hughes, A. K. and Evans, J. S. O., 'A synthetic route to size-controlled fcc and fct FePt nanoparticles', *J. Am. Chem. Soc.*, 2005, **127**, 10140–10141.

25. Schmittel, M., Kalsanl, V. and Kienle, L., 'Simple and supramolecular copper complexes as precursors in the HRTEM induced formation of crystalline copper nanoparticles', *Chem. Commun.*, 2004, 1534–1535.

26. Bowers, M. J., McBride, J. R. and Rosenthal, S. J., 'White-light emission from magic-sized cadmium selenide nanocrystals', *J. Am. Chem. Soc.*, 2005, **127**, 15378–15379.

27. Beer, P. D., Cormode, D. P. and Davis, J. J., 'Zinc metalloporphyrin-functionalized nanoparticle anion sensors', *Chem. Commun.*, 2004, 414–415.

28. Davis, J. J., 'Interfacial sensing: surface-assembled molecular receptors', *Chem. Commun.*, 2005, 3509–3513.

29. Makha, M., Purich, A., Raston, C. L. and Sobolev, A. N., 'Structural diversity of host–guest and intercalation complexes of fullerene C_{60}', *Eur. J. Inorg. Chem.*, 2006, 507–517.

30. Atwood, J. L., Koutsantonis, G. A. and Raston, C. L., 'Purification of C_{60} and C_{70} by selective complexation with calixarenes', *Nature*, 1994, **368**, 229–231.

31. Diederich, F., Jonas, U., Gramlich, V., Herrmann, A., Ringsdorf, H. and Thilgen, C., 'Synthesis of a fullerene derivative of benzo[18]crown-6 by Diels–Alder reaction–complexation ability, amphiphilic properties and X-ray crystal structure of a dimethoxy-1,9-(methano 1,2 benzenomethano)fullerene 60 benzene clathrate', *Helv. Chim. Acta*, 1993, **76**, 2445–2453.

32. Meyer, R. R., Sloan, J., Dunin-Borkowski, R. E., Kirkland, A. I., Novotny, M. C., Bailey, S. R., Hutchison, J. L. and Green, M. L. H., 'Discrete atom imaging of one-dimensional crystals formed within single-walled carbon nanotubes', *Science*, 2000, **289**, 1324–1326.

33. Smith, D. K., 'Dendritic supermolecules–towards controllable nanomaterials', *Chem. Commun.*, 2006, 34–44.

34. Araki, K., Takasawa, R. and Yoshikawa, I., 'Design fabrication, and properties of macroscale supramolecular fibers consisted of fully hydrogen bonded pseudo-polymer chains', *Chem. Commun.*, 2001, 1826–1827.

35. Hanabusa, K., Miki, T., Taguchi, Y., Koyama, T. and Shirai, H., '2-component, small-molecule gelling agents', *J. Chem. Soc., Chem. Commun.*, 1993, 1382–1384.

36. Wei, Q. and James, S. L., 'A metal–organic gel used as a template for a porous organic polymer', *Chem. Commun.*, 2005, 1555–1556.

37. Greene, N. T., Morgan, S. L. and Shimizu, K. D., 'Molecularly imprinted polymer sensor arrays', *Chem. Commun.*, 2004, 1172–1173.

38. LaVan, D. A., Lynn, D. M. and Langer, R., 'Moving smaller in drug discovery and delivery', *Nature Rev. Drug Disc.*, 2002, **1**, 77–84.

39. Autumn, K., Sitti, M., Liang, Y. C. A., Peattie, A. M., Hansen, W. R., Sponberg, S., Kenny, T. W., Fearing, R., Israelachvili, J. N. and Full, R. J., 'Evidence for van der Waals adhesion in gecko setae', *Proc. Natl. Acad. Sci. USA.*, 2002, **99**, 12252–12256.

40. Autumn, K., 'How gecko toes stick', *Am. Sci.*, 2006, **94**, 124–132.

41. Yurdumakan, B., Raravikar, N. R., Ajayan, P. M. and Dhinojwala, A., 'Synthetic gecko foot-hairs from multiwalled carbon nanotubes', *Chem. Commun.*, 2005, 3799–3801.

42. Muller, A., Bogge, H. and Henry, M., 'Coordination chemistry under confined conditions: a simplified illustrative view', *C. R. Chim.*, 2005, **8**, 47–56.

43. Oliver, S., Kuperman, A., Coombs, N., Lough, A. and Ozin, G. A., 'Lamellar aluminophosphates with surface patterns that mimic diatom and radiolarian microskeletons', *Nature*, 1995, **378**, 47–50.

44. Yang, H., Coombs, N. and Ozin, G. A., 'Morphogenesis of shapes and surface patterns in mesoporous silica', *Nature*, 1997, **386**, 692–695.

Index

Printed and bound in the UK by
CPI Antony Rowe, Eastbourne

Printed and bound by CPI Group (UK) Ltd, Croydon, CR0 4YY

27/10/2024

14580377-0002